About Island Press

Island Press, a nonprofit organization, publishes, markets, and distributes the most advanced thinking on the conservation of our natural resources—books about soil, land, water, forests, wildlife, and hazardous and toxic wastes. These books are practical tools used by public officials, business and industry leaders, natural resource managers, and concerned citizens working to solve both local and global resource problems.

Founded in 1978, Island Press reorganized in 1984 to meet the increasing demand for substantive books on all resource-related issues. Island Press publishes and distributes under its own imprint and offers these services to other nonprofit organizations.

Support for Island Press is provided by the Geraldine R. Dodge Foundation, The Energy Foundation, The Charles Engelhard Foundation, The Ford Foundation, Glen Eagles Foundation, The George Gund Foundation, William and Flora Hewlett Foundation, The James Irvine Foundation, The John D. and Catherine T. MacArthur Foundation, The Andrew W. Mellon Foundation, The Joyce Mertz-Gilmore Foundation, The New-Land Foundation, The Pew Charitable Trusts, The Rockefeller Brothers Fund, The Tides Foundation, and individual donors.

Global Marine
Biological Diversity

Edited by Elliott A. Norse

Center for Marine Conservation
World Conservation Union (IUCN)
World Wildlife Fund
United Nations Environment Programme
World Bank

A Contribution to the *Global Biodiversity Strategy*

Illustrations by Jill Perry Townsend

GLOBAL
MARINE
BIOLOGICAL
DIVERSITY

A Strategy for
Building Conservation into
Decision Making

ISLAND PRESS
Washington, D.C.
Covelo, California

Library of Congress Cataloging-in-Publication Data

Global marine biological diversity : a strategy for building
    conservation into decision making / edited by Elliott A. Norse ;
    Center for Marine Conservation . . . [et al.] ; illustrations by Jill
    Perry Townsend.
        p.   cm.
    "A contribution to the Global biodiversity strategy."
    Includes bibliographical references (p.   ) and index.
    ISBN 1-55963-255-0 (cloth).
    ISBN 1-55963-256-9 (pbk.).
    1. Marine biology.   2. Biological diversity conservation.
    3. Marine resources conservation.   I. Norse, Elliott A.   II. Center
    for Marine Conservation.
    QH91.8.B6G58   1993
    333.95'216—dc20   93-25350
                                                              CIP

Printed on recycled, acid-free paper

♲

Manufactured in the United States of America

10 9 8 7 6 5 4 3 2

ABOUT THE EDITOR

Elliott A. Norse, Chief Scientist of the Center for Marine Conservation in Washington, D.C. (USA), and Affiliate Professor at the Institute for Environmental Studies of the University of Washington in Seattle, WA (USA), is a marine and forest conservation biologist. His Ph.D. and postdoctoral research in the 1970s examined the ecology of blue crabs (*Callinectes* spp.) in Jamaica, Curaçao, Mexico, Panama, and Colombia. Since then, he has devoted his career to incorporating conservation biology into environmental decision making as a staff member or consultant for US federal agencies, international governmental organizations, scientific professional societies, conservation organizations, and foundations. His writings include more than 50 publications on environmental policy, conservation biology, marine ecology, forest ecology, and human-caused climatic change. As Staff Ecologist at the White House Council on Environmental Quality during President Carter's Administration, Dr. Norse was senior author of the seminal 1980 CEQ *Annual Report* chapter that first defined the concept of biological diversity. His books include *Conserving Biological Diversity in Our National Forests* (The Wilderness Society, 1986) and *Ancient Forests of the Pacific Northwest* (Island Press, 1990).

ABOUT THE CO-SPONSORS

The **Center for Marine Conservation** is a Washington, D.C. (USA)–based non-governmental organization dedicated to maintaining the diversity, abundance, and integrity of life in oceans and coastal areas through science-based advocacy. Founded in 1972, CMC has over 60 scientists, attorneys, and other staff members working to prevent the overexploitation of marine species and the degradation of marine ecosystems, and to restore them where they have been diminished. Using research, education, and interaction with governments, CMC's long-standing programs emphasize sustainable fisheries, prevention of solid-waste pollution, conservation of protected species, and management of marine protected areas in US waters and, increasingly, in other nations. Cutting across these program areas is a biological diversity program that aims to establish the science of marine conservation biology. CMC administers the fast-growing annual International Coastal Cleanup Campaign, which, in 1992, brought 162,000 volunteers in 32 nations to clean up and document marine pollution from plastics and other solid wastes.

Founded in 1948, **IUCN—World Conservation Union** brings together states, government agencies, and diverse non-governmental organizations in a unique world partnership: more than 720 members in all, spread across 118 countries. IUCN exists to serve its members by representing their views and providing them with the concepts, strategies, and technical support they need to achieve their goals. Through its Commissions, IUCN draws together over 5,000 expert volunteers. A central secretariat coordinates the program and leads initiatives on the conservation and sustainable use of biological diversity, and the management of ecosystems and natural resources. IUCN has helped many countries prepare National Conservation Strategies, and demonstrates the application of its knowledge through its field projects. Operations are increasingly administered through an expanding network of regional and country offices, located principally in developing countries. IUCN works with its members to achieve development that is sustainable and improves the quality of life for people worldwide.

In the USA, **World Wildlife Fund** is the largest private organization working worldwide to protect endangered wildlife and wildlands. It is the US affiliate of the international WWF family, with national organizations or representatives in 40 countries. WWF works with hundreds of in-country conservation groups and foreign governments to ensure the success of its field projects and emergency assistance grants. The organization is committed to reversing the degradation of the natural environment while also meeting human needs. A top priority is the protection of tropical forests and wildlife in Latin America, Asia, and Africa. Over the past three decades, WWF has sponsored more than 3,000 conservation projects in 140 countries, resulting in the protection of thousands of rare plant and animal species and millions of acres of wildlands.

Conceived at the 1972 Stockholm Conference and created the same year, the **United Nations Environment Programme** is the environmental conscience of the UN. Its primary function is to motivate and inspire, and to raise the level of environmental action and awareness at all levels of society worldwide. UNEP coordinates the environmental activities of all the UN agencies and works to win the cooperation and participation of governments, the international scientific and professional communities, and non-governmental organizations. UNEP is staffed by nearly 200 professionals, and its activities include a program of global environmental quality monitoring and management, environmental law, public information, education and training, an International Register of Potentially Toxic Chemicals, a worldwide information network (Infoterra), the Regional Seas Programmes, and a network of environmental treaties and conventions negotiated under the auspices of UNEP (the Montreal Protocol, Basel Convention, Convention on International Trade in Endangered Species [CITES], Biodiversity Convention, and Climate Change Convention).

The **World Bank** is an international lending institution working to improve living conditions in developing nations. Founded in 1944, the Bank is owned by more than 174 member countries and functions as a large cooperative in which members are shareholders. Responding to the concerns of its members, the Bank began to integrate environmental concerns into its work in 1987. In 1989, the Bank initiated environmental assessments of its projects to detect and remedy environmental problems. That same year, the Bank began involving government officials, academic experts, non-governmental organizations, and international agencies in drawing up national environmental action plans for all borrowing countries. In fiscal year 1992, the Bank disbursed 19 environmental protection loans totaling $1.2 billion. Fully half of the Bank's projects now have environmental components. The Bank also administers the Global Environment Facility (GEF) in cooperation with UNEP and the United Nations Development Programme. GEF is a program to help developing nations manage environmental problems that transcend international boundaries.

Contents

Preface

As THE 20th century ends, the Cold War that drained so much of the world's resources is finally over. Although we are by no means free of political strife and there is deepening concern about nuclear proliferation, the specter of nuclear holocaust enveloping our entire planet has all but vanished. For this, all humankind can celebrate.

Nonetheless, the joy of celebration dims as we realize that the Cold War diverted attention from global threats even more grave than nuclear war. In trying to fulfill our needs, as individuals and nations, we have been ruining our home. The Earth's living systems are showing unmistakable signs of breaking down: Biological diversity is decreasing sharply and our planetary metabolism is being pushed ever further out of equilibrium. How to accommodate our material needs and growing desires while not degrading the life-support systems of a world that was not designed to accommodate billions of us is *the* greatest challenge facing the human species.

In 1989, the Center for Marine Conservation (CMC) joined a large group of international organizations under the leadership of the World Resources Institute, the World Conservation Union (IUCN), and the United Nations Environment Programme (UNEP), organizations with the vision and courage to assemble a plan to save life on Earth, a *Global Biodiversity Strategy* (WRI et al. 1992). It became clear, however, that the GBS could not possibly give enough treatment to the distinctive conservation needs of life in oceans, coastal waters, and estuaries to help leaders decide how to protect the sea's living systems. As a result, CMC, IUCN, World Wildlife Fund (WWF), UNEP, and the World Bank have assembled as its companion document *Global Marine Biological Diversity: A Strategy for*

Building Conservation into Decision Making (*The Strategy*), to focus on ways to save, study, and use the sea sustainably.

In assembling *The Strategy*, CMC staff members traveled to forums in Australia, the Federated States of Micronesia, the USA, the Dominican Republic, Costa Rica, Venezuela, and Brazil to consult with experts in the natural and social sciences, conservation, industry, and government on the outline and text of this document, and corresponded with people world-wide. More than 100 authors—marine biologists, oceanographers, econo-mists, attorneys, government officials, and environmentalists—drafted parts of *The Strategy*, and hundreds of people were asked to review the text, with participants from more than 40 nations. Rather than presenting the views of one profession, one sector, or one nation, CMC sought to present the most up-to-date information and something approaching a consensus of the wisdom of the world's best thinkers in marine conserva-tion on how to meet the challenge of conserving the living sea. Our intended audience is the decision makers in governments, industries, funding institutions, and environmental organizations who most directly influence the health of the oceans—in other words, the people with the greatest interest in, and responsibility for, saving, studying, and using the wealth of life in the sea. We also hope to provide a foundation in marine conservation for the people who are training to be the decision makers of tomorrow.

This document differs from the *Global Biodiversity Strategy* in that it contains more background information (Chapters 1 through 8) and less specific prescription (Chapter 9) because the need for conserving life in the sea and the principles for doing so are far less appreciated than for the land; marine conservation lags terrestrial conservation by roughly two decades. Even a book-length document, however, cannot delve very deeply into a topic as broad and complex as marine biodiversity. Further-more, although CMC has made every effort to make this a truly global *Strategy*, a disproportionate amount of information comes from the USA. Given a few more years, this document undoubtedly would be better . . . but the health of the marine environment would be even worse, and our options for curing the ills would be even fewer. *The Strategy*, therefore, is a first attempt to lay out basic principles and recommendations for people whose decisions affect the health of the seas.

Of course, understanding is of little value unless it is a prelude to action. As the UN Conference on Environment and Development showed, there is an urgent need for decision makers worldwide to build networks to put these principles into action. It is now clear that the threat to the biological integrity of our planet is an unequaled emergency: We have the whole world in our hands. We can destroy it or coexist on it.

If this document and others like it inspire our leaders to protect, study, and sustainably use biological diversity, people will continue to benefit from the products and services of life on Earth. There is an encouraging precedent for success: Our species managed to end the Cold War before we destroyed ourselves. Now we must work together to end the destruction of our planet, lest we follow a different path to the same fate.

*   *   *

This is Contribution #2 in the Center for Marine Conservation's Marine Conservation Biology Series.

ACKNOWLEDGMENTS

First, the Center for Marine Conservation owes appreciation to the partner institutions that are co-sponsoring this document: the World Conservation Union, World Wildlife Fund, the United Nations Environment Programme, and the World Bank. They are vitally important to meeting the challenge of saving, studying, and sustainably using the diversity of life in the sea.

Just as the sun provides the energy that sustains marine organisms and ecosystems, the following organizations believed enough in this project to provide the funding that sustained it: the Surdna Foundation, the W. Alton Jones Foundation, the Educational Foundation of America, the David and Lucile Packard Foundation, the Beneficia Foundation, the Heart of America Fund, the Curtis and Edith Munson Foundation, the C. S. Fund, the Marcia Brady Tucker Foundation, and the US National Oceanic and Atmospheric Administration. Their investment in a broader understanding of marine life will pay dividends worldwide for many years to come.

Five other institutions deserve special thanks for generously providing facilities or the wisdom of their staffs to this project: World Resources Institute, the National Museum of Natural History of the Smithsonian Institution, the Biodiversity Support Program, the Consultative Group on Biological Diversity, and the University of Washington's Friday Harbor Laboratories.

The task of conserving life in the world's oceans is far greater than any individual, any institution, any nation. A staggering number of people from more than 40 nations drafted or reviewed sections of this document or provided information, insights, or other assistance crucial to its completion. To them I owe my deepest gratitude. They include: Janet Abramovitz, Soenartono Adisoemarto, Suraya Afiff, Tundi Agardy, Martin Albert, Anderson Rocha de Albuquerque, Lewis Alexander, Melody Allen, Anders Alm, Patricia Almada-Villela, Lee Alverson, Orane Alves, Elmo da Silva Amador, Nathalie Ames, V. R. Anand, Ivonne Arias, Oscar Arias, William Archambault, Margarita Astrálaga, L. F. Awosika, William Baker, George Balazs, Charles Barber, Mary Barber, R. S. K. Barnes, Dinah Bear, Pierre Béland, Matt Berlin, Jody Berman, Nora Berwick, Charles Birkeland, Richard Bishop, Chris Bleakley, Patricia Bliss-Guest, Mark and Sharon Bloome, Dee Boersma, James Bohnsack, Christopher Bolch, Idelisa Bonnelly de Calventi, Ann Bowles, A. Bradbury, Per Brinck, James Broadus, Jon

Brodie, Wallace Broecker, Hooper Brooks, Gardner Brown, Robert Burg, William Burke, Richard Burroughs, Ian Burston, Dave Butcher, Faith Campbell, Tom Campbell, Pierre Campredon, Dan Cao, James Carlton, Chad Carpenter, John Carr, Jeffrey Carrier, Jodi Cassell, Gonzalo Castro, John Catena, Billy Causey, John Chapman, Aldo Chircop, Colin Clark, James Coe, Theo Colborn, Bruce Collette, Rita Colwell, David Cottingham, Don Cowie, Gordon Cragg, Wendy Craik, Michael Crosby, Larry Crowder, Gustavo Cruz, Flinn Curren, Boyd Curtis, Herman Daly, Paul Dayton, Carola de Boulloche, Carlos de Paco, William Denison, Megan Dethier, Nora Devoe, Antonio Carlos Diegues, Ian Dight, Zena Dineson, Baba Dioum, Ana Dittel, Peter Douglas, François Doumenge, Patrick Dugan, David Duggins, Donna Dwiggins, Bob Earll, Patricia Edgerton, James Edwards, Stephen Edwards, Charles Ehler, David Ehrenfeld, Mark Eiswerth, Danny Elder, Lucius Eldredge, Richard Emlet, Jack Engle, Kevin Erwin, Ron Etter, Kristian Fauchald, William Fenical, Gary Fields, Peggy Fischer, Kathy Fletcher, David Fluharty, Nancy Foster, Sarah Fowler, William Fox Jr., Robert Francis, Rodney Fujita, Paul Gabrielson, Rodrigo Gámez Lobo, Richard Gammon, S. M. Garcia, Emily Gardner, Gudrun Gaudian, Mike Gawel, Janet Gibson, Kristina Gjerde, Lynda Goff, Arturo Gómez-Pompa, Robert Goodland, William Gordon, Thomas Goreau, Judith Gradwohl, Gary Graham, J. Frederick Grassle, Andrew Griffel, Charles Griffiths, Richard Grigg, Ted Grosholz, Dennis Grossman, Samuel Gruber, Susan Gubbay, Richard Gustafson, Arlin Hackman, Scott Hajost, Lynne Hale, Andrew Hamilton, Christopher Haney, Arthur Hanson, Richard Harbison, Larry Harris, Leslie Harroun, Gary Hartshorn, Carl Haub, Jon Havenhand, Mark Hay, Raymond Hayes, Joel Hedgpeth, C. Wolcott Henry, Sailas Henry, Marc Hershman, Robert Hessler, Charles Higginson, Elaine Hoagland, Robert Hofman, Martin Holdgate, Marjorie Holland, Paul Holthus, Nancy Hotchkiss, A. C. Ibe, Sixto Inchaustegui, Mohamed Isahakia, David Jablonski, Jeremy Jackson, Marcia Jacobs, Daniel Janzen, Alain Jeudy de Grissac, Robert Johannes, Connelly Johnson, Dick Johnson, Natalie Johnson, Carol Adaire Jones, Martin Jones, Peter Jutro, John Karau, James Karr, Leslie Kaufman, Graeme Kelleher, Janet Kelly, Richard Kenchington, Fred Kern, Gene Kersey, Susan Kidwell, Cecily Kihn, Jiro Kikkawa, Yumi Kikuchi, Lee Kimball, Aaron King, Ron Kipp, Jean Kirby, Robert Knecht, Roger Kohn, Joan Koven, Chadwick Kumpe, Rosa Lamelas, Justin Lancaster, Wilson Laney, Edward LaRoe, Pierre Lasserre, James Lawless, Gerry Leape, Stephen Leatherman, Bruce Leighty, Pedro León, Erkki Leppäkoski, Anne-Cathérine Lescrauwaet, Robert Lester, Colin Limpus, David Livengood, Paul Loiselle, Ira Lowenthal, Jane Lubchenco, Tashiro Ludwig, Carl Lundin, Indrani Lutchman, Richard Lutz, Patricia Mace, Rajeshwari Mahalingam, Linda Mantel, Albert Manville, James Maragos, Karen Martin, Joan Martin-Brown, Merri Martz, Don E. McAllister, Tim McClanahan, Maxine McCloskey, John McCosker, Margaret McMillan, Jeffrey McNeely, Pam McVety, Gary Meffe, Richard Meganck, Gray Merriam, Marshall Meyers, Lori

Michaelson, Antonio Mignucci Giannoni, Edward Miles, Julie Miller, Kenton Miller, Marc Miller, Richard Miller, Chris Morganroth, Peter Moyle, Jim Muldoon, Bruce Mundy, Kirk Munro, Katherine Muzik, J. Peterson Myers, A. Moses Nelson, Russell Nelson, Robert Netting, David Newman, Canice Nolan, James Norris, David Norriss, Brad Northrup, Bryan Norton, Steve Norton, Reed Noss, Hideo Obara, Craig O'Connor, Mary O'Donnell, Waafas Ofosu-Amaah, John Ogden, Reuben Olembo, Stephen Olsen, Molly Olson, Robin O'Malley, Makoto Omori, Jelili Omotolla, Daisuke Onuki, Suzanne Orenstein, Gordon Orians, Richard Orr, Rosario Ortiz Quijano, Julie Packard, Robert Paine, Andrew Palmer, Steve Parcells, Dave Parker, Susan Payne, Sam Pearsall, David Penrose, Salas Peters, Charles Peterson, Melvin Peterson, Ellen Pikitch, Stuart Pimm, Phil Pister, Stephen Polasky, Jan Post, Thomas Quinn, George Rabb, Gilbert Radonski, Katherine Ralls, Omar Ramirez, John Randall, Carleton Ray, Bill Raynor, Colin Rees, Walter Reid, Peter Reijnders, Renate Rennie, Artemas Richardson, Geoff Rigby, Richard Riskin, John Robertson, Robbie Robinson, Alejandro Robles, Caroline Rogers, Holmes Rolston III, Perran Ross, James Rote, Janis Roze, Laura Rubin, Dennis Russell, John Ryan, Robert Ryan, Andres Marcelo Sada, Carl Safina, Saul Saila, Peter Sale, Rodney Salm, Rudolfo Sambajon, Vicente Sánchez, Eleanor Savage, Lori Scarpa, Rudolf Scheltema, Peter Schröder, Rick Schwabacher, Jeff Schweitzer, Tucker Scully, Jeanne Sedgwick, Barbara Shapiro, Steve Sheinkin, John Shores, Brett Shorthouse, Caroly Shumway, Gail Siani, Steven Sillett, François Simard, Carl Sindermann, Fred Sklar, Edward Skloot, Michael Slimak, Cliff Smith, Theodore Smith, Tom Smith, S. H. Sohmer, Otto Solbrig, Michael Soulé, Frances Spivy-Weber, Ted Stanley, Craig Staude, Bruce Stein, William Stephenson, Carolyn Stewart, Michael Stewartt, Gregory Stone, Richard and Megumi Strathmann, Richard Strickland, Maurice Strong, Kathleen Sullivan, Tim Sullivan, Michael Sutton, Judith Swan, Timothy Swanson, Steve Swartz, Frank Talbot, Charles Tambiah, Mervyn Tano, David Tarnas, Phil Taylor, Martin Teitel, Patricia Tester, Janice Thompson, Boyce Thorne-Miller, Ramona Tibando, Clem Tisdell, Timothy Titus, Roxanne Turnage, Monica Turner, John Twiss, Peter Tyack, Joseph Uravitch, Georgia Valaoras, Sally Valdes-Cogliano, Dan van R. Claasen, Jack Vanderryn, Usha Varanasi, Michael Vecchione, Kristin Vehrs, Ana Marie Vera, Philomène Verlaan, Geerat Vermeij, Vance Vicente, Estelle Viguet, Tom Wathen, Michael Weber, Miranda Wecker, James Weinberg, Judith Weis, Donna Weiting, Katharine Wellman, Susan Wells, Tim Werner, John West, Colburn Wilbur, Dean Wilkinson, Paul Williams, Dennis Willows, Rob Wolotira, Robert Worrest, Gregory Wray, Alexei Yablokov, Oran Young, Douglas Yurick, Hamdallah Zedan, and Edward Zillioux. Undoubtedly there are others who helped this project whom I have unintentionally omitted. To them, too, I owe my appreciation.

The present and former staff members, board members, and volunteers at the Center for Marine Conservation made this document come to life. I owe special thanks to Deborah Crouse, Anne Dettelbach, Minette Johnson, Cynthia Sarthou, Jan Sechrist, and Chantal Stevens for their long hours of research, discussion, writing, editing, and exceptional teamwork; to David Allison, Natasha Atkins, George Barley, Rose Bierce, Amie Bräutigam, Andrea Brock, William Brown, Ray Burgess, David Challinor, Heather Dine, Cathy Dirksen, Marydele Donnelly, Sylvia Earle, Naomi Echental, Tim Eichenberg, Kirsten Evans, Sonja Fordham, Suzie Fowle, Michael Frankel, Nancy Goodman, Laura Hamilton, Burr Heneman, Suzanne Iudicello, Heidi Knapp, David Knight, Gary Magnuson, Linda Maraniss, Jennifer McCann, Alison Merow, Thomas Miller, William Mott, Irene and Ketzel Norse, Kathy O'Hara, Feodor Pitcairn, Edward Proffitt, Michael Rucker, Cameron Sanders, Rachel Saunders, Michael Smith, Tricia Snell, Jack Sobel, Roberta Ross Tisch, Jill Townsend, Harry Upton, Craig Vaniman, Ronald Vogel, Sharon Wiley, George Woodwell, Cynthia Yothers, Arthur Young, and Nina Young for serving as authors, reviewers, or sources of information; and, most of all, to CMC President Roger McManus, who conceived this project and whose extraordinary vision, insights, encouragement, and hands-on participation have been integral to it from the beginning.

Finally, I want to thank the people of Island Press for their commitment to helping people live sustainably in harmony with our planet, and for transforming the manuscript into this book with astounding speed, accuracy, and graciousness.

Elliott Norse
Center for Marine Conservation

CONSULTATIONS FOR *THE STRATEGY*

To develop the outline and the text of *The Strategy*, Center for Marine Conservation staff members consulted with experts at forums including:

Scientific Workshop on Marine Biological Diversity, National Museum of Natural History of the Smithsonian Institution, Washington, D.C. (USA), October–November 1990

IUCN General Assembly and Species Survival Commission Triennial Meeting, Perth, Western Australia (Australia), November–December 1990

Congress on Biodiversity in the Caribbean, Santo Domingo (Dominican Republic), January 1991

Second Woods Hole Workshop on Marine Biological Diversity, Woods Hole, Massachusetts (USA), February 1991

Meetings with Federated States of Micronesia and Pohnpei marine resources officials, Palikir and Kolonia, Pohnpei (Federated States of Micronesia), March 1991

US National Report Roundtable for UNCED, San Francisco, California (USA), June 1991

Annual Meeting of the American Society of Ichthyologists and Herpetologists, New York, New York (USA), June 1991

Workshop on Information for Decision Makers: How to Mobilize a Developing Nation's Biotic Wealth, National Institute for Biodiversity (InBio)/World Resources Institute, Heredia (Costa Rica), June 1991

North American Regional Biodiversity Consultation, World Resources Institute, Keystone, Colorado (USA), July 1991

Globescope Conference on Latin America and the Caribbean, Miami, Florida (USA), October–November 1991

Town Meeting and Public Hearing on the Earth Summit (Sustainable Development: A Northwest Perspective), United Nations Association,

Seattle Metropolitan Chapter, Seattle, Washington (USA), November 1991

National Forum on Ocean Conservation, Smithsonian Institution, Washington, D.C. (USA), November 1991

IUCN IVth World Congress on National Parks and Protected Areas, Caracas (Venezuela), February 1992.

Fourth Preparatory Committee meeting for the UN Conference on Environment and Development, New York, New York (USA), March 1992

Global Forum, United Nations Conference on Environment and Development, Rio de Janeiro (Brazil), June 1992

Symposium on Biodiversity in Managed Landscapes: Theory and Practice, Sacramento, California (USA), June 1992

Meetings with Great Barrier Reef Marine Park Authority officials, Townsville, Queensland (Australia), October 1992

Workshop on Nonindigenous Estuarine and Marine Organisms, Seattle, Washington (USA), April 1993

Executive
Summary

SINCE the UN Conference on the Human Environment in Stockholm in 1972, it has become increasingly clear that economic advances cannot be sustained unless we maintain the health of our environment. The cumulative activities of billions of people are now affecting the entire planet, including the 71 percent of it covered by oceans, coastal waters, and estuaries. As a result, the sea, like the land, is increasingly losing biological diversity. Stopping this loss is crucial to our well-being and survival.

Hundreds of experts—including more than 100 authors—from more than 40 nations have contributed to *Global Marine Biological Diversity: A Strategy for Building Conservation into Decision Making (The Strategy)*. Written as a companion document to the *Global Biodiversity Strategy* (WRI et al. 1992), its intended audience is decision makers in coastal countries, the people who have the greatest influence on the health of the sea. Its purpose is to provide the most up-to-date information and the wisdom of world experts to leaders and managers in governments, industries, international governmental organizations, and non-governmental organizations who are responsible for saving, studying, and using the wealth of marine life. *The Strategy's recommendations are given in Chapter 9.*

Conserving biological diversity in the sea has been even more neglected than that on land, yet the sea is rich in genetic, species, and ecosystem diversity. Recent research suggests that diversity in the marine realm, as with that on land, has been greatly underestimated. In fact, the sea is far richer in major groupings (phyla) of animals than the land; nearly half of animal phyla occur only in the sea.

The marine realm provides products and services that are important or essential to human existence. A large share (in some countries more than half) of the animal protein that people eat comes from the sea. The sea is the most promising source of new antiviral and antitumor medicines. Coral reefs and mangrove forests form bulwarks that protect coastal communities along many low-lying tropical coasts against storm surges. An indispensable marine ecosystem service is the marine "biological pump," by which the living ocean decreases the atmospheric concentration of the most important greenhouse gas, carbon dioxide. Many seminal ecological ideas, such as the keystone species concept, have come from the marine environment and have profound implications for conservation and sound resource management.

Despite their importance to us, humankind is destroying marine populations, species, and ecosystems. Leading marine scientists have concluded that *the entire marine realm, from estuaries and coastal waters to the open ocean and the deep sea, is at risk.* The damage is greatest near dense concentrations of humans, where our activities are most intense, but no place in the ocean is so remote that it has not been touched by human activities.

In an immediate sense, we harm marine biodiversity in five major ways:

1) overexploiting living things;
2) altering the physical environment;
3) polluting the sea;
4) introducing alien species; and
5) adding substances to the atmosphere that increase ultraviolet radiation and alter climate.

The effects of overexploitation and pollution are far better known than those of the other threats. Most marine species that have been driven to extinction in modern times were victims of overexploitation, and most fisheries are locked in boom-and-bust cycles. Toxic substances and excessive nutrients contaminate or overnourish many of the world's coastal ecosystems, especially estuaries, lagoons, and bays. But even experts have usually overlooked the other major threats. The physical effects of trawling, dredging, and coastal construction damage or destroy marine ecosystems; activities (such as dam-building, freshwater diversion, farming, and logging) that do not occur in the sea but which alter the movement of species, fresh water, sediments, and nutrients between the land and sea, are other underappreciated physical threats to marine biodiversity. Alien species that people have transported into regions where they do not historically occur, whether intentionally or in the ballast tanks of ships, are a growing plague that upsets ecological relationships that evolved over aeons. Additions of certain trace gases to the atmosphere leading to increased UV radiation at the sea surface and the many effects of global climatic change loom as major threats in the 21st century. Together, such human activities are dramatically increasing the intensity, pace, and kinds of environmental change, placing severe stresses on living things. We have already ruined once-bountiful fisheries, eliminated vital ecosystem services, and diminished the abundance or caused the extinction of species ranging from great whales to humble invertebrates.

Biological impoverishment is the inevitable consequence of the ways in which our species has used and misused the environment during our rise to dominance, and has extended from the land and fresh waters into the sea. Five root causes underlie the threats to marine life:

1) There are too many people.
2) We consume too much.
3) Our institutions degrade, rather than conserve biodiversity.
4) We do not have the knowledge we need.
5) We do not value nature enough.

Unless these underlying causes are addressed, efforts to diminish overfishing, stop pollution, or to maintain the composition of the atmosphere can only delay the inevitable.

Conserving marine biological diversity requires us to understand that the sea has properties that distinguish it from the terrestrial realm, including fluid boundaries that shift on all time scales, buoyancy, vastly larger size, three-dimensionality, and the prevalence of planktonic dispersal. Although there is a growing movement to protect, understand, and sustainably use the sea, there are also significant obstacles to marine conservation. Some are scientific or technical (our ignorance of the sea's value and vulnerability to us and the fact that what is known is not available to all who need it). Other impediments are cultural (the replacement of diverse human cultures adapted to living sustainably in diverse coastal ecosystems by a wasteful, consumption-oriented world culture), economic (the "Tragedy of the Commons," intergenerational inequities, undervaluing of life in the sea), political (North/South friction, national sovereignty, fragmented decision making), and legal (gaps and overlaps in jurisdiction, placing the burden of proof on those who would conserve marine life).

There are alternatives to degrading the sea, but they require changes in the ways that we think and act. Focusing solely on species—whether it is on maximizing biomass yield or preventing extinction—has proven to be insufficient in the sea, as it is on land. Ecosystem protection and management are essential complements to species protection and management. Critical marine areas that merit the highest priority for protection include ones with high diversity, high endemism, or high productivity, spawning areas, nursery areas, and migration stopovers and bottlenecks. Large marine ecosystems based on biogeographic provinces offer a promising way to manage the ocean holistically.

Stopping the human-caused loss of genes, species, and ecosystems is essential but not sufficient. Rather, the goal should be to ensure that living things do not become endangered, that is, to *maintain the integrity of life*. This means keeping not only the *parts* (genes, species, ecosystems), but also the *processes* that generate and maintain the parts, the ecological connections among living things.

Saving our planet is not a luxury that can be left to someone else. It is an imperative that requires us to make a fundamental change in our course by building conservation into the decision-making process. Decision makers need to have options that permit sustainability and rewards for choosing them. At present, individuals and institutions are generally free to act until it is proven that their actions are overwhelmingly harmful. The burden of proof needs to be shifted onto those whose acts would diminish marine biodiversity. Under the do-no-harm or precautionary principle, polluters, developers, and other users must demonstrate that their activities are *not* harmful to the sea before engaging in them. By shifting the burden of proof, conservation and management become *proactive*, not *reactive*.

There are tools that decision makers can use to encourage protection and sustainable use of the sea. These include: expanding the knowledge base through biological inventories, research, monitoring, training of professionals, and public education; planning (environmental impact assessment, action plans, and integrated area management); regulating the threats to marine species and ecosystems; using powerful economic tools such as assigning usage rights, valuation, and market forces for conservation; establishing protected areas; actively manipulating (through restoration or mitigation) populations and areas; and ensuring active involvement of citizens in government decision making. Decision makers in governments, industries, conservation organizations, and funding institutions who can use these tools will lessen harm to marine species and ecosystems, thereby improving the well-being of all who depend on products and services from the sea.

Public education, in the classroom and through electronic and print media, undergirds all marine conservation; informed people opt for sustainability. Decision making needs firm bases in modern scientific information and—where it has not been lost—traditional knowledge. Taxonomy, biogeography, ecology, and other established but undersupported sciences need far greater support to inventory, research, and monitor marine species and ecosystems; long-term studies and networks that transcend national and North/South boundaries to share information rapidly are especially important. Support is vitally needed to develop disciplines that have not yet been born or are in their infancy: marine conservation biology, marine landscape ecology, marine restoration ecology, and ecological economics. IUCN and international scientific organizations can play a central role in helping governments to solve conservation problems, and merit substantially increased funding.

Integrated area management in coastal zones and offshore areas, especially where it reflects biological boundaries and operates at a large scale, is the most effective means of countering fragmented decision making, but it needs broad public participation and overriding legal authority to regulate activities that can harm the sea.

Existing institutions and instruments have had limited success in conserving marine life. The division of the world into competing geopolitical blocs, nations, and economic sectors, none of which reflect oceanographic and marine ecological patterns and processes, impedes efforts to maintain the health of the sea. To conserve species and ecosystems that transcend the borders we create requires much stronger commitment and cooperation at all levels, from localities and nations to regional and global organizations. Institutions at all these levels have crucial and complementary roles in marine conservation. Local and national governments are best able to deal with the particular needs of their people. Global institutions can best establish frameworks and deal with activities that are inherently worldwide in scope. But regional institutions whose coverage corresponds to marine species' distributions and biogeographic provinces are probably best able to deal with many human activities in the sea. Nations and local governments need to reexamine long-standing views about sovereignty and reconfigure themselves to cooperate effectively in the task of maintaining life on our planet. This, in turn, requires us to modify practices that were born at a time when far fewer people and far weaker technologies meant far less environmental impact. Rich nations have a special responsibility to help poor nations to find sustainable solutions to their problems. But international funders need to move beyond merely withholding support from environmentally destructive projects to actively supporting ones that benefit the marine environment. The Global Environment Facility, in particular, should be strengthened and extended.

Non-governmental organizations are integral to marine conservation because they help governments counterbalance the influence of economic sectors that otherwise dominate local and national government decision making and thereby undermine measures to protect the sea even at the regional and global level. Governments that can limit the influence of major economic sectors, and that can keep their citizens informed and actively involved in decision making, are most likely to fashion enduring systems of marine biodiversity protection and management.

Global Marine
Biological Diversity

CHAPTER ONE

Conserving the Living Sea

IN 1741, the crew of a Russian ship stranded on Bering Island in the cold North Pacific discovered huge sea creatures in the surrounding waters. Unlike the seals and whales that the sailors knew, these four- to ten-ton mammals grazed the abundant seaweeds like cattle. The hungry men devised a method of killing the beasts, and found the meat and fat delectable. On reaching safety, they told others of their good fortune. More ships came, taking advantage of this bounteous food source, until, in 1768, sailors killed the last Steller's sea cow (*Hydrodamalis gigas*) (Figure 1-1). From its discovery by Western civilization to its extinction took only 27 years (Reynolds and Odell 1991).

What was lost? Steller's sea cow was a magnificent species, possibly the largest herbivore in the world, one shaped by the same forces that shaped our own species. It was one of only two members of its family; the other, the dugong (*Dugong dugon*), a smaller sea cow of the tropical Indian and Pacific Oceans, is now endangered in most of its range. Steller's sea cow apparently was ecologically important. In prehistoric times, it had ranged widely in the North Pacific, and its grazing probably played a major role in kelp forest ecosystems (Dayton 1975). It was a species that could have benefited humans as a resource. Its gut microorganisms could well have been used to generate fuel gas using seaweed as a feedstock. And, as a source of food for humans, Steller's sea cow was unique: Not only was it delicious, but it grazed marine pastures that cattle and sheep cannot. Whether our interests are ethical, ecological, or economic, the extinction of Steller's sea cow was a tragedy.

Although this drama unfolded two and a half centuries ago, it cannot be dismissed as the kind of error humankind made only in the past, before the genesis of modern conservation. Rather, as the 20th century ends, we

Figure 1-1. Steller's sea cow. European hunters drove this gigantic seaweed-grazing mammal to extinction 27 years after its discovery in the North Pacific Ocean. Loss of genes, species, and ecosystems is a rapidly worsening problem in the sea, as on land.

are playing countless variations on the same theme, in the tropics and the cold regions, in countries rich and poor, on land, in fresh waters, and in the sea. The wealth of life on Earth is now being—or is about to be—lost at a rate exceeding any in 65 million years, since a global disaster (probably the impact of a comet or asteroid) killed off the dinosaurs and vast numbers of other species.

This present-day global mass extinction event differs from those of the past: It is not due to the inevitable momentum of a mindless mass of rock. Rather, it is because an intelligent species threatens life on our planet, a species able to recognize its impact and change its course.

The ethical implications of this mass extinction are enormous and must not be ignored. Religions, philosophies, and laws honored throughout the world make humans responsible for actions affecting other living things. Humankind, the other animals, plants, and microorganisms share a common ancestry on the tree of life that goes back 3.5 billion years; the same processes that created us created them as well. But in a world in which hunger, disease, ethnic strife, social upheaval, political oppression, and extreme disparity in wealth plague our species, some people might think that the loss of biological diversity is less important than pressing human needs, such as economic development and national security.

This attitude ignores one fundamental fact: Humans are utterly dependent on other living things. Every breath we take, every bite of food we eat, and every drop of water we drink comes from the diversity of life. Living things are our resources and the life-support systems that maintain conditions in which we can survive and prosper. Living things are the basis of economic development and national security. Without them, we would all soon be starving, thirsting, roasting, choking for breath, and drowning in our wastes.

Our species now faces its greatest challenge: how to provide for the needs of ever-growing numbers of people without degrading the biological, chemical, and physical systems essential to our survival and well-being. Since the first United Nations (UN) Conference on the Human Environment in Stockholm in 1972, it has become increasingly clear that economic advances cannot be sustained unless we maintain the health of our environment. Stopping the loss of biological diversity is absolutely essential if we are to meet human needs.

For the most part, attention to the loss of biological diversity has focused on the land, particularly the world's tropical rainforests. These ecosystems are exceptionally rich in species and are indeed imperiled, but they are far from the only ones in desperate need of attention. Other terrestrial ecosystems, including tropical dry forests, temperate coniferous forests, grasslands, and mediterranean shrublands, merit far more attention from scientists, citizen activists, and decision makers. Freshwater wetlands, the great floodplain rivers, and the great African Rift lakes do as well. And so does the sea.

At first, the need for marine conservation might seem strange. Humans do not yet live beneath the waves, and the vastness of the marine realm makes it appear inexhaustible and invulnerable. But that is how we once regarded the land. Since the Stockholm Conference, however, we have increasingly seen that the activities of billions of us are affecting not only the land, but the entire planet, including the 71 percent of it covered by oceans, coastal waters, and estuaries.

Reports from newspapers, radio, and television scarcely hint at the magnitude of the problem. They tell of growing numbers of shellfish beds closed due to pollution, of North Atlantic bluefin tunas (*Thunnus thynnus*) being fished to commercial extinction, of massive oil spills from tanker accidents. In these cases, the immediate causes are clear. But they also tell of Bangladeshis suffering from lowland flooding, thousands of dead seals washing up on North Sea beaches, blooms of toxic phytoplankton, and episodes of coral bleaching on the increase throughout the world's oceans. Are these natural phenomena, or do they reflect the impact of humans on the sea? Our inability to provide definitive answers highlights how much is *not* known about marine species and ecosystems, including how we affect them.

Scientists are continually astounded by what is not known. Until 1938, coelacanths, a group of fishes related to the ancestor of amphibians, reptiles, birds, and mammals, were thought to have died out 70 to 80 million years ago. The discovery of one species (*Latimeria chalumnae*) off South Africa and the Comoro Islands in the Indian Ocean was one of the great biological finds of the century (Thomson 1991). Until 1977, no one suspected the existence of the most unusual ecosystems on Earth, the deepsea hydrothermal vents, whose spectacular animal communities are the only ones known not to depend on plant photosynthesis (Box 1-1). Not until 1989 did scientists realize that viruses play a far greater role in marine food webs than previously thought (Colwell and Hill 1992). And it was only last year that scientists (Grassle and Maciolek 1992) estimated that the deep sea could harbor 10 million species that have not yet been described and named, a diversity of species roughly comparable to that of tropical forests. The sea is far richer biologically than anyone had thought, and we still have much to learn about its patterns and processes.

Although humankind depends on the oceans in many ways, there have been few intensive studies of marine life. We know far less than is necessary to protect and use marine resources sustainably. Nonetheless, there *is* enough knowledge about threats to the sea and ways to diminish them to make major improvements now. The key question is whether individuals, groups, and nations can share this knowledge and use it locally, nationally, and internationally in ways that will benefit people on a sustainable basis.

As a companion document to the *Global Biodiversity Strategy* (WRI et al. 1992), *Global Marine Biological Diversity: A Strategy for Building Conservation into Decision Making* (*The Strategy*) offers a comprehensive summary of necessary background information, by examining the diversity of life in the sea and its importance to humankind, the similarities and differences between conservation on land and in the sea, the threats to marine biodiversity, and reasons why marine conservation and management efforts have not stemmed the loss of biodiversity. It then offers a goal

BOX 1-1. Biodiversity at deep-sea hydrothermal vents

On an expedition near the Galápagos Islands in 1977, scientists discovered hot water issuing from deepsea springs. Surrounding these hydrothermal vents were remarkable communities of previously unknown animal species that live in a way unlike any that scientists had ever seen.

Scientists have since learned that hydrothermal vents occur along tectonic spreading ridges on the seafloor in all the oceans. Since the initial discovery of the vents at 2,500 meters (8,200 feet) in the East Pacific, dense communities of animals have been reported from undersea hot springs on both sides of the Pacific and on the Mid-Atlantic Ridge (Grassle 1986). Of 236 vent species listed by Tunnicliffe (1991), 223 are new to science. These belong to upward of 100 new genera and at least 22 new families.

Perhaps the most remarkable thing about the vents' animals is their source of energy. Independent of sunlight and plant photosynthesis, they depend on chemosynthesis by bacteria. The bacteria make carbon compounds using energy from reduced sulfur compounds in fluid issuing from fissures and chimneys on the seafloor. This primary production supports the large biomass of invertebrates and fishes at hydrothermal vents. Within invertebrates such as the tube-worm *Riftia pachyptila*, the mussel *Bathymodiolus thermophilus*, and the large white clam *Calyptogena magnifica*, these bacteria live as mutualists inside their hosts' cells and provide their hosts with food.

The animals of the hydrothermal vents have a high degree of endemism. At least 158 of the vent species are known from only a single hydrothermal vent field, and the overwhelming majority of the others are endemic to a single spreading ridge. In general, the closer the vent fields, the more similar their faunas.

Hydrothermal vents are isolated and short-lived—issuing heated water for perhaps a few decades—hence local vent populations appear to be short-lived, disappearing when vents stop discharging warm water and colonizing new vents as they begin discharging. Although population densities and endemism are high at hydrothermal vents, species diversity is low compared with many other deepsea ecosystems (Tunnicliffe 1991; Grassle 1989), but consistent with other deepsea ecosystems that are disturbed often.

---

whose achievement would allow humankind to save, study, and use the seas sustainably, a strategy for meeting that goal, an examination of tools available for conservation and management, and a candid look at the strengths and weaknesses of some existing efforts. Finally, it provides a set of concrete recommendations for actions that can be accomplished in the coming decade at local, national, and international levels that would markedly improve marine conservation and management.

The fossil record suggests that humans originated in Africa a few million years ago, perhaps some distance from the coast, and our dependence on the sea might have been limited to breathing the oxygen created by marine organisms and living in climates shaped, in large part, by marine ecosystems. Even if some humans foraged along the shore and our wastes entered the sea, our limited numbers and technologies prevented significant effect on the sea for our first few million years.

However, since the advent of agriculture, writing, metallurgy, and cities in the last 10,000 years, our populations and technologies have been increasing dramatically. Now billions of us crowd the Earth, mainly in the coastal zone or in drainages that empty into the sea. Moreover, technologies that either intentionally or inadvertently change the environment are increasing at an astounding rate. The oceans are the primary sink for substances we discharge into the atmosphere and waterways. The well-being of humankind increasingly depends on the well-being of the sea. And this connection goes both ways: The health of the sea now depends on what our species does.

Life in the sea is roughly 1,000 times older than the genus *Homo*. Since its origin, marine life has responded to changes: global geochemical changes and drifting continents, cataclysmic volcanic eruptions and asteroid impacts, changes in climate and the evolution of new enemies. Now the sea—indeed, the entire biosphere—faces an unprecedented threat: exponentially increasing, unrelenting stresses from human activities. Virtually all of humankind already suffers from the environmental damage we cause, and great numbers of people die from it, especially in the poorer countries. The acceleration in environmental impacts cannot fail to harm us more and more. If our species can understand both the vulnerability and the resilience of the living sea and the rest of the biosphere, we have an opportunity to act in our own interest and benefit from life's products and services in perpetuity. If we do not—if we fail to relieve the stresses that we impose on the biosphere—the relatively brief story of the human species will almost certainly come to an inglorious end. *The Strategy* is written in the hope that we can act effectively and soon enough to prevent that from happening.

CHAPTER TWO

Marine Biological Diversity: Definition and Importance

B Y EXAMINING the molecular basis of life, scientists have concluded that all known living things share a common origin; there is a fundamental unity of life on Earth. No less remarkable, however, is the diversity of life, that is, biological diversity.

What Is Biological Diversity?

The central idea in the definition of biological diversity most used by conservation biologists and decision makers is that the diversity of life occurs at several hierarchical levels of biological organization. Although biologists have been interested in diversity for decades, the term "biological diversity" (which is sometimes shortened to "biodiversity") appeared in conservation publications only about 1980, and its originators either did not define it (Lovejoy 1980) or defined it inadequately (Norse and McManus 1980). The most widely used definition of biological diversity (Norse et al. 1986) considers three levels: genetic, species, and ecosystem diversity. An alternative scheme (Soulé 1991) defines five levels, splitting genetic and ecosystem diversity into two each. The first is simpler conceptually; the second offers finer resolution. (There could be useful schemes in which genetic and species diversity are divided still further.) The concept that there are different hierarchical levels of biological diversity is so useful as an organizing principle in conservation that it has quickly been adopted worldwide.

Species Diversity

In the three-level definition, the most obvious level is the middle level, species diversity, which nonexperts often incorrectly equate with biological diversity. The number of species varies greatly among higher taxonomic groups, such as families or classes, and among geographic areas. For example, there are far more species of snails (Gastropoda) than of chitons (Polyplacophora). As on land, the sea has far more small species (e.g., diatoms and snails) than large ones (e.g., mangrove trees and sharks). Similarly, there are far more species of marine animals than "plants" (including vascular plants, algae, and photosynthetic cyanobacteria). And, as on land, the diversity of small organisms is much less known than diversity in groups of larger organisms.

Marine species diversity varies enormously at different localities. There are two clear geographic gradients (Briggs 1974). First, as on land, for many groups (e.g., seagrasses, corals, snails, crabs, fishes), although not all (Clarke 1992), diversity is considerably higher in tropical regions than in cooler regions. Exceptions include starfishes and kelps (brown algae in the order Laminariales), which are most diverse in cold temperate Northeast Pacific waters off Canada and the USA. Second, within the tropics, many taxa reach highest species diversity in the Indo-West Pacific Ocean (especially in the area between the Philippines, Indonesia, and northeast Australia); intermediate diversity in the East Pacific and West Atlantic, and lowest diversity in the East Atlantic (for example, see Table 2-1).

Ecosystem Diversity

Ecosystem and genetic diversity—the highest and lowest levels of biological diversity—are less understood by decision makers and the public, but are no less important than species diversity. Ecosystem diversity is not difficult to see. Different physical settings favor very different communities of species. In practice, the physical conditions in ecosystems are so important to the organisms in them that the concept of a community is not a very useful one unless it is considered in the context of ecosystems. Ecosystems differ not only in the species composition of their communities, but also in their physical structures (including the structures created by organisms) and in what the species in their communities do.

The composition, structure, and function of estuarine salt marshes,

TABLE 2-1. Species diversity and endemism of penaeid shrimp in four marine regions

Biogeographic Region	# of Species	# Endemic
Indo-West Pacific	125	124
East Pacific	16	16
West Atlantic	21	18
East Atlantic	16	3

Source: Dall, W., B. J. Hill, P. C. Rothlisberg, and D. J. Sharples. 1990. The biology of the Penaeidae. *Advances in Marine Biology* 27:1–484.

coral reefs, and the sediment plains of the deep sea are very different (Figure 2-1). For example, salt marshes in many regions have high primary production—production of organic carbon-containing compounds from carbon dioxide through photosynthesis—based on high availability of nutrient elements such as nitrogen. Coral reefs, in contrast, maintain high primary production using far scarcer nutrients by cycling nutrients efficiently. Deepsea ecosystems, other than hydrothermal vents, have no primary production at all because they lack sunlight. The pathways of energy flow and the proportions of organisms performing particular functions (such as burrowing) differ markedly among them as well.

As Soulé (1991) points out, ecosystem diversity actually includes several levels of diversity because it occurs not only among different types of ecosystems, but also within types. Estuaries, for example, are very different in Peru, Norway, Saudi Arabia, and Borneo. Even within a single geographic region, estuarine ecosystems having different physical conditions (such as differing amounts or temporal patterns of freshwater input) can have marked differences in composition, structure, and function. There is no universally agreed upon classification system for marine ecosystems, and the population, community, and trophic (food-web) dynamics of most marine ecosystems are scarcely understood. Still, ecosystem diversity in the sea is high, probably higher than on land.

Genetic Diversity

The least visible and least studied level of biological diversity is the lowest level, genetic diversity within species. Each species consists of one or more populations of individuals. Within each population of a sexually reproducing species, individuals are more likely to breed with one another

Figure 2-1. Marine ecosystem diversity. *The sea probably hosts a higher diversity of ecosystems than the land. Salt marshes (top), coral reefs (center), and the deep seabed (bottom) differ markedly in their composition, structure, and function.*

than with individuals in different populations. This often happens because breeding occurs in different locations. For example, green sea turtles (*Chelonia mydas*) from the coastal waters of Brazil and Madagascar have virtually no likelihood of interbreeding.

But individuals need not live far apart to belong to different populations. Indeed, populations can be separated by the timing, rather than the location, of breeding. For instance, chinook salmon (*Oncorhynchus tshawytscha*) might intermingle on their feeding grounds in the North Pacific, but populations segregate and return to particular streams to breed at particular times. There can be spring, summer, and fall populations breeding in the same stream.

Because different populations have limited genetic mixing, they tend to diverge genetically because of mutation, natural selection, and genetic drift. Thus, some populations have specific versions of genes (alleles) that are absent in others. Or alleles that are very rare in one population might be abundant in another. Some of these genetic differences are adaptations that make organisms more likely to reproduce successfully under the specific conditions of their local environment.

Like ecosystem diversity, genetic diversity actually includes more than one level: There is not only genetic diversity among populations but also within them. In a given population, some individuals possess particular versions of genes that others do not. This genetic diversity within populations is the raw material for evolution: Populations with higher genetic diversity are more likely to have at least some individuals that can withstand environmental change and pass on their genes.

The certainty of accelerated environmental change makes conservation of genetic diversity an important conservation goal, both within and among populations, even in widespread species. Genetic differences are also very important in maricultured species being bred for desirable traits, and they provide the basis for the rapidly growing biotechnology industry.

One of the most significant scientific findings about marine biological diversity in recent decades is that organisms that had been considered single species on the basis of form have been found to be clusters of similar-looking "cryptic" or "sibling" species (Grassle and Grassle 1976; Weinberg et al. 1990; Knowlton et al. 1992). On closer examination, it is found that these species have important genetic, behavioral, and even morphological differences. What appear to be discrete populations can, in fact, be distinct species.

Two Other Kinds of Biological Diversity

Although conserving biological diversity at all hierarchical levels is a major advance over protecting only the diversity of species, there are two other useful ways to look at biological diversity that complement the hierarchical approach. One is to consider the diversity of *higher taxonomic groups*. Some people have believed that marine conservation deserves less priority than terrestrial conservation because, until very recently (Grassle and Maciolek 1992), the sea was thought to host fewer species than the land. But at a more basic taxonomic level—the level of fundamental types of body plans or phyla—marine animals display much greater diversity (or disparity, as Gould [1989] and others are now calling this more fundamental kind of diversity) than terrestrial ones (Ray 1988).

Most of the phenomenal species diversity on land comes from the phylum Arthropoda. Current estimates suggest that the millions of species in the arthropod class Insecta outnumber the species in all other land animal phyla. Insects, however, have fared poorly in the sea, whereas many taxa absent from the land or fresh waters occur in the sea. Of the 33 animal phyla listed in Margulis and Schwartz (1988), 32 occur in the sea. Fifteen, including comb jellies, lamp shells, and echinoderms, are exclusively marine (Table 2-2); five more, including sponges, coelenterates, and bryozoans, are nearly so (more than 95 percent of the species). Thus, the sea hosts almost the entire extant variety of basic animal body plans, whereas land and freshwater animals comprise myriad variations on the theme of insectdom and just a smattering of other body plans.

Still another useful way to look at biological diversity is to examine *function*. For example, in all but the least diverse ecosystems, species can be grouped on the basis of similarities in what they do. For example, a hectare (2.47 acres) of seabed might have 500 species of polychaete worms, each of which functions in a unique way but which can be aggregated in a smaller number of functional groups, or guilds. If feeding is considered, the worms of some guilds eat other worms, the members of others filter bacteria borne by the currents, the species in still other guilds consume organic material in sediments, and so on. If relation to the seabed is considered, the members of some guilds construct permanent burrows below the surface of soft sediments, the species in others hide under hard objects, or construct tubes that project above the sediment surface, and so on. Still another basis for dividing species into functional groups concerns

TABLE 2-2. Animal phyla in marine and nonmarine ecosystems

Exclusively marine	Marine & nonmarine	Exclusively nonmarine

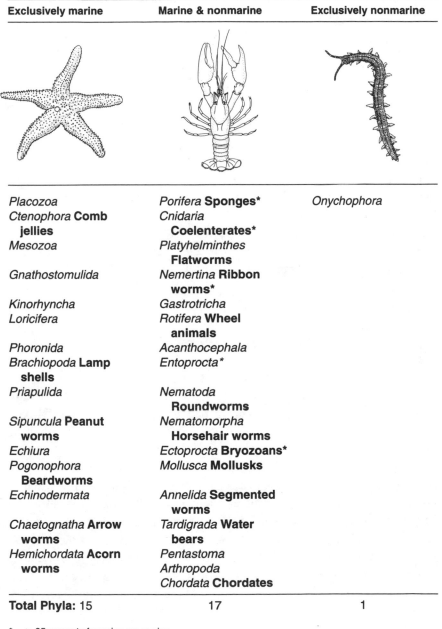

Exclusively marine	Marine & nonmarine	Exclusively nonmarine
Placozoa	Porifera **Sponges***	Onychophora
Ctenophora **Comb jellies**	Cnidaria **Coelenterates***	
Mesozoa	Platyhelminthes **Flatworms**	
Gnathostomulida	Nemertina **Ribbon worms***	
Kinorhyncha	Gastrotricha	
Loricifera	Rotifera **Wheel animals**	
Phoronida	Acanthocephala	
Brachiopoda **Lamp shells**	Entoprocta*	
Priapulida	Nematoda **Roundworms**	
Sipuncula **Peanut worms**	Nematomorpha **Horsehair worms**	
Echiura	Ectoprocta **Bryozoans***	
Pogonophora **Beardworms**	Mollusca **Mollusks**	
Echinodermata	Annelida **Segmented worms**	
Chaetognatha **Arrow worms**	Tardigrada **Water bears**	
Hemichordata **Acorn worms**	Pentastoma	
	Arthropoda	
	Chordata **Chordates**	
Total Phyla: 15	17	1

* = > 95 percent of species are marine

reproduction: whether they reach maturity quickly or slowly, produce tens of thousands of eggs annually, or settle from the plankton in a particular season or throughout the year. The 10 or 20 polychaete species that attach to hard surfaces, filter-feed and reproduce in spring have more in common ecologically than they do with members of other guilds, and can therefore be grouped for some purposes.

Another kind of functional diversity is biochemical diversity. Ecosystems with high disparity in sessile organisms and high pressure from predators, such as coral reefs, are likely to have high diversity of defensive chemicals, something of considerable interest to the pharmaceutical industry.

Because so many marine species are virtually unknown to scientists, grouping species by function in guilds, as Fagerstrom (1991) has for reef-builders, can be a useful way of looking at biological diversity. As important as it is to conserve individual species, it is even more important not to lose any broad functional category of species, because much of the ecological significance of species and many of their benefits to humankind concern their functions.

Importance of Marine Biological Diversity

The loss of biological diversity might be no more than a subject of academic interest if other living things were not so important to humankind. But our species is inextricably tied to other living things. Many people feel a special kinship with other living things, and from these feelings come a special responsibility not to harm them.

Our concern about other species is not only our recognition of their intrinsic importance, however. We depend on other organisms for essential products, including foods, medicines, and raw materials. Because living things create the conditions essential to human survival, we are utterly dependent on the services that ecosystems provide as they function. We have profited greatly from ideas that come from observations of biota. Many humans find living things beautiful, and are willing to make considerable sacrifice to have such beauty around them. Finally, the diversity of life provides an insurance policy for a future in which change is inevitable. All these things make biological diversity—in the sea as well as on land and in fresh waters—vitally important to humankind.

Products from Marine Life

The common idea that the sea is an inexhaustible cornucopia for human-kind masks a subtler truth. The marine realm provides a great abundance and diversity of foods, medicines and raw materials, and will undoubtedly provide important new ones as we learn more, but the wealth of the sea is finite. In a world where human numbers and demand for products are increasing rapidly, it behooves us to know how much we can reasonably expect from living marine resources.

Foods

Although *Homo sapiens* is not a marine species, we are very much a part of marine food webs, for we consume huge amounts of marine fishes, inver-tebrates, and algae. The global catch of fishes ("finfish" hereafter) and crustaceans and mollusks ("shellfish" hereafter) in 1989 totaled 99.5 million metric tons. The marine catch was 85.8 million metric tons, with the rest caught in inland waters (FAO 1991). Although annual increases have been low in recent years, the catch of finfish and shellfish is the world's largest single source of animal protein, exceeding production of beef, sheep, poultry, or eggs. Indeed, they are an essential source of animal protein for much of the developing world (Table 2-3).

Because of population growth, the UN Food and Agriculture Organi-zation (FAO) estimates that another 19 million metric tons will be needed by the year 2000 simply to maintain consumption at the current annual rate of about 12 kilograms per capita. An increase in consumption could boost demand by another 10 million metric tons, suggesting that demand could increase in this decade by 30 percent.

How can the increasing demand for finfish and shellfish be met? Current fishing activity exceeds sustainable yields on a large number of traditional fishing grounds, especially in coastal areas and on conti-nental shelves. The take of most currently exploited stocks of commer-cially valuable species has reached or exceeded levels of sustained rational exploitation. This is occurring when many coastal nations are accelerating efforts to increase domestic food supplies, generate foreign exchange, or diversify their economies. Unfortunately, it is unlikely that increasing demand for marine finfish and shellfish can be met without dramatic changes. These might include effective management and en-forcement, reversal of habitat destruction, reduction of processing waste,

TABLE 2-3. Seafood as percentage of animal protein consumed in various countries

Country	%	Source	Country	%	Source
Australia	6	2	Japan	51	2
Canada	10	2	Madagascar	15	1
China	19	2	Morocco	24	1
Ghana	50	1	Philippines	50	2
India	13	2	Senegal	38	1
Indonesia	60	2	United Kingdom	9	2
Italy	10	2	United States	6	2
Ivory Coast	31	1			

Sources:
1) Bonzon, Alain and Benoit Horemans. (1988). *Socio Economic Data Base on African Fisheries.* FAO Fisheries Circular No. 810. All figures are for 1986.
2) Laureti, Edmondo. (1991). *Fish and fishery products: world apparent consumption statistics based on food balance sheets (1961–1989).* FAO Fishery Circular No. 821(1). All figures are for 1989.

and improvement of post-catch technologies, exploitation of a wider variety of marine organisms (especially at lower trophic levels), and significant advances in combining mariculture with capture fisheries in the form of sea ranching or culture-based fisheries, perhaps employing genetic engineering.

At present, the largest marine catches come from the temperate and subpolar continental shelves, and from upwelling areas. Although there are more than 15,000 species of marine fishes, only a tiny fraction are of major commercial importance. For example, during the 1980s, small pelagic schooling fishes provided most of the observed increase in fish production. Actually only three pelagic species (Peruvian anchoveta [*Engraulis ringens*], South American sardine [*Sardinops sagax*], and Japanese sardine [*S. melanostictus*]), and one semi-pelagic species (walleye pollock [*Theragra chalcogramma*]) constituted about half of the increase in landings.

The former USSR, China, and Japan have been the leading fishing nations, each taking about 11 percent of the total catch, followed by Peru, Chile, and the USA. Not all of the world's catch is used directly for food. Approximately 30 percent becomes fish meal, which is used as animal feed.

Because most fishes, mollusks, and crustaceans are not taken commercially, there is a widespread belief that the sea is a cornucopia that harbors "untapped resources" that could lead to expanded food production. However, this is unlikely to happen. For example, perhaps the world's most numerous and widely distributed fishes are bristlemouths, genus *Cyclothone*, which occur in the open ocean at depths from 500 to 2,000 meters (1,650 to 6,500 feet). But these fishes are unlikely to become a food source for hungry people. Exploiting fishes as small (to 6 centimeters [2.4 inches]) as *Cyclothone* spp. poses serious, probably insurmountable problems. At present, efficient technologies to capture and process them do not exist. Even if they did, *Cyclothone*'s value as a resource is doubtful: These species have a very high water content, they are relatively slow growing, and their density in the sea is low. Their numbers are huge only because the volume of the bathypelagic zone where they live is so great. There are very good technological and economic reasons why people capture sardines instead of *Cyclothone*.

Mariculture (both intensive and extensive) will likely contribute significantly more to total marine fishery production in the near future. Mariculture production is growing at the rate of 5 percent annually, much faster than the take from capture fisheries. By the year 2000, up to one-third of all fish production may be derived from mariculture-aquaculture activities. The 1988 aquaculture production (including marine species) was over 14.6 million metric tons, with 80 percent of the reported yield from Asia. Freshwater culture of carps and related fishes was by far the largest component (4.5 million metric tons), followed by brown seaweeds, mussels, and oysters. Culture of anadromous salmon, trout, and smelts, and of shrimps and prawns, each amounted to about 0.5 million metric tons.

Although *in situ* mariculture might seem promising, it is far from trouble-free. Nutrient loading and antibiotics in maricultural wastes, accidental releases of alien or genetically altered organisms, disease transmission to native species, and destruction of coastal ecosystems to create mariculture ponds are all substantial problems that could diminish future mariculture yields.

Scaweeds have been collected for food for hundreds, even thousands of years (Tseng 1981) in some parts of the world, and are important food items in coastal areas of many tropical developing countries. Use in East Asia far exceeds that in other areas. Some algae, including Irish moss (*Chondrus* spp.), agarweed (*Gelidium* spp.), and giant kelp (*Macrocystis pyrifera*), are taken from natural populations. Others, such as *Eucheuma* spp., *Gracilaria* spp., nori or laver (*Porphyra* spp.), and kombu (*Laminaria* spp.), are cultivated *in situ* on artificial substrates such as poles and nets. Japan, Korea, and China have extensive cultivation of red and brown seaweeds for domestic consumption and export.

Consumption of seaweeds (about 2 million metric tons annually) is increasing as human population growth places more demands on both historically important species and previously unexploited species for food, animal feed, and industrial applications. The scientific community is uncertain about the effects of seaweed gathering on marine ecosystems (Lindstrom and Gabrielson 1990), but seaweed farming can provide a sustainable livelihood for fishers affected by declining fishery stocks.

The sea is a very important source of food for humankind. We are already exploiting existing major marine capture fisheries at very high levels. The potential for increases from these fisheries is limited without dramatic changes in management and policies. Indeed, sustaining current levels is contingent on our ability to stop and reverse the mounting threats to the marine ecosystems on which fisheries depend.

Medicines and Tools for Biomedical Research

Long before the birth of modern science, shamans and herbalists used diverse living things as medicines. Most medicines come from land plants and fungi, which, sessile and unable to flee, have evolved chemical compounds that deter predators, parasites, or competitors.

The sea has its sessile fungi and plants, but differs from the land by also having sessile animals. Further, the sea is home to many animal phyla that are not found on land, so its biochemical diversity is greater. Together, these facts suggest that the sea is a major source of living things that defend themselves chemically and that could produce chemicals of interest to the pharmaceutical industry. The existence of poisonous and venomous invertebrates and fishes has been known for millennia, but the chemical identities of their toxins are only now being determined. Diligent investigation of the pharmaceutical potential of the sea began only with the advent of reliable scuba gear a few decades ago. As in land plants, the diversity of chemical defenses in marine organisms, and hence the potential for pharmaceutical development, is sharply higher in the tropics.

Medicinal use of marine organisms has a far richer tradition in the East than in the West. But even in the West, cod and shark liver oils were used as sources of vitamins D and A before other sources were developed.

Although all areas of medicine are open to new substances, medicines to cure cancers and viral diseases have been most intensively sought in the past two decades, probably due to the potential rewards (more than US$1 billion per year) for a successful product and to funds made available in the last 35 years by governmental agencies (particularly in the USA) for large-scale screening of natural products for antitumor agents and, in recent years, anti-HIV agents. The potential from marine sources is high enough to have spawned commercial ventures to obtain marine-derived substances in Spain, Japan, Australia, and the USA. A number of pharmaceutical companies worldwide are supporting research on marine-derived compounds to be used directly as medicines and as novel chemical structures that could be modified to become medicines.

Modern interest in obtaining medicines from the sea dates from the discovery of naturally occurring compounds called arabinosides in extracts of the sponge *Tethya crypta* in 1950. These led to development of the antiviral pharmaceuticals Ara-A and Ara-C, which are still used to treat herpes infections, with annual sales of US$50 to $100 million. Indian scientists have recently reported that a number of India's mangrove trees, seagrasses, and seaweeds exhibit antiviral activity (Premnatha et al. 1992).

More and more antitumor agents have marine origins. In 1978, extracts of a Caribbean tunicate ascidian of the genus *Trididemnum* were found to be strongly toxic to cells. The active agents, didemnins, belong to a new family of compounds, some of which show promise as anticancer medicines. Didemnin B is now in advanced clinical testing at the US National Cancer Institute. A modified didemnin B recently isolated from the Mediterranean tunicate *Aplidium albicans* has significant antitumor activity (Rinehart 1991), and is being considered for clinical trials. Tunicates are the source of another family of very active antitumor compounds, the ecteinascidins.

A second, very promising marine antitumor agent in clinical trials is bryostatin, from the bryozoan *Bugula neritina*. There are now reports of compounds in preclinical study that are active against solid tumors, which have not generally been responsive to chemotherapy. Dolastatins, first isolated from sea hares in the genus *Dolabella*, include one of the most active antitumor agents yet identified. Chemicals with antitumor activity have also been purified from marine microorganisms. Cyanobacteria produce scytophycins and tolytoxin, which seem to have a unique mechanism of action (Patterson and Smith 1991).

In addition to providing agents that are of direct use as medicines, marine organisms are prolific sources of chemicals whose pharmacologic actions permit scientists to study cellular communication. For example, the 500 or so species of venomous tropical cone snails (*Conus* spp.) produce diverse toxins (Olivera et al. 1990) that are exciting new tools for the study of nerve cells. Other substances that are vitally important to neurobiology include tetrodotoxin and saxitoxin, which were first identified from puffer fish or fugu (family Tetraodontidae), and the dinoflagellate *Protogonyaulax tamarensis*, respectively. Many neurological effects were studied using giant nerve cells from marine invertebrates, namely squid (*Loligo* spp.) and sea hares of the genus *Aplysia*.

Moreover, biological materials from the sea are finding intriguing applications in medicine. For example, since 1985, French researchers have experimented with using pieces of New Caledonian reef coral (*Porites* and other genera) skeleton as bone grafts for people requiring maxillofacial and cranial surgery (Roux et al. 1988). Calcareous pieces of gorgonian corals (*Corallium johnsoni*) and molluscan mother-of-pearl have also been used successfully (Lopez et al. 1989). The calcium carbonate in these implanted fragments is resorbed and is replaced by bone.

The sea is a rich source of materials of use to medicine, and seems as rich in novel pharmaceuticals as the land. Although few compounds from marine sources have yet become commercial medicines, potential sales for some are very large. Indeed, the demand for natural products from marine organisms can become a serious problem if collectors overexploit the typically slow-growing sessile organisms that are most likely to be chemically defended. Sustainable collection of pharmaceutically interesting species requires the development of appropriate management techniques which exist in few, if any, nations.

Because of the complexities of the testing systems, chemical isolation, and clinical studies, major marine-based pharmaceutical operations will probably continue to originate in industrialized nations in the near future, although other countries might be important sources of supply. Both developing and industrialized nations have recognized the problems inherent in this pattern. The biodiversity treaty drafted in preparation for the

Earth Summit in Rio de Janeiro in June 1992 was intended to ensure continued opportunities for pharmaceutical companies and adequate compensation for countries that provide raw materials. The USA, however, did not sign the biodiversity treaty, dealing a serious blow to production of medicines from the sea.

Raw Materials

Some of the products from marine species are used with little or no modification. But others serve as raw materials that are further modified before being used by various industries. The benefits that are derived from marine foods and medicines are better known than those from marine raw materials, but a sizable fraction of humankind uses raw materials from marine organisms every day, and many people's lives depend on income from these products.

Algae and Cyanobacteria: Marine algae and cyanobacteria are the essential primary producers at the base of the food pyramid in most marine ecosystems and are important sources of raw materials. They range in size from less than 1-μm (1/2,500th of an inch) cyanobacteria to 50-meter (164-foot) kelps. They number more than 15,000 species, yet, despite their ecological importance, very little is known of species' ranges or their basic ecological requirements.

Among the better-known species are seaweeds (red, brown, and green marine algae) that provide raw materials, principally polysaccharides, for human uses. The brown algae, about 1,500 species, are most important for alginate, which has many food, technological, and medical applications. The kelps (e.g., *Macrocystis pyrifera*) and fucoids (e.g., *Ascophyllum* spp.) are harvested from wild and commercially grown populations for a net annual value of US$150 million. In Ireland, Norway, and France, the seaweeds cast ashore are used as livestock feed or composted for agricultural soil conditioner and fertilizer. Brown algae are collected along the coasts of India, Australia, Mexico, Chile, Argentina, Iceland, and South Africa, as well as in the major alginate-producing countries of China, Canada, the USA, the UK, Norway, and France (Lewis et al. 1988). Nations vary widely in their regulations regarding collecting wild seaweeds.

The marine red algae, about 5,000 species, are important as sources of carrageenans and agars, which, like alginates from the brown algae, are cell wall polysaccharides whose texture makes them useful in food manufacture. The annual value of carrageenan is approximately US$110 million and that of agar about US$140 million. Cultivation of the carrageenanweeds *Kappaphycus* spp. and *Eucheuma* spp. is a major source of income for thousands of families in China, Thailand, Malaysia, and Indonesia, as well as in the Philippines, where carrageenanweed farming can bring a family about US$3,000 per year, more than the average annual salary for a university instructor. Similarly, cultivation of agarweeds such as *Gracilaria* spp. and *Gelidium* spp. is of increasing economic benefit to Brazil, Chile, and other South American and Asian countries. Natural populations of the agarweeds *Gelidium* spp. and *Pterocladia* spp., and the carrageenanweeds *Iridaea* spp., *Chondrus* spp., *Eucheuma* spp., and *Gigartina* spp. are collected in Canada, Mexico, Spain, and Asia. Some populations of these species are presently overexploited and have declined significantly, but the species are not known to be endangered.

Some countries, such as Japan and the Philippines, have made preliminary efforts to establish gene banks for many cultivated species to maintain the wild parent stocks, but conservation of seaweeds has long been neglected. Both the method of collecting and the kinds of interactions between targeted species and associated species determine the impact that seaweed collecting has on an ecosystem (Foster and Barilotti 1990).

Single-celled algae are less frequently used than seaweeds as raw materials. Marine green flagellates in the genus *Dunaliella* are cultivated widely in salt ponds in Israel, China, Australia, and the USA for production of ß-carotene used in vitamin capsules and as a food coloring. Cyanobacteria in the genus *Spirulina* are cultivated as a health-food supplement and fish food color supplement in saline ponds. No other marine phytoplankton are collected for direct use as raw material in human food products, although many species are cultivated in tanks and ponds as food for larvae in commercial shrimp culture systems.

Mangrove Trees: Mangrove trees such as *Rhizophora mangle* are a source of tannins used by the leather industry, and of wood for construction, fuel, and charcoal. They are also used as as a raw material for making rayon. Although these uses could be sustainable, overexploitation of mangrove trees is depleting mangrove forests throughout the tropics. An even greater threat is the destruction of mangrove ecosystems for agriculture, housing, and commercial mariculture ponds. In Ecuador alone, more than 150,000 hectares (371,000 acres) have been lost to shrimp mariculture operations (Lugo 1990).

Marine Animals: Marine animals are also sources of raw materials. Living corals, coral rock, and coral sand, whether used unmodified, burned for lime, or incorporated into concrete, are important building materials in India and island nations including Sri Lanka, the Philippines, Indonesia, and French Polynesia. And yet, this mining can cause enormous structural damage to reefs (UNEP and IUCN 1988/89). In the Maldives, for example, reef flats have been completely denuded of living corals and showed no sign of regeneration after 20 years (Brown and Dunne 1988).

A more environmentally benign product comes from marine crustaceans. Chitin, derived largely from the processing of shrimp and crab shell waste, is used extensively in agriculture, biotechnology, wastewater clarification in industry, human and animal food supplements, and medical, dental, and cosmetic applications. The total value exceeds US$50 million per year (Skjak-Braek et al. 1989).

The biochemical diversity of marine plants, microorganisms, and animals suggests that humankind will discover many new uses for them as raw materials, foods, and medicines. But we will reap the benefits of these organisms only if we conserve them and the ecosystems in which they live.

Ecosystem Services from the Sea

When astronauts leave the Earth, their survival depends on sophisticated life-support systems created by engineers. The remaining billions of us depend on life-support systems that are vastly more intricate, being composed of countless evolving organisms that, by serving their own needs, create the conditions in which people can live and prosper. Until about 425 million years ago, *all* life was marine (Figure 2-2), and marine ecosystems still provide a large share of the services on which we depend today.

Coastal Processes

The services from marine ecosystems directly benefit coastal areas. Mangrove forests form a bulwark against storm surges and waves along many low-lying tropical coasts. Without them, human settlements, agriculture, and infrastructure are far more vulnerable to cyclones and tsunamis. Each year, some 25 billion tons of nutrient-rich sediments are eroded from the

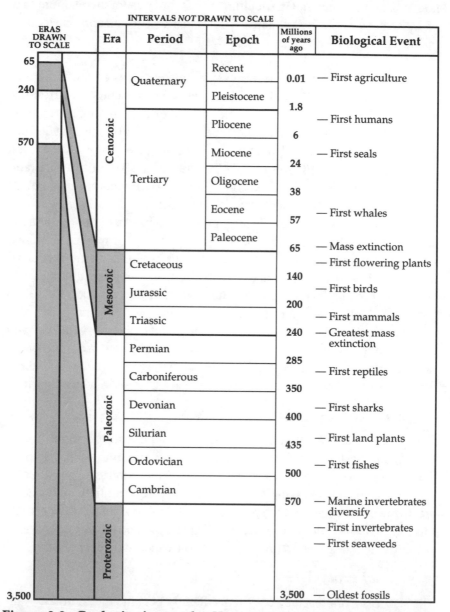

INTERVALS *NOT* DRAWN TO SCALE

ERAS DRAWN TO SCALE	Era	Period	Epoch	Millions of years ago	Biological Event
65	Cenozoic	Quaternary	Recent	0.01	— First agriculture
240			Pleistocene	1.8	
570		Tertiary	Pliocene	6	— First humans
			Miocene	24	— First seals
			Oligocene	38	
			Eocene	57	— First whales
			Paleocene	65	— Mass extinction
	Mesozoic	Cretaceous		140	— First flowering plants
		Jurassic		200	— First birds
		Triassic		240	— First mammals
	Paleozoic	Permian		285	— Greatest mass extinction
		Carboniferous		350	— First reptiles
		Devonian		400	— First sharks
		Silurian		435	— First land plants
		Ordovician		500	— First fishes
		Cambrian		570	— Marine invertebrates diversify
	Proterozoic				— First invertebrates
					— First seaweeds
3,500				3,500	— Oldest fossils

Figure 2-2. Geologic time scale. *Humans are newcomers on Earth. Life began in the sea at least 3,500 million years ago. Many marine invertebrate phyla appeared 570 million years ago. But the genus* Homo *arose around 3 million years ago; agriculture, cities, metallurgy, and writing date only from the last 10,000 years, almost an instant in the history of life.*

land (Brown et al. 1989), and more yet would enter coastal marine ecosystems but for marine wetlands. The roots of trees in mangrove forests and the shoots of plants in salt marshes and seagrass beds act as baffles. When moving water encounters these structures, it is slowed, causing suspended sediment particles to settle. These same three-dimensional biological structures also help to prevent resuspension of the particles. Ultimately, this accumulation of sediments and nutrients allows the colonization of land plants, which stabilizes the sediments. By amassing sediments and preventing their erosion, mangrove, salt marsh, and seagrass ecosystems build some of the world's most fertile coastal lands. In doing so, they not only trap the particles that would otherwise smother benthic species and diminish light penetration and productivity in coastal waters, but they also remove dissolved nutrients, thereby improving water clarity.

Coral reef ecosystems also create land, as first detailed by Charles Darwin (1842). In the tropics, the fringes of new volcanic islands are colonized by reef corals, massive coralline algae, and sediment-forming algae such as *Halimeda* spp., which produce calcium carbonate rock (limestone) and sand. As the volcano subsides, accumulation of these carbonates can continue until what remains is a lagoon ringed by a necklace of coral islands, that is, an atoll. Many Indo-West Pacific island groups, including the Maldives, Tuamotus, and Marshalls, owe their existence to this biogeochemical process. In many more tropical nations, coral reefs provide the defense against storm waves that would otherwise inundate coastal lands.

Global Climate

Ecosystem services are not limited to coastal waters, but rather affect the entire ocean-atmosphere system through their role in stabilizing global climate. The most important of these global-scale marine ecosystem services is the marine "biological pump," by which the living ocean controls the atmospheric concentration of carbon dioxide (CO_2). Second only to water vapor in its importance as a heat-trapping or "greenhouse" gas, CO_2 in the atmosphere (700 billion metric tons of carbon) is largely maintained by exchanges with the much larger oceanic reservoir (35,000 billion metric tons of carbon).

The surface waters of the world's oceans contain less dissolved carbon than the deep waters. This vertical gradient is produced by phytoplankton in the sunlit photic zone (the upper 100 meters [330 feet]). These organisms, including diatoms, coccolithophorids, and dinoflagellates, take the dissolved carbon out of solution. After dying or being eaten by zooplankton, their organic tissue (and in some species calcium carbonate shells) sinks into deeper waters or to the ocean floor, where it decomposes. The processes of photosynthesis and decay pump carbon from the surface into the deep ocean.

The total flow of fixed carbon downward from the surface waters of the global ocean is still poorly known, but it might be comparable to the global annual combustion of fossil fuels, about six billion tons of carbon per year. Although fossil fuel burning and deforestation that have raised atmospheric CO_2 have thereby increased the concentration of dissolved carbon in surface waters by about 2 percent, the ocean's phytoplankton have probably not been affected because they are not limited by CO_2 scarcity but by nutrient scarcity.

However, any major trauma to the marine biota, such as might be expected from climatically induced changes in ocean circulation and/or increases in UV-B radiation at the sea surface, could have a dramatic impact on the efficiency of the marine biological carbon pump, and hence on the level of atmospheric CO_2. Imagine, for example, that the surface ocean phytoplankton were to die completely but that the ocean circulation were to remain unchanged. Within a rather short time, a few centuries at most, the atmospheric CO_2 level would rise dramatically to two or three times its present value, as deep ocean waters recirculate to the surface and release CO_2 to the atmosphere. It is only the action of the biological "pump" that prevents this from occurring today.

Some of the most productive regions in the sea, where the "pump" is working hardest, include the upwelling areas of continental shelves and slopes, and the upwelling areas in the open ocean associated with wind-driven divergences (e.g., equatorial and subpolar zones). Phytoplankton blooms in such areas are often dominated by diatoms, which have silicious shells that are not made from dissolved carbon, and by coccolithophores, which have carbonate shells that are. The rain of organic matter and coccolith shells in these upwelling areas are very important parts of the "pump."

Although the oceans are very unlikely to die, the greenhouse ocean of the future is likely to be less productive than today's ocean, just as today's ocean is known to be less productive than the ocean during glacial periods.

The reason is that the wind's stirring of nutrients into the sunlit waters (which fuels the productivity) was higher in glacial times than it is today, and it is predicted that it will be weaker still in the less vigorously circulating ocean-atmosphere system of the next century.

Marine phytoplankton might play an important role in the heat balance of the Earth by controlling the extent and whiteness of clouds over the ocean, a key component determining how much sunlight is reflected from our planet (Charlson et al. 1990). It is hypothesized that certain types of phytoplankton emit a volatile, reactive form of sulfur (dimethyl sulfide, or DMS), which escapes to the atmosphere and is rapidly oxidized to sulfuric acid. The acid, in turn, forms the condensation nuclei necessary for water vapor to coalesce to form cloud droplets over the sea. This is an active area of research; while still a controversial hypothesis, many links in the chain of argument have already been validated, and none have been repudiated.

Anything that brings significant changes in the abundance and species mix of phytoplankton could, therefore, affect marine cloud cover and the Earth's albedo (reflectivity). Paleoclimatic records of the forms of sulfur that exist in Antarctic ice (Legrande et al. 1988) suggest that this mechanism was more efficient in the more productive glacial ocean. If, indeed, the greenhouse ocean will be less productive, it will not only have a weaker biological "pump," but will be less able to cool itself by the DMS/cloud mechanism. Both of these biotic responses to global climatic change are positive feedbacks that would exacerbate the damage due to increasing atmospheric CO_2.

It has long been clear that humankind depends to a considerable degree on products from life in the sea. But it is becoming increasingly clear that we depend even more on the services that healthy marine ecosystems provide.

The Sea as a Source of Ideas

As important as it is in its own right and for its material benefits to humankind, the diversity of marine life has provided the raw material for ideas that apply far beyond marine biology. Among the many examples that illustrate this point are ideas about the ecology of uncommon species, past global climates, and the evolution of predation and anti-predation mechanisms.

Until midcentury, most ideas about the science of marine ecology came from observational studies. In the 1940s, British ecologists began conducting experiments in the rocky intertidal zone, and were joined by American ecologists in the 1950s. Their experimental methods and results added profound new insights to the understanding of species interactions. For example, scientists had often assumed that the importance of species in their ecosystems is roughly proportional to their abundance. But Paine (1969) found that the process of feeding by the ochre sea star, *Pisaster ochraceus*, a Northeast Pacific starfish that is far from the most abundant species in its rocky shore community, nonetheless affects the physical structure and species composition of the entire community. The starfish eats mussels that would otherwise eliminate other species by occupying all available primary space—the rock surface—in the community. The gaps it creates in the mussel beds allow other species to colonize the rock surface. This starfish inspired the term "keystone species," one that plays a crucial role in its community.

The keystone species concept has vital implications for conservation: Some species—even uncommon ones—are so important in their ecosystems that they merit special attention in conservation efforts. Removal of these species might not even be noticed, but would have profound effects on the composition, structure, and functioning of their entire community. For example, the removal of crabs decreased the productivity of an Australian mangrove forest, because crab burrowing aerates soils (Smith et al. 1991). While scientists have only begun to identify keystone species in the sea, it is clear that they are especially important priorities for conservation.

Studies of mangroves (Rabinowitz 1978) and coral reef fishes (Sale 1991) were important sources of another idea with great importance for the management and protection of living things. This is the concept of "supply-side ecology," the idea that differences in the amount and timing of recruitment—the addition of new individuals to the population in a particular place—can be crucial in determining species' distributions and abundance.

The fossil record of marine organisms is indispensable for studying the world's climatic history because it is far more complete than that of land and freshwater realms. We know, for example, that reef-building corals in the genus *Pocillopora* are more or less restricted to waters warmer than 18°C (64°F). The discovery of *Pocillopora* from the Miocene epoch (Figure 2-2) in the Portuguese island of Madeira provides strong evidence that northwest Africa was substantially warmer than it is today (Boekschoten and Wijsman-Best 1981). Similarly, during the middle Miocene (about 15 million years ago), many tropical or warm-temperate invertebrates lived in Kamchatka, Russia, where the climate is now sub-Arctic (Fot'yanova and Serova 1987). And tiny shell-bearing protozoans called foraminifera have played a major role in deciphering the Earth's climatic history.

Biologists have long studied the co-evolution of predators and prey (Ehrlich and Raven 1964), but this "arms race" is especially clear in shallow-water tropical mollusks. Shell characteristics that inhibit predation are most developed in the Indo-West Pacific, moderately developed in the eastern Pacific, and least developed in the Atlantic Ocean. The predators' offensive weapons (e.g., the robustness of crab claws) for catching and breaking molluscan shells show a similar pattern. This observation, subsequent experiments, and fossil evidence of predation on marine mollusks have helped evolutionary biologists to understand the conditions under which adaptations and counteradaptations occur (Vermeij 1987).

These examples merely scratch the surface. Their main lesson is that far-reaching ideas often come from life in the sea. By harming marine ecosystems and species, we are drying up the fountains of new thought.

Aesthetic/Recreational Resources

More than two centuries ago, the British writer Samuel Johnson observed, "The joy of life is variety. . . ." He probably was not contemplating a choice between kippered herring or smoked salmon when he penned these words, but it is clear that aesthetic appreciation of diversity is a powerful motivating force for conservation.

At first, this idea might seem odd. Economics so dominates our world that even some conservationists think species and ecosystems must "pay their way" if they are to survive. But diverse cultures have a special connection with marine life. Dolphin and octopus motifs are common in ancient Greek pottery and mosaics. Saltwater crocodiles (*Crocodilus porosus*) and the large estuarine fish called barramundi (*Lates calcarifer*) play major roles in the art and spiritual life of northern Australian aborigines. Bald eagles (*Haliaeetus leucocephalus*) and orcas (*Orcinus orca*) are similarly valued by British Columbian native peoples. Indeed, the millions of moviegoers who have delighted in being horrified by giant octopuses and "man-eating" great white sharks (*Carcharodon carcharias*) show the grip of marine organisms on human psyches.

The oceans, with their powerful storms, their shimmering palette of colors, and their varied mysterious sea life, have inspired some of the world's finest painting, poetry, stories, and music. For many people, however, the main aesthetic experience of marine organisms is gustatory. Peoples as varied as Ghanaians, Seychellois, Japanese, Venezuelans, and Norwegians have made seafood a significant part of their culture. The legions of people worldwide who also enjoy whale watching, scuba diving, sport fishing, shell collecting, aquarium keeping, and visiting public aquaria attest to human enjoyment of marine organisms. The fact that these activities produce large economic returns suggests how important they are as aesthetic and recreational experiences. The number of certified scuba divers worldwide exceeds 5 million, and nearly half of those who visit the Cayman Islands scuba dive on the islands' remarkable coral reefs. Each year the National Aquarium in Baltimore (USA) attracts 1.4 million marine biodiversity enthusiasts, who bring some US$80 million into the local economy. Some major aquaria in the USA draw more than 4 million people annually. Some hobbyists will go to great lengths, and spend significant sums of money, to display fishes such as the Indo-West Pacific clown triggerfish (*Balistoides conspicillum*), which can retail for over US$250.

The aesthetic affinity for marine organisms and ecosystems is a mixed blessing, for it is all too easy to love them to death. A number of marine animals, including precious corals (*Corallium rubrum* in the Mediterranean, and others) and hawksbill sea turtles (*Eretmochelys imbricata*), have become rare because their hard body parts are considered beautiful and are fashioned into artifacts. In many places, people pay a premium to live near beautiful barrier beach sand dunes, salt marshes, or coral reefs, not realizing that they are bringing problems such as oily runoff from roads and sewage pollution, thereby degrading the ecosystems that attracted them.

However, aesthetic feelings also stimulate the formation of conservation movements, which play a crucial role in stopping or slowing damaging practices. Commercial whaling did not end because of a resource management decision that whales were overexploited but because so many people in so many countries feel profound emotional kinship with whales. Outrage over the drowning of dolphins and other marine mammals in purse seines and driftnets has led to controls that also benefit marine food webs by diminishing deaths of nontarget fish species and seabirds. Opposition to pollution that degrades the beauty of the Mediterranean has led public-interest organizations and governments in Spain, France, Italy, and Greece to acknowledge the need to eliminate sources of marine pollution, benefiting not only people's aesthetic sensibilities but also the entire ecosystem. This stewardship would not have happened if people did not care about the sea and its creatures.

Many decisions affecting marine biological diversity are based mostly on traditional market economics. That economics is not the sole factor, however, attests largely to the aesthetic value of marine species and ecosystems. Not all people are touched by the beauty of nature, but for those who are—and they are many, in both industrialized and developing nations—love of nature is a motivating force of unequaled potency.

Adapting to a Changing World

The Earth is always changing. Over aeons, marine organisms have repeatedly been confronted by abiotic challenges such as changing currents and climates, and biotic challenges such as diseases and competition. Those that met the challenge either managed to avoid or adapt to change. Those that did not became extinct.

Human activities have dramatically increased the intensity, pace, and kinds of environmental change, posing severe adaptive challenges to living things. Anthropogenic introduction of species, release of toxic pollutants, siltation, fragmentation of habitats, and climatic change are now more rapid, intensive, and widespread than natural changes. Responses to anthropogenic changes have included severe declines of many fisheries, contraction of species' ranges, and extinction of some large vertebrates (Thorne-Miller and Catena 1991). It has often been assumed that marine organisms are more resistant to extinction than terrestrial ones, but this assumption could reflect the fact that the oceans are difficult for people to study, so people are far more ignorant about the marine realm than they are about the land. The fossil record provides ample evidence of mass extinctions in the sea, and recent reports of the extinction of a mollusk (Carlton et al. 1991) and the near extinction of corals (Glynn and de Weerdt 1991; Glynn and Feingold 1992) suggest that modern marine species are not immune to extinction.

Organisms can avoid extinction only by finding a refuge from change within which they can reproduce enough to maintain their populations, or by becoming better adapted to new conditions. Human activities are destroying the habitats of many species, diminishing habitat quality for others, and confining still others to isolated refuges with suitable conditions. These changes either eliminate a species or reduce its abundance.

The alternative, adaptation, results from the differential reproductive success of the fittest individuals within a population, a process that Charles Darwin termed "natural selection." In most organisms, differences in fitness result mainly from differences in their genes. Hence, without genetic diversity, natural selection cannot operate and adaptation cannot occur (Futuyma 1979). Genetic diversity is therefore essential for the survival of species in a changing world.

Reducing populations can reduce genetic diversity in two ways. First, if whatever reduces population size selects some genotypes in favor of others, those selected against will become rarer and eventually disappear. But even if no selection occurs, decreasing populations lose genetic diversity when the individuals possessing rare genes, purely by chance, do not breed. Hence, sexually reproducing species are at risk when their habitats are fragmented, in part because there is decreased interchange of genes between isolated populations. They are also at risk when population densities are very low and the chance of mating successfully is reduced (Lande 1988).

Another problem in marine conservation is that, in some cases, what looks to scientists like one "species" is really a number of separate populations of "sibling" species that cannot interbreed (Grassle and Grassle 1976; Weinberg et al. 1990). Marine invertebrates are less cosmopolitan than once thought, and are, therefore, likely to be more vulnerable to localized events.

Biological diversity serves as insurance in an unpredictably changing world. Maintaining genetic diversity within species is the best strategy for maximizing the ability of individual species to adapt to rapid change. Maintaining the existing species diversity in marine ecosystems is the best strategy for maximizing the long-term survival of these systems and minimizing the harmful effects of change.

The Value of Life for Its Own Sake

"Man is the measure of all things," said the ancient Greek philosopher Protagoras; the modern Western version is that all values are anthropocentric, that is, centered on human interests. Perhaps anthropocentric values will be the most persuasive ones in environmental policy. But the arrogance of thinking that only humans count (Ehrenfeld 1978) is nowhere clearer than in the sea, where we are out of our element and life flourished for several billion years with no attention from us.

Because life originated in the sea, every living thing is descended from marine life. Indeed, we carry the ancient sea inside us, for the ionic composition of the blood of vertebrates on land bears an unmistakable resemblance to seawater. In this, as in many other things, our place in the natural world is evident. But humans are also different from other living things. Of all the many millions of species, *Homo sapiens* stands alone as both the one most capable of affecting other species and the one with the unquestionable ability to care about the fate of the others.

Does our unequaled power give us the right to do anything we want to other living things? Or do other species have value unto themselves? Conscience and science both go astray if we value the welfare of only one species—our species—as important in itself, with all others valued only for their utility to us. We need a deeper ethic.

From one perspective, the activities of each organism, whether kelp, crab, or human, are just the whir and buzz of enzymes. But from an equally valid—and objective—perspective, when an organism acquires food, defends itself, or reproduces, it is protecting something of value, its self and thus its genes. People who think, "Kelps and crabs don't care, so why should I?" should recognize that they act very much as if they *do* care. The behaviors and physiologies of organisms equip them to survive and perpetuate themselves, just as ours do. It is obvious to anyone who has been pinched by a crab or watched seaweeds compete for space that other species act in their own interests without doing so for our benefit, however much their activities might benefit us.

Individual organisms conserve themselves and their genes, but conservation is even more important at the population and species levels. Species and sub-specific populations are dynamic, evolving entities whose behaviors allow them to perpetuate their genes for millions of years. Their significance transcends that of individual organisms, and they are more appropriate for moral concern. Extinction shuts down the evolutionary process. To kill a particular fish is to stop a life of a few years, while other such lives continue unabated; to eliminate a fish population or species is to end a far older story and leave no future possibilities.

Although our lives are tied to those of all other species, however strongly or tenuously, people know little of just which species' activities are vital to us. A conservation ethic that goes beyond short-term self-interest is a form of life insurance, because it also favors the long-term survival of our genes and cultural elements. By treating other things as if they matter, we not only demonstrate commendable humility, but also benefit our own self-interest. Thus, conserving biological diversity unites the interests of religious leaders and deep ecologists with those of practical people whose major concern is ensuring the survival and well-being of themselves, their loved ones, and their descendants.

CHAPTER THREE

Marine and Terrestrial Conservation

H U M A N S are terrestrial creatures, and often assume that we comprehend the whole planet, but at best, only some of what we know applies in the realm beneath the waves. To save, study, and use the seas sustainably, we must recognize which lessons we can transfer from the land, and which must be modified or fashioned anew.

For example, the 1980s saw the flowering of a new scientific discipline, conservation biology, the science of conserving biological diversity. Conservation biology has become an increasingly powerful influence in many natural resource discussions. But its central ideas come from the terrestrial realm, and their extrapolation to the sea is risky. Effective conservation and management of life in the sea, then, requires a new science of marine conservation biology that reflects the sea's distinctive nature.

Distinctions Between Marine and Terrestrial Conservation

Because some attributes of organisms transcend the land-sea interface, some aspects of terrestrial conservation are applicable in the sea. But differences in media, dimensionality, and scale between terrestrial and marine realms have major implications for marine conservation. As a result, principles of marine conservation can be very different from the ones derived from experience on land.

A Buoyant Liquid Medium in Motion

Perhaps the most obvious difference between the sea and land involves the fluid media above the geological substrate. The water overlying the seabed is more viscous and has greater surface tension than the air overlying the land. It is about 850 times denser, which provides increased buoyancy and thus allows organisms to survive without needing powerful supporting structures. This allows for very different kinds of organisms to exist in the sea.

For example, giant kelp (Figure 3-1) can reach a height of 40 meters (131 feet) while devoting very little energy to supporting its weight, unlike trees on land. The kelp uses the buoyancy of its gas-filled bladders, rather than strong but heavy wood, to loft its photosynthetic structures to the light.

Indeed, buoyancy allows the existence of the most prominent group of sea-dwellers—the thousands of species of nekton (active swimmers, including penguins, whales, fishes, and squids that have density near that of seawater)—to spend much or all of their lives *within* the water column without having to rest *on* the seabed. In contrast, terrestrial dragonflies, birds, and bats are, at most, temporary aero-nekton. They have to expend substantial amounts of energy to remain airborne, and seldom or never leave the ground for more than hours at a time.

Seawater's viscosity is about 60 times greater than that of air. This dramatically slows the sinking of particles compared with their sinking rates in air. Combined with increased buoyancy, slower sinking rates allow the existence of another major group of marine organisms, the plankton. Phytoplankton, including a diversity of single-celled algae such as diatoms, dinoflagellates, coccolithophorids, and zooplankton, which are animals ranging from tiny copepod crustaceans to bathtub-sized jellyfishes, are drifters in the water column that do not have to rest on the bottom. In contrast, aero-plankton—fungal spores, wind-dispersed seeds, ballooning spiderlings, and flying insects—are, at most, temporary air-dwellers, and far less abundant.

The buoyancy and surface tension of seawater support still other lifestyles. Dwelling at or just beneath the sea-air interface, an environment without analogue in the terrestrial realm, are the pleuston and neuston, including water-striding insects (*Halobates* spp.), the Portuguese man-of-war (*Physalia physalis*), and the larvae of many bottom- and water-

Figure 3-1. Giant kelp. *A giant kelp 40 meters (131 feet) tall can weigh about 50 kilograms (110 pounds). The aboveground portion of an equally tall tree weighs perhaps 140 times as much. The difference comes mainly from the tree's heavy support tissues (wood). The sea is so buoyant that some kelps can hold their photosynthetic tissues near the surface with small, light, gas-filled bladders.*

column-dwellers. In the Arctic and Antarctic, the upper and lower surfaces of floating ice and the polynyas (open waters surrounded by sea ice) provide habitats unlike any on land. Indeed, under-ice communities provide a significant share of the primary production in polar regions. Thus, the water column—the pelagic realm of the sea—supports a great diversity of nektonic, planktonic, pleustic, neustic, and under-ice organisms that have few or no counterparts on land.

Seawater's buoyancy and viscosity keep food particles suspended, allowing marine organisms to feed by filtering the particles. This is a rare mode of obtaining food on land, the main example being orb-weaving spiders. But in the sea, thousands of species, from microscopic protozoa to colossal baleen whales, filter water as they move through it, eating the particles. Thousands of others, including benthic and intertidal invertebrates, rest on or in the bottom and filter food borne to them by currents.

For some organisms, the distinctive physical characteristics of the sea have left no obvious mark on life history patterns. Like many land organisms, whales and most sharks give birth to small numbers of large young, and entire groups of animals, such as amphipod, isopod, and mysid crustaceans, brood relatively small numbers of eggs and release young resembling tiny versions of their parents.

In the sea, however, the buoyancy, viscosity, and currents that favor permanent plankton (holoplankton) are also vitally important to species that shed their reproductive products into the currents to become temporary plankton (meroplankton) (Strathmann 1990). *Most* species in *most* marine groups in *most* marine ecosystems, whether nekton (e.g., herrings, squids) or benthos (e.g., seaweeds, starfishes), release large numbers of small planktonic eggs, sperm, spores, or larvae. For example, tropical conchs in the genus *Strombus* can release hundreds of thousands of eggs at a time; *Crassostrea virginica*, an estuarine oyster from the temperate northwestern Atlantic, can produce up to 70 million eggs in a single spawning. This life history strategy allows the young of even sessile species to disperse over long distances and reach less densely populated habitats.

On land, although fungi, plants, and spiders disperse mainly during early life history stages, many species, including insects, birds, and mammals, disperse mainly after they attain adult form. A much larger proportion of marine species disperse before they attain adult form. Some meroplankton are completely passive dispersers. Others, such as coelenterate, mollusk, and echinoderm larvae, can move, but their lateral movement is not strong enough to prevent them from being transported by currents. Still others, including some estuarine fish and crustacean larvae, have complex behaviors for migrating vertically among layers of water that move predictably in different directions, raising the chance that they will be carried to suitable habitats. The time spent in the plankton ranges from minutes to more than a year, but is most often hours to weeks. Even closely related species can differ in the amount of time their larvae spend drifting in the water column. The prevalence of planktonic dispersal has important implications for species' conservation and management.

As a habitat, the sea differs from the land by being far more three-dimensional. One reason is that the inhabited part of the sea is divided into discrete layers. Because seawater is a much better absorber of light than air, sunlight penetrates only a few hundred meters at most. This has at least two important consequences for life in the sea.

First, compared with the terrestrial realm, a far greater fraction of the marine realm is light-limited. In the vast majority of the sea's volume, there is no sunlight and therefore no photosynthesis. With the exception of the hydrothermal vents (see Box 1-1), the unlighted depths depend on the productivity of the surface layer, which rains down into the darkness as sinking whale carcasses and waterlogged wood, phytoplankton cell walls, and zooplankton fecal pellets.

Second, essentially all life on land occurs within the lowest layer of the atmosphere, the troposphere. But in the sea, sunlight directly warms the upper few percent of the water column, forming a "mixed layer" of warmer water that sits upon cold layers of progressively denser water masses. Because there is much more mixing within each layer than between adjacent ones, water masses tend to maintain distinct chemical characteristics and biological communities. Life in the sea occurs not only on the seabed, but in all of the distinct layers above it.

Both the land and sea have vertical gradients, but they differ. From the sea surface downward, illumination, temperature (other than in polar waters), salinity (in some places), and wave motion change rapidly, so biological zones tend to be very compressed near the surface but progressively less so as it gets deeper, a zonation pattern quite different from the land's.

Because seawater has a much greater heat capacity than air, temperatures change more slowly than on land (Steele 1985). As a result, organisms experience much smaller and slower temperature changes.

Moreover, many tropical marine organisms live very close to their upper thermal limit. For both reasons, marine ecosystems could be more susceptible to both local (thermal pollution) and global temperature changes.

Ocean currents impelled by winds, temperature, and salinity gradients, in combination with the Coriolis force, move water masses both horizontally—the Benguela, Kuroshio, Humboldt, and other surface currents—and vertically—the downwelling of cold polar water, upwelling from offshore winds, and the entrainment process in estuaries. Many organisms have special behaviors that allow them to stay within suitable habitats, rather than being swept away by these currents.

Both the air above the land and the water above the seabed are fluids, but, in general, the movements of water are far more important to the sea than are air currents to the land. Water movements strongly influence species' evolution and distributions (Denny 1988).

Another significant difference is that the sea and many of its ecosystems are far larger than those of the land. The Pacific Ocean alone could enclose all of the continents even if there were two Australias. But the difference in the volumes of marine and terrestrial habitats is actually much greater than the difference in area. Mt. Everest (Qomolangma Feng), the tallest mountain on land, measures more than 8,800 meters (29,000 feet), but is frozen and lifeless on its upper kilometers. Below the zone of perpetual snow and ice, nearly all terrestrial life occupies a thin, relatively two-dimensional zone from a few meters below the soil surface to 125 meters (410 feet), at very most, above it. Omitting the air above the land and sea, the average thickness of the terrestrial portion of the biosphere is probably less than 20 meters (66 feet). In contrast, marine life ranges from a few meters above sea level in the splash zone along wave-swept coasts, to 11,700 meters (more than 38,000 feet) deep in the Marianas Trench in the western Pacific, a realm averaging nearly 4,000 meters (13,000 feet) deep. Thus, the sea's habitable volume is hundreds of times greater than the land's.

Not only are the oceans much larger than landmasses; they are all connected. Species, on average, have larger ranges. Nonetheless, because the sea is full of boundaries, very few marine species are cosmopolitan. Unlike those of the land, however, many marine boundaries move on a time scale of days or even hours. These movements may be invisible to a casual observer on land, but they are vibrantly illustrated in photographs from Earth orbit showing swirls of phytoplankton concentrations and the animals they support shifting many kilometers from day to day. Biological changes are rapid when shifting currents bring different water masses and different pelagic communities to a given spot. In some years, for example, tropical water masses shift poleward and many tropical species temporarily penetrate normally temperate waters. Such short-term biotic shifts are far more common than on land, and bedevil attempts to draw fixed boundaries of marine biogeographic provinces.

The sea's boundaries are created by depth, salinity, temperature and light gradients, winds, currents, and upwelling. Although the boundaries move, within water masses conditions are remarkably constant. When swirls (eddies) of water break off from a warm current, forming huge (100 to 300 kilometers [62 to 186 miles]), lens-shaped masses of water surrounded by cooler waters, these warm-core rings maintain their distinctive physical characteristics and biological communities for many months.

Key to understanding life in the sea, far more than on the land, are the characteristics of the fluid medium, especially its motion. Unfortunately, the same currents that transport nutrients and organisms also transport pollutants. To avoid catastrophic changes in their internal fluids, organisms had to evolve ways to avoid desiccation before they could colonize the land, including having outer layers that are relatively impermeable. Marine algae and invertebrates are far more permeable than their terrestrial counterparts, so they are generally in greater physiological continuity with their surrounding medium (Ray 1976), which increases their susceptibility to chemical pollutants.

Marine Systems Are Biogeochemical Sinks

The oceans are salty, but they were not always so. Their salts have come mainly from the land. Over the aeons, some salts have issued from subsea volcanoes, some have come through the air as dust blown from barren land, but most have probably been leached from eroding soils and carried by streams into the oceans.

Salts continue to accumulate in the sea, a one-way and irreversible process. What has changed is the advent of a powerful new biogeochemical force: Humans have become a dominating influence on the cycling and movement of biologically important chemical elements within and among the land, atmosphere, and sea. We have greatly accelerated scarification of the land, thereby speeding soil erosion by water and wind. Humans are also adding untold kinds and amounts of chemicals to the biosphere. The same processes that have made the oceans salty are carrying those substances that we have released or mobilized into the oceans. The sea is biogeo-chemically downstream from the land.

Some of the most powerful evidence that this is happening has come from the circulation of radioactive materials released by nuclear bomb tests during the 1950s and early 1960s. Radioactive dust was washed by rain from the atmosphere over the land and sea, and was carried from the land to the sea, where it accumulated. Some entered marine food webs immediately, contaminating tunas and other fishes in Japanese markets. All traveled complicated biogeochemical paths and appeared again and again in the waters, sediments, and life of the oceans. By looking at an isotope of carbon (C^{14}) and one of hydrogen (H^3) as tracers, scientists learned details of oceanic circulation, including pathways and rates of movement (Woodwell 1967). Similarly, in the aftermath of the Chernobyl accident, intertidal seaweeds contaminated by air-borne iodine[131] from the nuclear reactor were found in Washington (USA) and British Columbia (Canada) (Druehl et al. 1988).

These experiences offer a cautionary tale. Virtually any substance loosed into the biosphere is carried seaward: The sea is biogeochemically "downstream" from the land. Insecticides or fertilizers sprayed on crops in Kazakhstan or Kansas might appear in rain over the Sea of Japan or Gulf of Maine. Polychlorinated biphenyls (PCBs) loosed from industries in São Paulo or Dakar become universal contaminants in marine systems, even the organisms of the deep sea.

Just as the oceans accumulate salt, they also accumulate those wastes of human enterprise that remain long enough to circulate. Introduction of long-lived substances into the sea therefore commits us to long-term contamination. For DDT, PCBs, and other chlorinated hydrocarbons that are highly soluble in fat and slow to decay, the commitment is long, indeed. This guarantees the contamination of life, including the human food web, until these substances are removed from circulation or broken down and rendered harmless. There is no reversing this process any more than there is the possibility of reversing the salinization of the oceans.

Tiny Producers, Larger Consumers

One important distinction between the land and sea concerns differences in primary producers and the animals that consume them (Ray 1976). In terrestrial ecosystems, primary producers (mainly vascular plants, such as grasses and trees) constitute the great majority of biomass and are often large. Except where seaweeds and marine vascular plants (e.g., seagrasses) occur along the coasts, primary producers in the sea are generally tiny and tend to be outweighed by the animals that eat them. Such animals are not disobeying the laws of thermodynamics; rather, their food organisms are microscopic, short-lived phytoplankton that can reproduce rapidly. In other words, the sea's primary producers have high turnover rates.

As a result, phytoplankton respond quickly to environmental changes. A pulse of nutrient enrichment might trigger a phytoplankton bloom, thereby affecting growth and reproduction of the zooplankton that eat them and of higher trophic (feeding) levels of consumers with progressively longer turnover times. In the sea, many of the longest-lived organisms are large fishes, seabirds, and whales, which occupy the highest trophic levels. But on land, the longest-lived organisms are trees (Steele et al. 1989; Steele 1991), which are primary producers. Their populations respond much more slowly to environmental fluctuations than the sea's primary producers. Because marine primary producers have such high turnover rates, they and the animals that depend on them can be vulnerable to severe disturbances of even relatively short duration.

Research and Monitoring Are More Difficult

Humans are terrestrial, warm-blooded, air-breathing animals who depend mainly on sight, so the sea is an alien environment for us. Our lungs cannot extract the scant amounts of oxygen in it. Below 1,000 meters (3,280 feet), almost all of the sea's volume is frigid, even in the tropics, and completely devoid of sunlight. Morcover, a human at 10 meters (33 feet) below the surface experiences twice the pressure as one at sea level, and the pressure increases by one atmosphere with each additional 10 meters of depth. These hostile physical conditions inhibit exploration and dramatically increase the cost of obtaining data, which explains why people know so little about the oceans, especially the deep sea.

Until quite recently, most research in the sea has used indirect means such as nets, grabs, bottles, and instruments lowered into the depths from a surface platform. But these methods provide a very incomplete picture of the sea and its life; little would be known of rainforests, deserts, or agroecosystems if standard oceanographic techniques were used to explore them. The sea surface is relatively easy to study using well-known, although not necessarily inexpensive, techniques. Ships, buoys, aircraft, and satellites allow researchers to collect data without having to immerse themselves. The invention of practical scuba equipment in the 1940s has dramatically increased the amount of research conducted in nearshore waters. But the effects of pressure on human physiology generally limit divers to the uppermost 30 or 40 meters (about 100 to 130 feet), or, in exceptional cases using exotic mixes of gases, down to about 700 meters (2,300 feet). The time divers spend underwater is obviously far shorter than researchers can spend on land.

Frustration with these approaches has stimulated scientists and engineers to develop a growing array of manned and remotely operated vehicles equipped with cameras, lights, and sensory devices. Application of such technology has barely begun, however. The deepest part of the ocean has been glimpsed directly only once when, in 1960, two men ventured there in the bathyscaphe *Trieste*. More than 30 years later, only five manned systems are capable of visiting even half the ocean's depth (Walsh 1990), the deepest-operating being the Japanese submersible *Shinkai 6500*. However, there are now hundreds of remotely operated vehicles that can send television images of the depths to operators and researchers on the surface or on land.

Although the cost of these machines is a small fraction of those used for space exploration, they are not inexpensive to build or operate. These costs will be considered affordable only when people realize that a much improved knowledge base is essential to sustainable use and protection of the 71 percent of our planet covered by the ocean.

Similarities Between Marine and Terrestrial Conservation

Although major differences between the sea and the land affect strategies for protecting and sustainably using their biological diversity, some lessons from terrestrial conservation and resource management apply to the sea. The sea, like the land, is a complex patchwork of physical environments that are occupied by different communities of species. As on land, some marine species and ecosystems are of special concern to humans. Marine biological diversity is being reduced at all levels, through selection against certain genetic attributes, elimination of populations, extinction of entire species, and destruction, fragmentation, and simplification of ecosystems, the same processes that are happening on land.

The Sea Is Not Homogeneous

From the edge of the polar sea ice to the tropics, the sea looks homogeneous to anyone viewing it from a low angle. Only differences in color hint at the diversity of marine topography and overlying waters. This heterogeneity is crucial, however, because different physical settings favor different kinds of organisms, that is, ecosystem diversity. Ecosystem diversity in the marine realm very likely exceeds that of the terrestrial realm, yet because marine ecosystems are mostly hidden from view, humans are less aware of it.

Factors Creating Diversity in the Sea

The diversity of marine life reflects the physical factors affecting organisms and their interactions with other organisms. What factors create the heterogeneity that, in turn, creates opportunities for different genotypes, species, and communities?

Continents are masses of rocks that sit above and are separated by ocean basins made of different kinds of rocks. The continents' edges, known as continental shelves, are submerged beneath shallow marine waters. Deeper areas include steeper continental slopes, flat abyssal plains, steep-sided trenches, and submerged mountains. Topographic relief is high in the sea; the Earth's tallest mountain, longest mountain range, and deepest canyon are all in the ocean. Much seabed topography, however, is smoothed by thick accumulations of sediments. Different communities of organisms live in these different ecosystems.

Along most shorelines, tides expose the edge of the seabed to the air daily or monthly. These intertidal zones receive strong sunlight, drying winds, high temperatures, sub-freezing cold, fresh rainwater, the crashing force of waves and ice floes, and, on some coasts, even air pollution. Despite these factors, many intertidal zones around the world are remarkably diverse, although severely threatened by human activities. The influence of these factors on the seabed disappears or decreases sharply with depth, affecting the composition of intertidal and shallow subtidal communities.

Surface waves and currents can suspend sediment particles of various sizes, so areas experiencing different wave forces have substrates ranging from boulders to fine muds. Other substrates are solid rock rather than sediment. Patches of these different substrates have different communities of organisms. Light is another factor that determines the communities on the seabed: Only the shallower areas receive enough light to support attached plants and reef corals, which need light to support the algae that live within their tissues.

Like the continents, the seas are affected by plate tectonics, the movement of the Earth's crustal plates. Until the Jurassic period (Figure 2-2), the continents were joined, surrounded by one world ocean. The Atlantic Ocean began forming when plate movements separated the Americas from Eurasia and Africa about 170 million years ago. The Sea of Cortez was born only 6 million years ago. When and where these bodies of water formed and what has happened since their formation determine which species occupy them now. For example, the Caribbean has been separated from the tropical East Pacific by the Isthmus of Panama for only a few million years. Although many identical or very similar species do occur on both sides of the Isthmus, many others have no close relatives on the other side. The differing communities in different regions reflect their separate histories of extinction, colonization, and evolution.

Added to the heterogeneity of the seabed is the heterogeneity of overlying waters. The amount of sunlight and fresh water reaching the sea surface determines seawater density, which increases as temperature falls or salinity rises. Masses of seawater of different densities mix slowly and tend to keep their identities both horizontally and vertically. They have distinctive biotic communities, even when they are moved long distances by currents. In the open ocean, the lighted upper layer—the euphotic zone—is very different from deeper and usually colder layers that lack photosynthesizers and host very different animals. Indeed, two places on the surface 1,000 kilometers apart are likely to be far more similar to each other biologically than either is with the waters 1 kilometer below.

Coastal waters differ markedly from oceanic waters. The physical conditions in estuaries, bays, and the neritic waters that overlie the continental shelf are more variable than those in the open ocean—they can be much warmer or colder, fresher or saltier—and are more turbid because they carry far more suspended sediments, nutrients, and phytoplankton. Indeed, these differences often manifest themselves as contrasting colors: neritic waters are usually brownish or greenish, while oceanic waters are almost always blue. For these reasons, coastal communities are very different from those of the open ocean.

Nutrient availability is very important in the sea. When the shallows are well lighted, phytoplankton absorb nutrients. Zooplankton that eat phytoplankton egest nutrient-rich fecal pellets, which sink quickly into dark, cold water masses, thereby depleting the shallows of nutrients. In some places, however, upwelling currents carry deep, nutrient-rich water back into the shallows, which stimulates the growth of phytoplankton and changes the color of the water. The sea's gloriously complex shifting patterns of nutrient-poor and nutrient-rich water masses, with their different plankton communities, are easily seen from space.

Disturbances in the sea harm some species but can create opportunities for others. Disturbances vary widely in space and time, from millennial glacial advances and retreats that affect an entire hemisphere, to cyclones, typhoons, and hurricanes that affect hundreds of thousands of square kilometers for decades, to the disturbances caused by the foraging of individual crabs, which affect a few square centimeters for weeks. This varied array of natural perturbations greatly increases biological diversity in the sea. That is why areas with similar physical conditions now but whose histories are different can differ biologically.

The sea's layered mosaic of physical factors is mirrored by a dynamic patchwork of distinct biological communities. The sea appears homogeneous to a casual observer only because our terrestrial vantage does not allow us to see its great diversity.

Biogeography of the Sea

The geographic distribution of life and the processes that shape it are the subjects of the science of biogeography. The basic concept of biogeography, whether on land, in fresh waters, or in the sea, is simple: Organisms occur someplace only if they can get there *and* survive there. In practice, biogeography is wonderfully intricate because our planet is so dynamic. Indeed, the process of inferring why species or communities of organisms are where they are requires the deductive reasoning of a detective because so many answers about modern distributions lie buried (literally) in the past (Vermeij 1978).

Biogeography and Conservation: An appreciation of biogeography is vitally important in marine conservation because so many policy decisions hinge on where things occur and how to treat particular places. It becomes even more important in a world where greenhouse warming is likely to bring very large changes in distributions within (as biogeographers reckon) a very short time.

In the sea, as in the other realms, the basic tool of the biogeographer is the distribution map. By examining distributions of different species, biogeographers can infer which factors shape their distributions. Indeed, much classical biogeography (e.g., Briggs 1974) is devoted to examining where distributional limits of large numbers of species coincide (which are the borders between biogeographic regions) or seeking clusters of endemic species (species that occur only within a limited area). But marine biogeographers also use information from many other disciplines, including physiology, ecology, taxonomy and systematics, evolution, oceanography, vulcanism and plate tectonics, and paleontology.

As complicated as terrestrial biogeography is, marine biogeography is even more so because the sea is more three-dimensional and dynamic than the land. Within a particular ocean basin, the distributions of intertidal species, species of the upper layer of the open ocean, and deepsea species are very different because geographic patterns of physical conditions in their respective realms are so different. Experts in biogeography of intertidal, open ocean, and deepsea species would have difficulty agreeing on the boundaries of biogeographic regions. Further, so few people are working in marine biogeography that major realms of the sea have been studied only minimally. As a result, there is no global marine biogeographic scheme that is widely agreed upon by experts, although the work of Hayden et al. (1984) is a step in the right direction.

Classical biogeography tends to assume that the modern distributions of organisms are limited by barriers to dispersal; for example, the cold waters of southwestern Africa make it very difficult for species to disperse between the tropical waters of West Africa and East Africa. But there is another "school" of biogeographers—vicariance biogeographers (e.g., Springer 1982)—who see present distributions more as reflections of past events, such as continental drift. For marine conservationists, the distinction between these two schools is not trivial because the actual factors that limit species' distributions could have important implications for the management of populations and areas.

The lack of a broadly accepted marine biogeographic scheme is a serious gap in marine conservation and management. Analogous terrestrial schemes by Bailey (1989), Küchler (1964), Udvardy (1975), and others have been useful in deciding what to protect and how to manage. If one biogeographic province has 20 protected areas and another has none, it makes sense to focus more attention on the neglected region. If it is known that an ecosystem within a biogeographic province functions in a certain way, managers dealing with a similar ecosystem in the same province can give this information more weight than information from an ecosystem in another biogeographic province. It is safer to extrapolate coral reef protection or management principles from the Australian portion of the Great Barrier Reef to the part that lies in Papua New Guinea than to those of the reefs of Panama.

Another gap is our lack of understanding about endemism in the sea. There is growing scientific evidence that many marine species are actually far less widely distributed than previously thought. Understanding the geographic patterns of endemism should be as valuable to conservation of biological diversity in the sea as it is starting to be on land and in fresh waters.

Applications of Biogeography to Management: Two applications of biogeography to the management of marine species and ecosystems merit special attention. The large marine ecosystem (LME) concept (Sherman et al. 1990) is based on the idea that the sea consists of fairly distinct regions within which physical conditions, biological communities, fish stocks, etc., are so closely linked that they are best managed as a regional ecological unit. LMEs are not based solely on biogeography, however; they are also economic and political. The idea of ecosystem management—which seeks to consider all the factors that affect the species within an ecosystem, rather than taking a narrow sectoral approach—is gaining increasing attention from decision makers. Ecosystem management has the potential to improve effectiveness of management across a wide spectrum of situations, from areas "untouched" by humans to those that are heavily used, from areas solely within the jurisdiction of one country to those shared among nations.

The concept of "hotspots" has gained increasing popularity for use in selecting areas to be protected. These hotspots (the term also has other uses) are areas rich in total numbers of species or numbers of a particular kind or category of species, such as seagrass, endemic, or endangered species. Although it has been applied mainly to the terrestrial and freshwater realms, it could be equally useful in the sea. For example, McAllister et al. (1986) showed that the Cumberland-Tennessee Plateau was richer in freshwater fish species than any other area in the United States and identified several hotspots of endemic species. Hotspots for different taxonomic groups might or might not coincide.

A powerful tool for finding hotspots is geographic information systems (GIS), which are computer software systems that can compile records of the occurrence of individual species to make maps of, say, species density. GIS can be used at global and regional levels to help select areas of importance in conservation and to map hotspots of human impacts. For example, the Coral Reef Fish Specialist Group of the Species Survival Commission of the World Conservation Union (IUCN) is using GIS software on a microcomputer to find hotspots for coral reef fish species, localized coral reef fish endemics, and human impacts.

Understanding life on the scale of biogeographic provinces is essential, but not sufficient. Managers and conservationists need to know how natural processes in the sea produce hierarchical spatial scales, from particular community types all the way up to biogeographic regions, because various human activities exert their effects at different scales. Although there are established fields of marine community ecology and marine biogeography, there is not yet a marine analogue to the emerging field of landscape ecology (Forman and Godron 1986; Urban et al. 1987), which is needed to fill the gap between them. Strengthening and applying the science of marine biogeography, and creating a science of marine landscape ecology (e.g., Ray 1991), will provide powerful tools for decision makers intent on improving protection and sustainable use of the seas.

Some Species Attract Special Concern

In one sense, all species are equally important: All are survivors of 3.5 billion years of change. But certain species loom large in people's mental landscapes. Some of this "importance" we accord species is subjective, coming from admirable traits people believe they possess: noble tarpon (*Megalops atlanticus*), amusing penguins, graceful albatrosses (*Diomedea* spp.), fierce sea eagles (*Haliaeetus* spp.), gentle baleen whales, intelligent dolphins, etc. Some species evoke special interest for their uniqueness, having survived aeons largely unchanged, such as coelacanths, or for defying our notions of what they should be, such as mudskipper (*Periophthalmus* spp.) fishes that climb trees. Others have special ecological significance, such as krill (*Euphausia superba*), which are shrimplike crustaceans that support most Antarctic squids, fishes, birds, and mammals, and sea otters (*Enhydra lutris*), a keystone species that encourages growth of kelp forests in the Northeast Pacific by eating kelp-grazing sea urchins. Still others are objects of interest because they are at risk of extinction. Whether the interest that people accord them is based in science, economics, or humane feelings of kinship, these species receive heavy emphasis in current efforts to protect and manage marine life.

Marine Mammals

One group considered of special concern as a whole is the marine mammals, the archetypal "charismatic megafauna" of the sea. Their large size and the fact that they spend time at the sea surface make them more likely to be noticed by people. Furthermore, marine mammals remind us of ourselves, being homeothermic, air-breathing creatures that bear few live young, have long parental-care periods and complex social structures, and exhibit behaviors sometimes attributed to intelligence and altruism. Another seldom articulated but ecologically more important reason is that marine mammals can serve as "strong interactors" or keystone species in ecosystems.

Marine mammals are among the more diverse groups of large mammals. There are more than 75 cetacean (whale and dolphin) species (Minisian et al. 1984), 35 pinniped (seal, sea lion, and walrus) species (King 1983), several species of sirenians (manatees and the dugong) and otters, and the polar bear (*Ursus maritimus*). Marine mammals are found throughout the world from the tropics to the poles, from the open ocean to coastal waters and estuaries; indeed, several dolphin species are strictly freshwater animals. And although all must surface to breathe, some can dive to depths of more than 1.5 kilometers (0.93 miles) for more than an hour.

Marine mammals have provided subsistence and commodity goods (meat for food, skins for leather, blubber for oil, baleen for corsets) for coastal peoples worldwide. Although some populations have withstood hunting reasonably well, the fact that they bear few young and, in general, are slow to reach reproductive age makes marine mammals especially vulnerable to overexploitation. As a result, *all* cetaceans are currently listed on Appendices I or II under the Convention on International Trade in Endangered Species of Wild Fauna and Flora (CITES), along with 11 species of pinnipeds, two of marine otters, three of manatees, and the dugong. Thus, for a variety of reasons, marine mammals receive special attention from marine conservationists and managers.

Sea Turtles

Another group of special concern is sea turtles, an ancient lineage of reptiles with a fossil record of more than 200 million years (Pritchard 1979). Although there are currently only seven species recognized, most of them are widespread and had very large population sizes until this century, suggesting their successful adaptation to the marine environment. Like marine mammals, sea turtles are large (a leatherback sea turtle, *Dermochelys coriacea*, weighing 916 kilograms [2,020 pounds] washed up in Wales in 1988) and breathe air. Indeed, the leatherback, like marine mammals, is warm-blooded. However, sea turtles lay eggs that require them to emerge periodically from the ocean to nest on beaches, a time when both adults and eggs are extremely vulnerable to exploitation.

Like marine mammals, sea turtles have been hunted around the world for both subsistence and commercial trade. The eggs are used as a protein source for both people and livestock in many coastal regions, the meat is consumed by humans, the cartilage of the green sea turtle is the source of green turtle soup, and the scutes of the carapace, particularly those of the hawksbill, are the only natural source of "tortoise shell," used for jewelry, eyeglass frames, Japanese ceremonial combs, and other items. However, as with marine mammals such as whales and manatees, sea turtles are becoming increasingly valuable alive, in a growing tourist industry based on viewing their nesting and hatching.

Although sea turtles lay numerous eggs, they provide no parental care and they reach reproductive maturity even later than marine mammals; Balazs (1981) estimates that some green turtles may not reproduce until they reach 50 to 60 years of age. The number of eggs is not large enough to compensate for the tremendous mortality of turtles of all sizes caused by directed hunting, incidental deaths in fishing gear, and losses of nesting habitat. Populations have plummeted worldwide (King 1981; Ross 1981), with many nesting colonies becoming completely extirpated. Moreover, sea turtles are susceptible to global warming because their sex is determined by the temperature at which the eggs are incubated. All seven species are now listed on CITES Appendix I; six are listed as endangered or threatened by IUCN. Thus, sea turtles are species of special concern by virtue of their evolutionary persistence, their values to humans, and their vulnerability and current rareness.

Endangered and Threatened Species

Efforts to assess marine species in jeopardy (Table 3-1) are in their infancy. The process of listing marine mammals, birds, and reptiles that are at risk is well under way, but these constitute only a tiny fraction of marine species likely to need protection. In contrast, the process of listing fishes, invertebrates, algae, and the various groups of microorganisms that might need special protection has barely begun. There is ample justification for extending protection to species that are large, beautiful, and appealing to people, and that are, indeed, in trouble, but there is no justification for ignoring things that people consider unappealing or too small to notice. The integrity of the biosphere could depend far more on a phytoplankton species or benthic polychaete worm lacking a common name than on a charismatic mammalian species that attracts the lion's share of conservationists' attention.

Some Areas Are Especially Important

Decision makers faced with difficult choices often ask biologists to provide a simple quantitative measure of an area's biological importance. Unfortunately, an appreciation of the complexities of living systems makes such requests impossible to fulfill. No index that measures a single biological attribute of geographical areas can convey intricacies (such as linkages among species) any better than such indices as intelligence quotient (IQ) and gross national product (GNP) measure intelligence or economic robustness. Nonetheless, even the most impartial biologist will admit that some areas merit special attention. Because there might not be enough resources to protect all of them at once, decisions will be made about which marine areas are most critical. Understanding which biological attributes make certain areas especially important is essential to decision makers wrestling with the question of what to protect.

Areas of High Diversity

Because species diversity is widely and incorrectly equated with biological diversity, it is tempting to make it *the* criterion for ranking areas. Species diversity is conceptually simple and "mappable," characteristics that lend nicely to pinpointing hotspots. It has been used for deciding what areas merit conservation in the terrestrial realm (especially tropical forests), and has recently been proposed for use in marine conservation as

TABLE 3-1. Some marine animal species in jeopardy (endangered, threatened, vulnerable, or similar listings).
CITES = Convention on International Trade in Endangered Species of Wild Fauna and Flora; IUCN = International Union for Conservation of Nature and Natural Resources Red Data Books; ESA = U.S. Endangered Species Act.

Name	List	Range	Cause
precious corals: *Corallium* spp.	IUCN	Mediterranean Sea, Central Pacific	overexploitation (jewelry, novelty items)
giant clam: *Tridacna gigas*	CITES, IUCN	Indo-West Pacific	overexploitation (shell and meat)
Triton's trumpet snail: *Charonia tritonis*	IUCN	Indo-West Pacific	overexploitation (shell and meat)
great white shark: *Carcharodon carcharias*	IUCN	Cosmopolitan, mainly cool temperate	overexploitation (persecution)
Key silverside: *Menidia conchorum*	IUCN	Lower Florida Keys (USA)	physical ecosystem alteration, pollution
Knysna seahorse: *Hippocampus capensis*	IUCN	South Africa	physical ecosystem alteration/ overexploitation (aquarium collection)
Cuban crocodile: *Crocodilus rhombifer*	All	Cuba	overexploitation (hides), hybridization
Peruvian penguin: *Spheniscus humboldti*	CITES, IUCN	Peru, Chile	El Niño (food reduction), physical ecosystem alteration (guano mining)
dark-rumped petrel: *Pterodroma phaeopygia*	IUCN	Galápagos (Ecuador) and Hawaiian (USA) islands	alien species (rats, mongooses)
Madagascar heron: *Ardea humbloti*	IUCN	Madagascar, Comoro Islands	overexploitation (meat and eggs)
Steller's sea lion: *Eumetopias jubata*	ESA	Japan, eastern Russia, northwestern USA	overexploitation (competition from trawlers)
marine otter: *Lutra felina*	All	Peru, Chile S of Strait of Magellan	overexploitation (fur hunting, persecution)
La Plata dolphin: *Pontoporia blainvillei*	CITES, IUCN	Argentina, Uruguay, Brazil	overexploitation (bycatch in shark fishery)

well. If well-studied taxa are used, species diversity is the easiest way to rank areas, because species are relatively easy to count, particularly on land. But how valid is it as a measure of conservation priority?

Species diversity is probably best viewed as one among a number of useful criteria. All things being equal, it is probably better to conserve an area with 500 species than one with 300, assuming that the 500 include the 300. But all things are seldom equal. What if the 300 (or at least some of them) are *not* among the 500? The smaller assemblage might contain species that people find more important to conserve for some reason such as economic importance, ecological importance, evolutionary significance, or endangerment. For example, an area that harbors the only remaining coelacanth species, the last surviving one of its subclass of fishes, might merit special priority even if the area is less diverse than adjacent ones. Or, an area with lower diversity might be better at providing ecological services important to people. Because marine ecologists have so little information on the importance of particular species and ecosystem services in each area, there will not always be a strong basis on which to make informed choices.

Furthermore, species diversity differs markedly on both ecological and biogeographic spatial scales. On an ecological scale, is it more important to conserve a coral reef with 60 fish species or a mangrove forest with 40 fish species, when this reef is unusually species-poor for a coral reef while this mangrove forest is unusually species-rich for its ecosystem type? Then again, bottom-dwelling invertebrates might be even more diverse in the deep sea than in coral reefs. Does this mean that coral reefs and other shallow-water ecosystems merit a lower priority for protection?

On a biogeographic scale, using species diversity as the sole criterion for priority would greatly simplify decision making because the Indo-West Pacific—especially the area between Indonesia, the Philippines, and northeastern Australia—is so much richer in species than any other shallow-water area that conserving the rest of the world could be ignored. Of course, people might not concede that the marine life of Haiti, Argentina, Angola, Turkey, Korea, or Antarctica is unworthy of attention.

If high species diversity is used as a measure of importance, it should be used *within* but not *among* different biogeographic regions. Similarly, within particular biogeographic regions, it should be used *within* ecosystem types but not *among* them. This approach would avoid scientifically dubious and politically divisive arguments about whether one nation's "best" coral reef is more important than another's "best" kelp bed.

Nonetheless, areas with high diversity do merit attention because they have the conditions necessary to maintain large numbers of potentially competing species in sympatry. Depending on what people consider important, however, areas of high diversity may not be most critical to the sea as a whole, for various reasons. Coral reefs generally have high species diversity, but tend to be low in another important biological attribute, endemism. Despite having lower species diversity than coral reefs, salt marshes, mangrove forests, and seagrass beds can have special importance because they serve as significant nursery areas and their productivity supports important food webs. Moreover, some areas are especially important seasonally because they are critical to key elements of marine biological diversity even if their diversity is low. These include courtship or spawning areas, nursery grounds, migration corridors, and stopover points.

Conserving biological diversity is not simply an exercise in counting. High species diversity is only one of a number of important factors to be considered in assigning priority in protection and management schemes.

Areas of High Endemism

In general, species with small geographic ranges are more susceptible to extinction than species with large ones. Species with smaller ranges are likely to have smaller populations. And, the smaller the range, the higher the likelihood that some adverse condition can overwhelm the entire population.

Endemic species are ones with relatively narrow distributions. For example, a murex snail (*Muracanthus nigritus*) and the Gulf opaleye fish (*Girella simplicidens*) are endemic to Mexico's Sea of Cortez (Brusca 1980; Thomson et al. 1987). Although endemics have a small geographic range, they are not necessarily rare within it. For example, the spotty (*Pseudolabrus celidotus*), an endemic wrasse of New Zealand, is the most abundant of the larger reef fishes there (Ayling and Cox 1987). The number of endemic species varies greatly from place to place in the sea, as it does on land. It is therefore important to identify marine areas that are rich in endemics and to establish whether some kinds of ecosystems are especially likely to have them.

Although much remains to be learned about patterns of endemism in benthic species, endemism seems especially common in temperate and marginally tropical regions where latitudinal temperature gradients are especially steep or where species have found a refuge from large-scale environmental fluctuations. For many groups of algae and invertebrates, examples of such regions are the northern (Senegal) and southern (Angola) limits of the West African marine province, southeastern Brazil and adjacent Uruguay and Argentina, the northern (Sea of Cortez) and southern (Ecuador and northern Peru) limits of the Panamic Province in the eastern Pacific, warm-temperate Japan and the adjacent coasts of Korea and China, the southwestern Cape of South Africa, and temperate Australia.

Other pockets of high endemism, mainly reflecting the isolation of islands or ocean basins, include the Okhotsk Sea and Kurile Islands in the northwestern Pacific, the continental coast of northern South America, the South Atlantic oceanic islands (especially Saint Helena, Ascension, and Fernando de Noronha), the Red Sea, the Coral Sea, the islands of Polynesia (especially Hawaii, the Marquesas, Easter Island, the Societies, and Tuamotus), and the Galápagos Islands. Endemism is low in many parts of the tropical Indo-West Pacific and Caribbean, but endemics occur even there. The coastal waters of Antarctica are very high in endemism compared with the Arctic Ocean.

Species in some ecological groups generally have smaller-than-average geographic ranges. This is notably true for high intertidal invertebrates. Rocky-shore fishes in the tropical East Pacific tend to have smaller ranges than do sand-bottom species (Rosenblatt 1967), perhaps because rocky habitats are patchier there than are the extensive soft-bottom ecosystems. A similar situation may well exist in West Africa. Mangrove-associated species, only a few of which cross significant oceanic barriers (see also Reid 1985), seem to have especially small ranges. Species associated with coral reefs, however, tend to have broad distributions, as do many holoplanktonic and deepsea species.

The fossil record strongly suggests that many species have undergone significant range contractions during the past several million years. The geographic refuges to which these species are now restricted often have high primary productivity and are near continents. These areas include the tropical eastern Pacific and eastern Atlantic, the north and east coasts of South America, the Indo-Malaysian region, the cool-temperate northwestern Atlantic, and the northwestern Pacific; there may well be others. Even for those species that have not undergone reductions in range, these refuges are probably important sources of recruits; that is, they export their young to other areas. Many of the restricted species have reasonably large ranges within these refuges and therefore would not usually be regarded as endemics, but ecosystem degradation in these regions might interfere with the recruitment of species and would bring many other species closer to extinction.

Our limited knowledge does permit identification of some regions where endemic species are especially common and of other areas that now serve as important refuges for previously more wide-ranging species. Both types of areas deserve special concern. However, a fundamental principle of conservation is that no population can persist if its members are overexploited or its resource base is destroyed throughout its range, no matter how large that range is.

Areas of High Productivity

To conserve what people value in the seas, scientists and decision makers need to talk about conserving not only the *components*—populations, species, and ecosystems—but also the *critical ecological processes* that maintain and are maintained by these components. Primary production is one such process. In shallow estuaries and coral reefs, productivity compares very favorably with intensively cultivated agricultural land. Gross productivity in these ecosystems can be as high as 20 grams of carbon per square meter per day. Indeed, the highest measured primary production in any ecosystem is in wave-beaten northeast Pacific intertidal kelp beds (Leigh et al. 1986). Areas of high primary productivity are usually good fishing areas and harbor the highest populations of fish-eating marine mammals and birds.

Two key factors affect marine productivity: the rate at which nutrient materials are renewed and the amount of light available to plants. Because these vary markedly in space and time, primary productivity in the ocean is far from uniform. Well-lighted shallows in the centers of ocean basins generally have low productivity because nutrient-rich materials (dead organisms, molts, and fecal pellets) sink below the photic zone and their nutrients are not rapidly replaced; they are nutrient-limited. Conversely, under heavy polar pack ice, nutrient levels can be high, but scarce light limits productivity. Nutrient concentrations are much higher on continental shelves and slopes, where there are closer sources of nutrients and vigorous vertical mixing, and in upwelling areas. Upwelling usually occurs where winds persistently drive surface water away from the coast, so that cold, nutrient-rich subsurface water rises to replace it. Upwelling returns nutrients to the photic zone as they are released by decomposition of organic material in deeper waters. Higher primary productivity results, five to ten times higher than in the open ocean. Most major fisheries are located in continental shelf waters or in areas of upwelling. Although upwelling areas cover only 0.1 percent of the ocean, they contribute one-third of the world catch (Cury 1991), making them exceptionally important areas to protect.

Upwelling areas are especially important to humankind because the nitrates and phosphates they lift into surface waters are nutrients essential for production of phytoplankton. In most marine ecosystems, phytoplankton are eaten by zooplankton, which in turn are eaten by small fishes. But in upwelling areas, the phytoplankton tend to aggregate into clumps or filaments, which can be consumed directly by small fishes such as anchovies. Not only is primary production high in upwellings, but their short, efficient food chains yield far more fish per unit of primary production, a fact that does not go unnoticed by seabirds, marine mammals, and humans that seek dense concentrations of fishes.

Upwelling occurs mainly off the western coasts of continents, particularly in the trade wind belts of the tropics and subtropics, and around Antarctica. The surface waters off Peru and northern Chile are colder than usual for their latitudes because of upwelling. Upwelling also occurs off the northwestern and southwestern coasts of Africa, in the Arabian Sea, and in the California Current system. All of these areas have high primary productivity and are locations of major commercial fisheries.

Every few years, a change in the ocean-atmosphere system called the Southern Oscillation alters wind and weather patterns all the way across the tropical Pacific or over an even larger area. Upwelling off Peru weakens or ceases—a phenomenon called El Niño—and primary productivity drops dramatically. El Niños can have disastrous consequences for Peru's anchovy fishery and for fish-eating seabirds. The frequency of El Niños seems to be increasing, perhaps due to the increase in atmospheric greenhouse gases.

Coastal and upwelling areas constitute only a small fraction of the marine realm, but they are disproportionately important to people and other animals that require dense concentrations of food fishes. For this reason, they merit special attention in marine conservation efforts.

Spawning Areas That Serve as Sources of Recruits

Most marine fishes and invertebrates have life cycles that include a plank-tonic larval phase. These phases last from hours in some invertebrates to many weeks, months, and even years in exceptional cases among both invertebrates and fishes. Although it is advantageous in several ways, this kind of dispersal has one important disadvantage: Larvae are carried to a different location from the one where they were born. This occurs because these tiny animals have little capability to resist transport by the water mass within which they are produced.

If this were to happen for generations, species' ranges would move progressively farther downcurrent. However, the adults of many species whose larval phases lasting several weeks or more overcome this problem by migrating to particular spawning sites. From these sites, currents carry their larvae to locations where the juvenile (and perhaps the adult) stages are spent. Thus, spawning migration by adults ensures that larvae are produced in a location from which, under usual conditions, they will be transported to suitable juvenile or adult habitat. This is called "life cycle closure" (Sinclair 1988).

The concept of life cycle closure is very important for protection and management of many marine species, on both local and broad geographic scales, because, without it, a population cannot persist. Because the areas where individuals reproduce are not the same as the ones where they live out their adult lives, these spawning areas are no less crucial to the species. Unfortunately, when adult animals migrate to them at predictable times to reproduce, their aggregations can be particularly susceptible to overex-ploitation.

Whereas the story of the various salmonid fishes that return from the ocean to home streams to reproduce is well known, many other commer-cial fish species have localized spawning grounds in specific oceanic or coastal locations (Harden Jones 1968). Sea turtles, many seabirds, and some cetaceans also return to the same sites to mate, lay eggs, or bear and rear their young.

Several species of tunas have broad ranges that extend from tropical waters into temperate waters, but all of them return to tropical waters to spawn. The most extreme example of this pattern is the southern bluefin tuna (*Thunnus maccoyi*), which is widely distributed between 30° and 50°S (Collette and Nauen 1984), but appears to spawn only in a relatively small area between northwestern Australia and Java (Caton et al. 1989). Similarly, bluefin tuna of the Northern Hemisphere can forage north of the Arctic Circle, but spawning in the Atlantic appears to be restricted to a small area in the Gulf of Mexico and another in the Mediterranean.

In coral reef ecosystems, many larger species exhibit spawning migrations to specific breeding sites (Johannes 1979). Among groupers (family Serranidae), in particular, spawning sites are strongly localized (a few thousand m^2 of reef edge) and are predictably occupied for short (5- to 15-day) periods at the same time each year (Figure 3-2). The spawning aggregations are dense with fish that have migrated from tens of kilometers away, and are attractive to fishers because these high-value species are more susceptible to capture.

Johannes (1978) has demonstrated how pre-contact Pacific civilizations managed fisheries for species that aggregate to spawn. Throughout Polynesia and Micronesia, local villages had tenure over adjacent reef resources, which they managed for long-term use. Micronesian fishers possessed knowledge unknown to scientists about spawning times and other fish behaviors, and their management could be considered "enlightened" compared with that of today's industrialized fisheries. Where spawning aggregations occurred within a village jurisdiction, fishing was sometimes banned during the period of aggregation.

This ban could either have been a religious injunction or taboo, or was a secular edict imposed by village leaders. In some places, such management has been lost with the influx of Western culture and fishing technologies that effectively broke down local control of sea resources. However, there are serious efforts under way in a number of Pacific island countries to reinvigorate these forms of marine resource management (Personal communication with Robert Johannes, Commonwealth Scientific and Industrial Research Organisation [CSIRO], Hobart, Tasmania, Australia). At present, in both the Pacific and Caribbean, fishers who know of spawning locations keep the information to themselves and fish to maximize short-term profit. A number of Caribbean spawning aggregations of grouper species are known to have been diminished or eliminated by fishing in recent years.

Migrations to breeding areas are not confined to fishes. The phenomenon has long been known (and exploited by hunters) in seals, baleen whales, seabirds, sea turtles, and penaeid shrimps, and was more recently discovered in spiny lobsters (Herrnkind 1980) and pelagic swimming

Figure 3-2. Spawning aggregation of Nassau groupers. *Like many tropical grouper species, these Caribbean fish come from long distances to spawn at certain places at the same time each year. In some traditional societies, people had taboos against fishing such aggregations. Many spawning aggregations are now specially targeted, and some grouper species have become rare.*

crabs (Norse and Fox-Norse 1977). There are strong indications that many squid species also concentrate in restricted areas for spawning and feeding. The best-known examples are several species of the commercially and ecologically important squids of the genus *Loligo*, which seasonally converge in huge numbers in specific bays at locations scattered around the world to mate and to lay clusters of benthic egg capsules. These spawning aggregations have long been the targets of heavy fishing pressure.

Not all aggregations are for spawning, however. The paralarvae of some squids and octopods concentrate in limited areas, sometimes in huge numbers (Vecchione 1987). This early life history stage is likely the most sensitive to anthropogenic disturbances. Because the paralarvae of many species (e.g., the long-finned squid, *L. pealei*) occur mainly in surface waters, human activities such as petroleum transportation in areas where paralarvae concentrate are of special concern.

Breeding aggregations and other aggregations in restricted areas make many marine organisms especially vulnerable to overexploitation and to anything—such as physical alteration or pollution—that might not be so worrisome if it occurred throughout the species' range. Such areas, therefore, require special protection.

Nursery Grounds

In some marine animal species, all life history stages share the same habitats. But juveniles in many mobile species have different ecological requirements than adults. They need large amounts of food that is small enough to ingest, and they need habitats in which they can avoid competition and predation from other species and (often) from adults of their own species. Ecosystems that meet these conditions host exceptionally high concentrations of juvenile fishes or invertebrates, and serve as nursery grounds or rearing habitats for much larger surrounding areas.

Estuaries are the most important nursery habitat for many marine species. For example, about 75 percent of the US commercial fishery landings are of estuarine-dependent species (Chambers 1991). Estuaries are usually highly productive because they have ample nutrients and light, the basic needs of photosynthesizers. Constant mixing of water by winds, riverine flow, and tidal flow resuspends nutrients in bottom sediments, ensuring high productivity. Much of the productivity in many estuaries depends on emergent or shallow-dwelling salt marsh grasses, seagrasses, and mangroves, which provide ideal habitat for juvenile fishes and crustaceans

(Fonseca et al. 1991). The abundance of decomposing detritus from these plants provides a rich source of food. Organisms growing on the submerged plant surfaces (aufwuchs or epibionts) are another abundant food source. And the complex three-dimensional tangles of stems, leaves, or roots create microhabitats that offer shelter from predators. Estuarine waters are often shallow and turbid, which further limits predators' access and their ability to locate and pursue juvenile shrimps, crabs, and fishes.

Nursery grounds are essential for many of the world's fisheries. Most or all of the warm temperate, subtropical, and tropical commercial shrimps (prawns) in the genus *Penaeus* depend on inshore nursery habitats in bays and estuaries. Young Florida spiny lobsters (*Panulirus argus*) usually inhabit soft sediments of shallow water seagrass beds in protected estuaries (Little 1977). First-year juvenile California spiny lobsters (*P. interruptus*) thrive in very shallow (0 to 4 meter [0 to 13 feet]) rocky shore surfgrass (*Phyllospadix* spp.) beds. The surfgrass provides shelter from predators as well as an abundant food source, which encourages high growth rates of juvenile lobsters in this warm-water zone (Engle 1979). After growing in such nursery habitats, commercial shrimp and spiny lobsters migrate to deeper continental shelf waters and reefs, respectively, where they are subject to intense, high-value fisheries.

Because many nursery areas are located in estuaries or near the shoreline, they are extremely vulnerable. Juvenile penaeid shrimps, for example, require specific salinity ranges for optimal growth and survival; irrigation projects that diminish the flow of fresh water to estuaries can alter salinity regimes in nursery grounds, and thus harm shrimp production. Similarly, the surfgrass nursery beds so vital to California lobster populations are sensitive to sewage discharge, oil spills, and shoreline alteration, and are very slow to recover from these major disturbances (Foster et al. 1988). Unfortunately, nursery grounds worldwide are being eliminated by the filling of estuaries and coastal lagoons to create land.

Nursery grounds that support dense populations of fishes, shrimps, and other marine invertebrates can be inviting targets. Fishers are often tempted to catch organisms before someone else gets them, but rarely does the short-term gain from taking juveniles increase the total value of the fishery; economic gains from increased growth more than compensate for losses from natural mortality. Moreover, some fishing techniques, such as trawling, are very damaging to nursery grounds. As a result, there is a growing movement to protect nursery areas from fishing to increase yields later, roller trawls have been prohibited for use in the southeastern USA to reduce damage, and, in Spain's Alacante Marine Reserve, artificial reefs have been used to snare nets and discourage illegal trawling in seagrass nursery habitats.

To the eyes of the untrained, nursery grounds might not be as scenic as some other marine ecosystems, but their value extends far beyond their borders. Their ecological and economic importance is so great, and their vulnerability to human disturbance is so high, that they merit high priority for protection.

Migration Stopover Points and Bottlenecks

Although many marine species are sessile or sedentary as adults and wait for currents to bring food to them, many others are highly migratory, with individuals covering hundreds or thousands of kilometers in annual or multiyear breeding cycles. Among them are some of the rarest and most threatened marine species, such as some baleen whales, sea turtles, tunas, and billfishes. Species migrate mainly to breed or in response to the appearance of high concentrations of food resources in different places at different times. When abundant resources appear predictably, migration routes are usually well defined and predictable. For example, on their annual migration, nearly all the world's California gray whales (*Eschrichtius robustus*) move through Unimak Pass in the Aleutian Islands, Alaska (USA) (Rugh 1984). The pass is only 18.5 kilometers (11.5 miles) at its narrowest point. This kind of concentration makes species especially vulnerable to purposeful human activities, such as hunting, and to accidents, such as oil spills.

A prime example of the importance of stopover points on migration routes involves migratory shorebirds (families Charadridae and Scolopacidae), with about 70 species in the Western Hemisphere. Some species are so abundant that they might not seem at risk, but they have low reproductive rates and undergo very energy-demanding migrations, which include stops at very specific places, such as certain coastal beaches and wetlands, which are generally being altered rapidly (Myers 1983; Myers et al. 1987).

Most New World shorebirds breed in the Canadian and Alaskan Arctic during the short summers, then migrate to spend the nonbreeding season in the southern USA, Mexico, and Central and South America. Some, including red knots (*Calidris canutus*), breed north of 70°N and migrate to Tierra del Fuego, thus covering more than 25,000 kilometers (15,500 miles) yearly. This makes stopovers, where the birds get the protein and fat reserves necessary for the long flights, crucial to their survival.

Concentrations of shorebirds can be staggering at stopovers. The Bay of Fundy in Canada, for example, hosts 80 percent of all semi-palmated sandpipers (*C. pusilla*) in the Western Hemisphere, providing the food resources needed to cover the metabolic demands of nonstop flight over the Atlantic Ocean to northern South America (Hicklin 1987). Similarly, during the spring, the Delaware Bay of the USA provides the food that fuels nonstop flights to the Canadian Arctic breeding grounds for 50 percent of all sanderlings (*C. alba*), 70 percent of all semi-palmateds, 75 percent of all ruddy turnstones (*Arenaria interpres*), and 80 percent of all red knots in the hemisphere (Manomet Bird Observatory 1986; Myers 1986; Castro et al. 1989).

Anything that reduces habitat quality at these stopover sites would therefore be disastrous to these species. The Western Hemisphere Shorebird Reserve Network is the first attempt to establish a network to protect critical stopovers, with 17 sites in six countries (through October of 1991) that protect a total of 30 million shorebirds. Similar mechanisms might be helpful for other species that migrate through narrow geographic bottlenecks.

Biological Diversity Being Reduced at All Three Levels

Scientists know very little about the recent loss of genetic, species, and ecosystem diversity in the world's oceans. A number of marine species have disappeared in historical times and this process is continuing. Ecosystems have been eliminated, and, from the loss of populations, one can assume that there has been great loss of genetic diversity. As a result, we are losing both products and services from the sea, losses no less important than they are on land.

Limitations to Our Understanding of Loss of Diversity

There are two great obstacles to understanding the extent to which humans have reduced marine biological diversity:

1) our ignorance of the scale and rate of such activity in pre-modern and even modern times; and
2) a catastrophic decline in scholarly work in systematics, natural history, and the biogeography of invertebrates, algae, and microorganisms.

Coastal zones of the world were devastated by human activity for centuries prior to the first extensive biological investigations, from over-fishing, pollution, and alteration of river flow into coastal waters (see Chapter 4). For example, more than 90 percent of marshes along some temperate coastlines have been destroyed. A modern tropical analogue is the destruction of mangrove ecosystems to create shrimp (*Penaeus* spp.) and milkfish (*Chanos chanos*) mariculture ponds. All types of coastal ecosystems around the world—rocky shores, mudflats, sandy beaches, dunes, seagrass beds, coral reefs, kelp beds, etc.—have been locally and selectively decimated, but no adequate analysis summarizes this destruction.

Largely anecdotal records show that many populations of inverte-brates, fishes, algae, and seagrasses have been eliminated from coastal areas. Lacking the necessary studies, at the end of the 20th century, we have no way to understand the scale of ecosystem destruction globally, although it appears to be much worse than the anecdotal and fragmentary published record would suggest. Museum records are available that could be used to document the loss of coastal species, especially marine snails, but very few scientists can find funding to do this research.

Even less is known about the concomitant destruction of genetic and species diversity in the sea. For most regions of the world, biologists do not know how the destruction of specific habitats may have led to species extinctions or loss of genetic diversity. In the past, biologists have sought consolation in the fact that although many populations have become locally extinct, the species survives elsewhere. It is now clear, however, from allozyme and molecular genetic (nuclear or mitochondrial DNA) studies, that enormous hidden diversity might exist within what were previously thought to be single marine species. The historic decimation of coastal populations thus makes it highly probable that much genetic, and thus perhaps species, diversity has been lost as well. Many museums contain specimens of marine invertebrates, particularly mollusks and crustaceans, some dating back to the 18th century. Specimens representing now demonstrably extinct populations may, in fact, represent extinct species. A means of testing this hypothesis, in part, for select taxa and regions would be by mitochondrial DNA analysis of these preserved specimens.

Moreover, the scientific literature and museum collections are replete with hundreds of marine species "known from a single specimen," often collected in the 18th or 19th century. Some people have assumed that such records reflect a failure to rediscover the microhabitat of a species, the rarity of collectors, or both. Of the many records of coastal marine organisms that have never been rediscovered, it is not known how many have been looked for again.

Unfortunately, our ability to acquire knowledge of the natural history, taxonomy, and biogeography of marine organisms continues to decline. Prospects for improving our ability to recognize extinction are low. The historical situation aside, there remain more undescribed than described marine macroinvertebrates and a staggering number of undescribed microinvertebrates and protists. At the close of the 20th century, fewer and fewer systematists and biogeographers are being trained, and young scientists are discouraged from investigating the natural history of marine organisms. We train no marine environmental historians; indeed, who would train them? To conserve the sea's biological diversity, these human resource trends must be reversed.

Known Extinctions

It should not be surprising that documented extinctions of marine organisms are rare, and are mainly of mammals and birds (Table 3-2); these taxa have also received the most attention on land. We know nothing of the possible extinction of recent marine fishes, algae, or seagrasses, nor, until very recently, of invertebrates.

In 1991, the first two historical extinctions of marine invertebrates were announced. Surprisingly, neither has yet been demonstrably linked to human activity. The highly specialized eelgrass limpet *Lottia alveus alveus* became extinct in the Atlantic Ocean about 1930 after the near-extinction of its plant host, the eelgrass *Zostera marina*, from a fungal disease (Carlton et al. 1991). A fire coral (*Millepora boschmai*) known only from a restricted region of Pacific Panama was thought to have become extinct following an exceptionally severe El Niño episode in 1982–83 (Glynn and de Weerdt 1991), but has since been rediscovered (Glynn and Feingold 1992). Whether the pathogen that nearly wiped out the eelgrass was imported by human agency, or if the frequency and severity of El Niños are increasing because of greenhouse warming, remains unclear.

Another possible, but less well-documented, extinction is the case of the mud snail *Cerithidea fuscata*, last collected in San Diego Bay, Cali-

TABLE 3-2. Marine birds and mammals known to have become extinct in historic times
(For a number of other species, evidence is suggestive but not as conclusive.)

Common name	Scientific name	Range	Ocean
Auckland Island merganser[1,2]	*Mergus australis*	Auckland Islands (New Zealand	Pacific
Guadalupe storm petrel[1,2]	*Oceanodroma macrodactyla*	Guadalupe Island (Mexico)	Pacific
Bonin night heron[1,2]	*Nycticorax caledonicus crassirostris*	Bonin Islands (Japan)	Pacific
Steller's spectacled cormorant[1,2]	*Phalacrocorax perspicillatus*	Komandorskiye Islands (Russia)	Pacific
Steller's sea cow[1]	*Hydrodamalis gigas*	Komandorskiye Islands (Russia)	Pacific
Labrador duck[1,2]	*Camptorhynchus labradorius*	Labrador (Canada) to New Jersey (Virginia?) (USA)	Atlantic
great auk[1,2]	*Pinguinus impennis*	Massachusetts (USA) to Norway and France	Atlantic
sea mink[1]	*Mustela macrodon*	Nova Scotia (Canada) to New England (USA)	Atlantic
Caribbean monk seal[1]	*Monachus tropicalis*	Caribbean Sea	Atlantic
Atlantic gray whale[3]	*Eschrichtius robustus*	North Atlantic Ocean	Atlantic

Sources:
1) Day, David. 1981. *The Doomsday Book of Animals—A Natural History of Vanished Species.* New York (USA): Viking Press.
2) Fuller, Errol. 1987. *Extinct Birds.* New York (USA): Facts on File Publications.
3) Mead, James G., and Edward D. Mitchell. 1984. Atlantic gray whales. In *The Gray Whale* Eschrichtius robustus, ed. Mary Lou Jones, Steven L. Swartz, and Stephen Leatherwood, pp 33–53. Orlando, Fl. (USA): Academic Press.

fornia (USA), in 1935 (Carlton et al. 1991). Human activities have dramatically altered San Diego Bay in the past century. If this taxon is distinct from the sibling *C. californica*, and if it proves to be extinct, then *C. fuscata* may be the earliest known human-caused marine invertebrate extinction. How many species preceded it on estuarine mudflats around the world is not known.

Small geographic range has not been an important factor in all historic extinctions. Pacific species that became extinct in historic times were, indeed, endemics, but those from the Atlantic did not have particularly small geographic ranges. The eelgrass limpet, for example, occurred from southern Labrador to Long Island (a span of about 12° of latitude) shortly before its extinction in the 1930s; the great auk (*Pinguinus impennis*) was found on both sides of the North Atlantic. Why endemism seems to have been a factor in extinctions of Pacific but not Atlantic species is a mystery.

Nonetheless, the limpet, the coral, and the mudsnail all had limited ranges or habitats. Changes in all their habitats within their geographic ranges led to extinction or near-extinction. This suggests what the future could hold for other marine organisms as the extent and intensity of environmental change increase worldwide.

Risk Is Higher for Certain Species and Ecosystems

As on land, some species and ecosystems in the sea are at much higher risk of disappearing than others. Small population size, low recruitment, extreme specialization, confinement to a small geographic range, value as a commodity (especially one for which there is no substitute), and exceptionally large size are all risk factors for marine as well as terrestrial populations and species. Proximity to humankind is risky for biological diversity at all levels.

Small Populations

All else being equal, four types of uncertainty make the risk of extinction higher in smaller populations than in larger ones (Shaffer 1981): demographic uncertainty, environmental uncertainty, natural or anthropogenic catastrophes (which are extreme environmental uncertainties), and genetic uncertainty. The best-studied of these, and perhaps the most important, is genetic uncertainty, although Lande (1988) and others suggest that demographic and environmental uncertainty are greater threats for very small populations.

Demographic Uncertainty: Small populations are subject to extinction by chance events affecting survival and reproduction. For example, sex ratios in small populations are likely to deviate randomly from 50:50; in the extreme there may be few or no members of one sex available for mating. Similarly, by chance, a smaller population is more likely than a larger population to contain a large proportion of individuals of an age class, such as post-reproductives or young individuals vulnerable to predation that do not reproduce. Thus, both unbalanced sex ratios and age structures, which are more likely in small populations, can lead to extinction.

Environmental Uncertainty: Unpredictable weather fluctuations, introduction of alien competitors or predators, disease epidemics, a change in food supply, or other biotic or abiotic changes can destroy small populations more readily than large populations. This heightened risk results from both the small numbers present (all are affected by the change) and low genetic variation in a small population, which makes it less likely that a population will have the genetic capacity to deal with the change (see "Genetic Uncertainty" section below).

Natural or Anthropogenic Catastrophes: Small populations are vulnerable to extinction through environmental catastrophes, whether natural or caused by humans. For the same reasons given above, an oil spill or a volcanic eruption is more likely to exterminate an entire population or species that is represented by few individuals than one that is more abundant and widespread.

Genetic Uncertainty: Small population size has genetic consequences that affect both short-term and long-term viability (Schonewald-Cox et al. 1983). The potential for genetic change (adaptability) in a population is directly proportional to the amount of available genetic variation. Because a small population contains less genetic variation than a similar large one, it cannot adapt as much or as quickly. Additionally, small populations suffer from random or directional genetic processes that may be maladaptive and lead to extinction. These include genetic drift (random and permanent losses of genetic variation from the gene pool through death or removal of individuals) and inbreeding (mating of close relatives, which typically exposes deleterious characteristics and reduces fitness, e.g., by increasing abnormalities in gametes). All of these genetic problems increase as population size decreases, making the population more prone to extinction through genetic deterioration and maladaptation.

The absolute or census size of a population is not nearly as relevant to conserving genetic diversity as the genetically effective population size, which measures the number of effective breeders and their relative genetic contributions to the next generation (Meffe 1986). The effective population size can be much smaller than the absolute population size, so a large population count does not mean that genetic diversity is secure. A census population of 300 seals, for example, might amount to a genetically effective population size of only 60.

The best-known example of the effects of small population size and loss of genetic variation on fitness and viability comes from a terrestrial mammal, the cheetah (*Acinonyx jubatus*). This species has no detectable genetic diversity, apparently as a result of historic population bottlenecks and inbreeding. Consequently, its reproductive capabilities have diminished, it is highly prone to disease, and young animals have poor survival. Evidence from many other species shows similar trends: Lower individual fitness may result from lower genetic diversity in small populations (Mitton and Grant 1984), leading to greater risk of extinction.

A marine species that might, like the cheetah, suffer from the loss of genetic diversity is the northern right whale (*Eubalaena glacialis*), which was subjected to more than 800 years of whaling. The North Atlantic population currently numbers approximately 350 animals. The effective population size might once have been as low as 50. Recent studies indicate that its genetic diversity is very low (Schaeff et al. 1991), which might explain why the population has shown little recovery despite more than 55 years of complete protection from whaling. This small population clearly is also vulnerable to the other forms of uncertainty listed previously.

The critically endangered vaquita, cochito, or Gulf of California harbor porpoise (*Phocoena sinus*), is restricted to the northern Sea of Cortez and now numbers a few hundred individuals. It is at great risk from drowning in fishing gill nets, which are estimated to remove about 35 animals annually. Pollution, as well as changes in salinity and water temperatures in the upper Gulf caused by flow reductions in the Colorado River (environmental uncertainty) could affect the entire vaquita population, either directly or through reduction in food supply. Also, such a small population could face genetic declines through inbreeding and genetic drift (genetic uncertainty).

The sea is so large that it is difficult to imagine that population sizes of some marine species are as low as those of terrestrial species that we consider endangered. Unfortunately, nothing about the sea safeguards marine species from extinction. There are two more important points about rarity and abundance. One is that naturally rare species are not necessarily vulnerable to extinction; it is likely that some naturally rare species long ago developed ways to survive at low densities. The other is that humans are capable of pushing even very abundant species to extinction. The most abundant bird (with a population of perhaps 5 billion) in the USA in the mid-1800s, the passenger pigeon (*Ectopistes migratorius*), was decimated by overexploitation and habitat destruction until its extinction in 1914. Its destruction took just one human lifetime. There is no reason to believe that the same could not happen to marine species whose populations are similarly reduced.

Populations/Species with Low Recruitment

Populations with low recruitment are especially vulnerable to overexploitation and disturbances. Recruitment is the addition of members to a population, but its definition varies depending on one's methods or goals. Recruitment can mean the number of individuals that reaches a size large enough to be useful in a fishery, the size that reproduces, or even the size that can be identified as belonging to a certain species.

Low recruitment can result from either low fecundity (number of young produced) or high juvenile mortality. When fecundity is low and growth is slow (some sturgeons and sharks need one to several decades to reach reproductive maturity), recruitment is always low. Discovering ways to predict, maintain, and enhance recruitment is a central problem in both fisheries biology and marine conservation biology.

Species that either protect their young or produce large young enhance early survival, but their large parental investments in each offspring impose a trade-off: low fecundity. Most sharks and rays have both low fecundity and slow growth, hence, low recruitment. Large blue sharks (*Prionace glauca*) and tiger sharks (*Galeocerdo cuvieri*) can have litters of up to 80 pups, but most sharks produce far smaller litters. Because the combination of low recruitment and late maturation severely limits the rate of sustainable exploitation, sharks are vulnerable to overfishing (Pratt and Casey 1990).

In contrast, sea turtles usually lay more than 100 eggs per nest but have low recruitment because their young have low survival rates and take a long time to reach sexual maturity. Mortality and late reproduction are far more limiting for sea turtles' population size than fecundity. In loggerhead turtles (*Caretta caretta*), in the southeastern USA, small reductions in mortality of large juveniles and adults, as occurs when turtle excluder devices (TEDs) eliminate drowning in shrimp trawls, are likely to be more effective in reversing population declines than a doubling of fecundity (Crouse et al. 1987). Although individual adult females contribute much to future generations, for some populations there are so few adult females that protection of adults alone is not a very effective management option. Although protection of nesting beaches is essential, the most effective strategy is to reduce the mortality of large juveniles, thereby magnifying recruitment to the adult reproductive population.

Even very high fecundity does not guarantee high recruitment. Many marine animals, including some commercial fishes and shellfish, produce more than one million eggs in a season yet have low recruitment because of lack of fertilization or very high larval or juvenile mortality. Their larvae are tiny (and therefore vulnerable to predators) and planktonic, so heavy losses can occur when currents carry them to unfavorable habitats. Even with astronomical starting numbers, when mortality is also high, recruitment can vary greatly, so adults that die are rapidly replaced in some places or years but not in others. Recruitment in species with planktonic larvae can be very sensitive to oceanographic variables. For example, differences in winds and water density can cause tenfold differences in recruitment of oyster (*Crassostrea virginica*) spat in adjacent estuaries by affecting transport of larvae (Boicourt 1982; Seliger et al. 1982). Recruitment of barnacles in southwest England is highly variable over periods of decades, perhaps owing to shifts in water masses, so the relative abundance of species changes (Southward 1991). Because recruitment in a species can vary in space and time, management policy must focus on the long-term patterns at a given locality.

Although the causes of low recruitment are diverse, a long adult life and repeated reproduction are common traits for populations of plants or animals with chronically low recruitment. These are the traits that allow such populations to persist, so anything that shortens life and reduces the number of times an organism reproduces endangers the population.

The geoduck clam (*Panope generosa*), for example, is a valuable food animal. Some geoducks in unfished areas in Puget Sound, USA, are more than 100 years old. Recruitment is low in undisturbed populations and even lower where geoducks have been taken recently (Goodwin and Shaul 1984). Recruitment is low naturally because young clams are eaten by crabs and snails. The few that survive long enough to grow large can bury themselves deep enough to avoid most predators (other than humans) and then can live more than a century. To ensure sustainability, the take is limited to 2 percent of the harvestable stock. Increasing recruitment by planting hatchery-reared juveniles has not been feasible because of the high cost of excluding their predators.

In contrast, organisms that mature quickly and reproduce only once can often withstand a high catch as long as their habitat is healthy. Whereas many marine species can return to their former abundance and biomass within 5 to 10 years of severe depletion of a population, populations with very low recruitment can take much longer. Indeed, this makes slow recruiters sensitive indicators of how well wild populations are being maintained.

Species with Very Specialized Requirements

Species with specialized requirements are especially vulnerable to environmental changes that diminish the resources upon which they depend because they have little ability to switch to other resources. These species might be specialized in habitat, diet, reproduction, or relationships with other species. For example, when a fungal disease devastated eelgrass beds in eastern North America during the 1930s, brant geese (*Branta bernicla*), which feed largely on eelgrass, almost disappeared. Eelgrass limpets (Figure 3-3), which lived *only* on eelgrass blades, were even less fortunate (Carlton et al. 1991). Similarly, about 500 species of tropical reef-building corals support some 4,000 species of reef fishes and an uncounted diversity of invertebrates, algae, and microorganisms, many of which are not found in other ecosystems. Loss of the important structure-forming corals or other keystone species can lead to cascading loss of the many other habitat specialists that depend on them.

Crustaceans called whale lice (*Cyamus* spp.) live on the skin of particular species of whales. Whether they or other species-specific symbionts disappeared with the extinction of the Atlantic gray whale will probably never be known. There are other cases of host specificity with rare species: At least three species of seaweeds are known only from carapaces of sea turtles that are themselves in danger of extinction (Personal communication with James Norris, Smithsonian Institution, Washington, D.C., USA).

A small (25-centimeter [10-inch]) seabird, the marbled murrelet (*Brachyrhamphus marmoratus*), exhibits extreme nesting specialization in the temperate Northeast Pacific part of its range (Marshall 1988). Between southeast Alaska and California (USA), these birds nest only in mossy depressions on large limbs of giant old-growth conifers. Because clearcut

Figure 3-3. Eelgrass limpet. *This snail disappeared from the Atlantic Ocean in the 1930s, when an epidemic devastated the eelgrass beds in which it lived. Marine invertebrates, fishes, seaweeds, and vascular plants can become extinct when their habitats are altered.*

logging has eliminated so much of the murrelet's ancient forest nesting habitat (up to 90 percent of the ancient forests in the northwestern USA), the bird is now absent from large sections of coast, and has been listed as threatened in Canada and the USA.

Some species are wide-ranging, yet, for reasons that are unclear, reproduce only in very few localities among a seemingly large array of similar habitats. Kemp's ridley sea turtles (*Lepidochelys kempi*) range along the USA's Atlantic coast and in the Gulf of Mexico, but nest almost exclusively on one small stretch of beach in eastern Mexico, where their breeding population has declined more than 99 percent since the 1940s. Waves from Hurricane Gilbert inundated the nesting beach late in the 1988 breeding season. Few nests were lost, but a hurricane one month earlier would have jeopardized the year's reproduction.

There are many highly specialized marine species whose conservation status is either not currently worrisome or is not known. But if their specializations tie them to other species or physical situations that are disappearing, these species are very likely to be in trouble.

Populations/Species with Limited Ranges

If a disaster of less than global scale happens someplace on the Earth, which species would most likely be eliminated? One clear answer is those whose entire ranges fall within the disaster's perimeter. In the sea, as on land, species that are widespread are less vulnerable to localized threats.

Marine species, on average, appear to have wider distributions than terrestrial species. Some widely distributed species are very mobile as adults; individual blue whales (*Balaenoptera musculus*) and blue sharks, for example, can range over thousands of kilometers. Both juvenile and adult gooseneck barnacles (*Lepas anatifera*) disperse (potentially across great distances) by rafting on floating pumice, seaweed, and logs (High-smith 1985; Jokiel 1990). But most widely distributed marine species disperse as planktonic larvae.

Some larvae (planktotrophs) feed while they are in the water column; others (lecithotrophs) do not. This distinction is important because planktotrophs tend to remain in the plankton longer than lecithotrophs, and therefore have the potential to move greater distances (Strathmann 1985; Emlet et al. 1987). Although there are few direct measurements of the distances that larvae travel, they are thought to range from a few meters to hundreds of kilometers (e.g., Strathmann 1974; Davis and Butler 1989).

Some have long enough larval lives—six months or more—that they can be transported several thousand kilometers even in moderate currents of only 0.5 kilometers (0.3 miles) per hour. Not all larvae of a particular species travel the maximum potential distance, but occasional ones that do can maintain or increase that species' geographic range.

Mobility at any life history stage of a species has the potential to increase the species' geographic range. For example, benthic organisms with larval dispersal tend to have greater geographic ranges than those that do not (e.g., Scheltema 1989). Further, in both modern and extinct species, those with planktotrophic larvae tend to have wider geographic ranges (Jablonski 1986). Greater geographic range, in turn, correlates with longer duration in the fossil record. But there are enough exceptions that the story is not completely clear. For example, high-latitude invertebrates without planktotrophic larvae are often widespread, and bivalve mollusks appear to show little correlation between developmental mode and geographic range or extinction rate (Jablonski and Lutz 1983).

It seems logical that planktonic dispersers should have fewer localized, genetically distinct populations, and they should be faster at recolonizing areas where local extinctions have occurred. Palumbi and Wilson (1990) found that two sea urchin species with dispersing larvae have smaller genetic differences at different localities than terrestrial species with comparable geographic ranges, which supports the first of these hypothesized effects. However, some species with pelagic larvae have genetically distinct local populations (Hedgecock 1986); indeed, this appears common. There are at least two possible explanations. One is that larvae with the potential to be carried long distances are not, but settle relatively close to their parents. Or, larvae of many genotypes within a species are present in the plankton, but only certain ones can survive under the conditions in a particular area; that is, natural selection is occurring.

Few hard data are available relating dispersal ability to recolonization following local extinction, the second hypothesized effect. Simon and Dauer (1977) found that species with planktotrophic larvae recolonized faster than those without, as might be expected, but many good colonizers do not have planktotrophic larvae. The size of the habitat patch in which a population has become extinct, its nearness to potential source populations, and whether currents bring larvae to it could all be important factors.

So, dispersal ability can influence geographic range, gene flow, and colonizing ability in benthic species; all of these factors can affect the risk of extinction. Theory and at least fragmentary evidence suggests that planktotrophs and other high-dispersal forms will be able to maintain gene flow among widespread—and even disjunct—populations, and can readily repopulate areas that have suffered local extinction, making these species more resistant to extinction from stresses that occur only in particular localities. In contrast, low-dispersal forms will tend to have smaller, more continuous geographic ranges and genetically subdivided populations that are more vulnerable to local, regional, or global threats; these species, then, will generally be more extinction-prone. Low-dispersal species will also be less able to maintain genetic connectedness among populations that live in marine reserves that are separated from one another by unfavorable or degraded habitat.

It does not necessarily follow that species with planktotrophic larvae are therefore extinction-proof to anything less than a global catastrophe. Not only must a species produce larvae; the larvae must have habitat to colonize. The alteration of marine ecosystems worldwide makes it likely that even good dispersers and colonizers will have increasing difficulty finding suitable habitats.

Species Valued as Commodities

Species that people use as commodities are inherently at risk of population reduction or elimination, and species for which there are no viable substitutes are at special risk. A basic principle of classic economics is that the price of a commodity increases with increasing demand if its supply remains constant, while at constant demand, its price will increase as supply dwindles. When the price increases, fewer people will be able to afford the commodity, so demand should drop. Moreover, if the price of one commodity rises beyond that of another that can serve as a substitute, demand for the first should slacken as people start to use the second in its place. But in the current world market, the interaction of culture, technology, and affluence complicates understanding of the economics of marine resources.

In theory, consumption of marine organisms is optional for human survival; our species can subsist on, say, rice and beans. But to coastal peoples, especially on islands with little arable land, relying on bountiful marine resources is necessary. For these peoples, seafood becomes not only an integral component of nutrition, but of culture as well.

Food items could more readily be substituted if cultural factors did not come into play. When technologies and affluence are limiting, people's only option when a species used as food becomes too rare is to switch to another species. If the preferred species has become rare due to over-exploitation, switching might allow the species to recover.

But culture, specifically cultural preferences, can also become a driving force for biotic impoverishment, especially as people advance technologically and economically. The progression from collecting seafood with wooden spears and nets made from plant fibers to using diesel engines and on-board freezers has allowed people to pursue and process preferred food species in increasingly remote locations. People who once caught a cherished type of fish from shore can now pursue it with the aid of satellites that transmit water temperature data to a fleet of ships in mid-ocean, thousands of kilometers from home. Although this might be incomprehensibly laborious and wasteful in terms of nutritional need when other fish or even non-seafoods might be far less expensive, it is wholly understandable in terms of cultural preference. When people want something, we devise technologies so we can get it. More powerful technologies can make commodities rarer in nature while actually *reducing* their prices by increasing their supply in the market. But even when more powerful technologies increase prices, affluent people can afford to pay them.

In Japan, for example, a country highly dependent on the sea for food, some species, such as giant bluefin tunas, are so prized in sushi that they are one of the most expensive food animals in the world and are affordable only by wealthier people. This demand has made giant tunas rare, and their continuing overexploitation by Americans, Canadians, and others to supply the Japanese market has become a contentious international conservation issue.

Some species are not put at risk by the cultural traditions of nations, but by hobbies that create subcultures with their own rules and traditions. The shells of many marine mollusks, for example, attract high prices on the specimen shell market. Many are rare deep-water species or endemics. There are few, if any, studies of their population sizes, distributions, or recruitment with which to assess the impact of their exploitation. But there are anecdotal reports of overexploitation of some "rare" species, such as the imperial harp (*Harpa costata*) and violet spider conch (*Lambis violacea*), both endemic to Mauritius; endemic cowries (Cypraeidae) in Mauritius, Australia, and Hawaii (USA); and endemic volutes (Volutidae) in New Caledonia (Wood and Wells 1988).

In some cases, the fate of species depends on the vagaries of fashion. Sea otter pelts that sold for US$2,000 in the early 1900s, for instance, fetched US$134 in 1971 (Novak et al. 1987). Throughout the late 1980s, demand increased for certain colors of precious corals, while prices for other colors fell, although they had not become any more common (Caddy and Savini 1989).

Unfortunately, as the number of technologically advanced and affluent societies increases, cultural preferences could drive an ever-greater number of species to depletion or even extinction. The seas are still biologically diverse, but are now crossed by a million wakes. Connoisseurs of giant bluefin or precious coral might have to make do with farm-bred catfish or colored plastic unless we take greater care.

Large Species

In many groups of organisms, the evolution of large size has conferred protection against predators. Until quite recently in evolutionary time, humans subsisted on fruits, roots, insects, and the occasional bonanza of a dead antelope that the vultures had not found, so large animals had little reason to fear us. For countless millennia, the great size and strength of mammoths (*Mammuthus* spp.) gave them the advantage in confrontations with humans until our increasingly effective technologies and social organization tipped the balance in our favor. Now there are 5.5 billion of us and no mammoths.

In the sea, advancing technologies brought improved gear that made hunting of large animals safe and profitable. The first Europeans to discover Steller's sea cow, which could weigh 100 times as much as a human being, found a way to kill them. Although their size once protected them, great whales, elephant seals (*Mirounga* spp.), great auks, giant bluefin tunas, giant clams (*Tridacna* spp.), and king crabs (*Paralithodes camtschatica*) became preferred targets, in part, because they offer higher economic yield per unit of hunting effort than smaller species. Giant kelp is far easier to cut than an equal weight of small algae with equally useful polysaccharides. Further, a given weight of a larger species is often easier to process for marketing and more acceptable to consumers; that is why giant mud crabs (*Scylla serrata*) are a prized meal from Mozambique to Micronesia, while hundreds of smaller portunid crab species are mostly ignored. Finally, the human preference for catching large organisms has also been shaped by an emotional need to "conquer the giant." Throughout our history, bringing home the biggest catch has conveyed social status and a sense of personal achievement for the hunter.

Large individuals of a species can be far more important sources of young than their increased length or weight might suggest. For example, one 61-centimeter (24-inch) red snapper (*Lutjanus campechanus*) produces the same number of eggs as *212* that are 42 centimeters (16.5 inches) long (Grimes 1987).

Large animal species, however, can be at greater risk because of their life histories. Within many taxa of organisms, larger species are often longer-lived and produce fewer offspring in their lifetimes (Peters 1983). Once they are reduced substantially, their late onset of sexual maturity and infrequent reproduction make for slow population growth. Thus, it can take many generations to return to their pristine population abundance. Furthermore, their generation time is longer than in smaller species. When they are reduced to very low densities, finding mates can be difficult or impossible (what conservation biologists call the Allee effect), further slowing recovery. And once large animals are reduced to rarity, smaller species that can increase faster might monopolize prey formerly used by the large species, possibly allowing smaller species to exclude the larger ones from the ecosystem.

Large animals are often wide-ranging and migrate across political borders, increasing their risk as they are subject to different economic, social, legal, and political systems. Their cross-boundary movements create the need for nations to cooperate in their management, something that does not always happen.

Large species are more likely than small ones to be top carnivores that eat other animals, but are seldom or never eaten. Unfortunately, seals, toothed whales, sea eagles, tunas, billfishes, thresher sharks (*Alopias* spp.), and other top carnivores that eat prey contaminated with persistent, fat-soluble organic compounds (e.g., PCBs, DDT, methyl mercury) can accumulate damaging concentrations of these substances, a process known as food-chain amplification, bioaccumulation, or biomagnification.

Because our increasing numbers and technologies have allowed us to exploit species of all sizes, large species, by virtue of their greater rewards to exploiters, slower recovery from depletion, wide-ranging habits, and position at the apex of food webs, are at heightened risk and therefore merit special attention.

Ecosystems in Proximity to Humans

Without a doubt, the greatest risk factor for marine ecosystems is having high concentrations of humans as neighbors. The open oceans, which are farther from dense human populations, are not yet severely contaminated by pollutants (GESAMP 1990), whereas estuaries have paid heavily for the towns, cities, and industries that border them. Estuaries receive materials from drainage basins that can be thousands of times as large as the estuaries themselves, so they collect and concentrate whatever washes off the land. Further, their rich abundance of fish and shellfish, the access they provide to upriver locations, and their protected anchorages have attracted most of the world's great cities and many smaller ones. Urban estuaries worldwide are in terrible condition, as are many non-estuarine coastal bays and lagoons.

Dense human populations have caused massive physical alteration. For example, wetlands bordering San Francisco Bay have been reduced more than 90 percent (Davis 1982) to provide for housing, industrial development, and harbor infrastructure. These stressors, combined with huge amounts of toxic materials and nutrients from millions of people, are devastating to urban estuaries. Nearly all are sites of petroleum refineries, and hydrocarbon sludge can form a high percentage of their bottom sediments. The constant parade of ships carrying vast numbers of organisms in their ballast tanks has made estuarine harbors the home of untold kinds of alien phytoplankton, animals, and pathogens.

The urbanization of estuaries has far-reaching effects. It destroys not only rich estuarine fisheries, but also coastal fisheries, a large fraction of which—often more than half—depend on estuaries, usually as nursery grounds.

The continental shelf, while generally less altered than estuaries and coastal lagoons, has suffered heavy impacts from humans in highly urbanized areas. Coastal waters are heavily polluted off some parts of Japan, Rio de Janeiro–Santos (Brazil), New York–New Jersey (USA), the western Baltic, North, and northern Mediterranean seas, among other places. For example, sediments in the western Baltic Sea contain high concentrations of heavy metals, including cadmium, lead, zinc, and copper. Its surface waters have elevated phosphate concentrations, and there are anoxic zones on the seabed, both of which are signs of eutrophication (Jickells et al. 1990). The dumping of sewage sludge can cause major changes to bottom communities, as it did off New York–New Jersey, where 9 million metric tons of sludge dumped per year have produced an area of 12,000 square kilometers (4,700 square miles) with oxygen deficiencies.

Oil spills are also most common where tankers and tankbarges navigate coastal waters, often those of densely populated industrialized countries. The great majority of the 84 large (more than 10,000-barrel [1.6-million-liter]) tanker spills from 1978 to 1988 occurred in neritic waters, especially in northwestern Europe, the Strait of Malacca (Malaysia and Indonesia), the northeastern USA, and the northern Gulf of Mexico (USA).

Coastal waters are harmed not only by what humans put into them, but by what we take away. Most of the world's fisheries are in shelf waters, and most of them are overexploited. For much of the 20th century, areas far from dense human habitation were free of our wastes. Technological and economic limitations made distant resource exploitation prohibitively expensive except for the most valuable products, such as whale oil. Now fisheries for products of low value per unit weight occur even off Antarctica, and pollutants are found even in the remotest surface waters and the deep sea. Indeed, the threats of stratospheric ozone depletion and climatic change will be manifested far from their sources as well as near them. Nonetheless, anthropogenic impacts such as overexploitation, physical alteration, pollution, and introduction of alien species are still greatest where our species is most abundant.

CHAPTER FOUR

Threats to Marine Biological Diversity

IN LATE 1990, 11 eminent marine biologists met at the Smithsonian Institution in Washington, D.C. (USA), to examine the basic dimensions of the marine biological diversity issue (Norse 1991). Perhaps their most startling conclusion was that *the entire marine realm, from estuaries and coastal waters to the open ocean and the deep sea, is at risk.*

While this risk is most obvious in estuaries and coastal waters, where human activities are most intense, no place in the ocean is so remote that it has not been touched by human activities. The vast expanses of open ocean are driven by the photosynthesis of microscopic phytoplankton whose rapid life cycles make them vulnerable to even temporary changes that affect large areas of sea surface. Such an event—probably the dramatic reduction of sunlight resulting from an asteroid's striking the Earth—occurred at the end of the Cretaceous era (see Figure 2-2). In the mass extinction that followed, planktonic organisms suffered even higher extinction rates than other marine species (McKinney 1987). The pelagic ecosystems of the open ocean could be threatened again. A major increase in ultraviolet radiation at the sea surface due to the depletion of stratospheric ozone could disrupt the sea's food webs on a scale dwarfing anything in human history.

The deep sea is at risk, in part, for the opposite reason. In the cold, lightless abyssal realm that predominates even in the tropics, life processes for many species appear to be very slow, and life spans are long. For example, it has been estimated that it takes the abyssal clam *Tindaria callistiformis* 100 years to reach the length of 8.4 millimeters (0.33 inches) (Turekian et al. 1975). Slow growth rates and the sluggish reproduction that accompanies this might make deepsea ecosystems especially slow to recover even after stresses have been removed. Trying to solve land-based problems, such as the depletion of metal ores and accumulation of wastes, by deepsea mining of manganese nodules and ocean dumping of toxic sewage sludge or radioactive wastes has the potential to harm vast areas of the biologically diverse deepsea realm.

Nearshore or offshore, the ways in which people threaten biological diversity can be grouped into classes of proximate or direct threats. These, in turn, are driven, in turn, by root causes within us.

Proximate Threats

Human activities damage life in the sea in ways that can be grouped into five broad classes: overexploitation, physical alteration, marine pollution, introduction of alien species, and global atmospheric change. This is the same as the list of proximate threats to biological diversity on land, although their relative importance in the sea differs somewhat.

In most cases, human activities have more than one effect on biological diversity. For example, the drilling muds discharged from offshore oil and gas operations affect marine ecosystems not only by adding toxic pollutants but also by adding fine particles to the seabed, which can alter larval settlement. Furthermore, the placement of some threats (e.g., noise pollution, treated here as physical alteration rather than pollution) in one class rather than another is somewhat arbitrary.

Overexploitation

Humans take a wide variety of marine vertebrates, invertebrates, and plants for use as foods, medicines, raw materials, pets, and curios. In theory, virtually any marine organism, even the slowest-maturing, least fecund species, could be exploited sustainably so long as no more were taken than reproduction could replace. Indeed, some tribal peoples took whales, dugongs, and sea turtles sustainably despite the fact that the life histories of these species make them highly vulnerable. But today's exploiters of marine organisms are far more numerous, their technologies are far more powerful, and the cultural inhibitions that prevented people from overexploiting have largely been eliminated. The alternative to overexploitation and losing resources that people need is to develop new cultural mechanisms to exploit marine organisms sustainably that are appropriate to modern societies with cash economies.

Before examining overexploitation, it is useful to discuss why *The Strategy* is sparing in its use of "harvest," a term that is widely used in fisheries, wildlife management, and forestry. In the sea, the term is applied to seaweeds, invertebrates, and vertebrates. Its use equates taking these organisms with gathering orchard-grown oranges or farmed potatoes. However, applying "harvest" to wild populations is misleading. Definitions of "harvest" in dictionaries say that the object of a harvest is a *crop*, that is, a cultivated plant or agricultural product that people breed or sow and grow before harvesting.

Agriculture sustains nearly all humankind, and we tend not to question it. In Western industrialized societies, at least, the killing of farmed animals is opposed mainly by people in the animal-rights and vegetarian movements, and nobody seems to object to killing farmed plants. In contrast, the killing of rare wild animals, and of trees in virgin forests, is a growing public policy issue worldwide. A large and increasing number of people are concerned that the spread of humankind is eliminating wild species, in large part due to overexploitation. Using "harvest" as a euphemism for killing wild lobsters, sardines, sea turtles, petrels, or dolphins lulls people into a false sense of security. That explains why lobbyists for fishing, whaling, hunting, and logging interests use the term.

There is no reason not to use "harvest" in discussing true farming operations in which humans control the organism's life cycle. Raft-cultivated nori, planted mangroves, maricultured abalone, cage-reared salmon, and captive-bred crocodiles are all crops that can be harvested. But killing members of wild populations is not harvesting.

Fisheries (Fishes and Shellfishes)

Most marine fishes and invertebrates that people catch can reproduce early enough and in sufficient numbers that they should be relatively easy to exploit sustainably. Their life histories are much more "forgiving" than those of species that reach maturity late and have few young. But this has not prevented humankind from overexploiting even these inherently less vulnerable species. Virtually all commercially valuable marine populations are now overexploited in at least part of their ranges, as are many others that are caught incidentally as bycatch. Yet there has been astoundingly little study of the effects of intense fishing pressure on sustainability. For example, Angel (1992), noting that 65 to 70 percent of the commercial-sized haddock are caught annually in the North Sea, wonders what the North Sea ecosystem would be like without fishing. It is becoming increasingly clear that overexploitation not only diminishes species' populations and reduces economic return, but also causes genetic changes in the exploited populations and alters ecological relationships with the species' predators, symbionts, competitors, and prey.

Targeted Species: As fishing effort continues to increase, fishing could, especially in combination with other threats, eliminate fish or invertebrate species, although no commercially fished marine species is known to have become extinct in modern times. But maintaining biological diversity goes beyond preventing extinction to maintaining the abundance of species and the functioning of ecosystems, including the production of exploitable populations. Fishing has profoundly decreased population sizes and productivity in many targeted fish and invertebrate species.

Two major changes have affected the world's fisheries in the last three decades. Since the 1970s, the process of enclosure—the change from open access to coastal jurisdiction—has placed most of the world's marine living resources under the control of coastal nations. And, after years of increasing catches, the total global landings have been stagnating at the 1989 level despite continuing increases in fishing effort. At the same time that enclosure has provided a potentially powerful tool for conservation, the end of growth in the world fish catch has created the need for fishery managers to devise ways to take food sustainably from a finite sea.

The oceans and their living resources are administered mainly by the nearest coastal nations. The degree to which nations actually manage depends on how much they have invested in fishery research, technology development, training, and management mechanisms.

Targeted fisheries, when managed at all, are often managed in a three-stage process. First, fisheries biologists study the targeted populations and try to determine how their populations respond to fishing, often using fishery models. Second, these researchers provide the information to decision makers who formulate regulations. Third, the fishers fish, either following the rules or not, depending on the level of enforcement and on the potential payoff.

The most common tool of fisheries biologists is a single-species sustainable yield curve, in which the catch depends on the biomass of the species and the amount of fishing effort. As long as the fishery operates in the left half of the curve, increasing effort results in a larger catch. But on the right half of the curve, more fishing effort causes the catch to *decrease*. The point of inflection between the left and right halves is called the maximum sustainable yield (MSY).

Fisheries managers have used estimates of the MSY point to set allowable catches since the 1940s and are still doing so today despite scientific criticism of the process dating back more than a decade (Larkin 1977). The Law of the Sea, among others, asks coastal nations to "maintain or restore populations of harvested species at levels which can produce the maximum sustainable yield" (Article 61) and to give other nations access to the surplus of the allowable catch (Article 62). This surplus is the difference between the nation's catch and the MSY, although the Law of the Sea allows the coastal nation to take into consideration economic and social factors before allocating its surplus to other nations.

As Anderson (1986) notes, among the many drawbacks of using MSY as a fishery management tool is the fact that the curve itself depends largely on a set of numbers and constants that quantify recruitment (number of fish entering the "catchable" population), individual growth (increase in weight of individual fish in the population), and natural mortality. Recruitment to fish stocks is usually highly variable (Shepherd 1988), making the determination of this very important parameter difficult and controversial. Natural mortality can also be difficult to assess.

When New Zealand initiated a commercial fishery for orange roughy (*Hoplostethus atlanticus*) in 1979, biologists had to estimate MSY for a species about which they knew very little. Employing sophisticated techniques to assess the population size and estimating mortality from that of a closely related species about which more was known, the scientists recommended a catch level of 47,000 metric tons in 1986. But further research uncovered a major mistake in the mortality estimate. After correcting it, the estimate of sustainable yield was lowered to 7,500 metric tons; their original estimate was six times too high (Personal communication with Ellen Pikitch, Fishery Research Institute, Seattle, Washington, USA). Estimating MSY is subject to serious error even in nations with substantial scientific and technological resources.

Fish stocks are easier to assess on a species-by-species basis, so stock assessment models work best in temperate or cold waters, where single-species fisheries are most common. In the waters of tropical countries such as Mauritania, where one trawl haul can catch up to 50 species (Gulland and Garcia 1984), multispecies modeling is more appropriate but has been used only for a decade or so. Multispecies fisheries are very difficult to study, evaluate, and manage. The web of competitive and predator-prey relationships is exceedingly complex and usually poorly understood. Production models for tropical multispecies fisheries are constructed by summing up the individual yield curves of the component species. These models give a rough estimate of total catch, but are not designed to deal with changes in species composition (Panayotou 1982). The inherent weakness in these models makes it possible that some targeted species could become commercially extinct—that is, they would provide negligible return for fishing effort—or even biologically extinct.

The second stage of fishery management is in the hands of decision makers who use the recommendations of biologists to decide whether or not to take action. If they do, they must weigh the long-term cost of not following scientists' advice against the short-term need to please their constituency. In the orange roughy fishery, after biologists recommended an 84 percent decrease in catch, decision makers, who had to consider the reaction of the fishing community, chose not to reduce the catch for the next two years. For the 1989–90 season, they ruled that the catch be lowered by only 20 percent despite being told that there was a 45 percent risk that the orange roughy fishery would collapse within five years unless the catch was sharply reduced.

Unfortunately, there are time lags between actual changes in biomass of fish stocks, when scientists realize that this has occurred, and when decision makers accept the scientists' advice (Sætersdal 1980). In most cases when fisheries suffered sharp declines, decision makers had failed to heed earlier scientific advice to reduce fishing effort (Glantz 1983).

Even if models are accurate in their predictions and decision makers mandate only sustainable catch levels of target species, the success of fishery management depends on whether fishers cooperate with the regulatory agency and catch only what is allowed, release without harm fish that are undersized, of the wrong sex, or at the wrong stage of their life cycle, and report accurately what they catch. In the real world, fishers, not unlike motorists who speed or taxpayers who do not declare all their gains, are profit-seeking individuals who sometimes break the law if the perceived cost of doing so is lower than the cost of getting caught (Sutinen and Gauvin 1987). Fishers in other nations' exclusive economic zones (EEZs) sometimes underreport their catches to lower the fees they pay to coastal states. Or they sometimes magnify the previous year's catch in the hope of being allocated a larger share of the fishery the next year. Given the inherent weaknesses in a system in which scientists do not know all that they would like, decision makers tend to ignore bad news, and fishers seek to evade regulation, it is little wonder that even managed fisheries can decline and collapse.

Incidental Take: Although poorly documented, the unintentional take of marine animals by fisheries is extremely widespread. Because most methods used to catch targeted marine fish and invertebrate species are nonselective, large numbers of invertebrates, fishes, sea turtles, seabirds, and marine mammals are caught, killed, and then discarded. Many more suffer nonlethal encounters that eventually reduce their survival or market value.

Two of the most common industrial fishing technologies, gill netting and bottom trawling, discriminate poorly between desired and undesired species. These and other techniques (e.g., purse seining for tunas associated with dolphins, longlining, dynamite fishing) have generated widespread concern over their effects on biological diversity.

Gill netting is a passive fish-entangling method that has been used for millennia, but until recent years its impact tended to be localized. Modern synthetic materials such as nylon have made it possible for a vessel to set up to 65 kilometers (40 miles) of gill net daily. In the North Pacific, the use of drifting gill nets (driftnets) to catch squid and tunas has involved the use of up to 3.5 million kilometers (2.2 million miles) of monofilament netting yearly, enough to circle the Earth 88 times.

In these fisheries, a gill net is set in the afternoon and allowed to drift overnight, hanging from the surface to 10 meters (about 32 feet) deep. Fish, squids, and other species become tangled when they contact the net. Observers report bycatch of nearly 200 nontarget species, or 40 percent of total numbers, including sea turtles, seabirds, and marine mammals (Fisheries Agency of Japan et al. 1991).

The effects of bycatch on North Pacific biological diversity have not been studied adequately. However, preliminary results suggest that the Japanese squid driftnet fishery is reducing northern right whale dolphin (*Lissodelphis borealis*), Pacific white-sided dolphin (*Lagenorhynchus obliquidens*), and albacore tuna (*Thunnus alalunga*) populations (Canada Department of Fisheries and Oceans 1991). International concern over the impacts of driftnetting prompted the UN General Assembly to declare a moratorium on all large-scale pelagic driftnet fisheries starting December 31, 1992.

A recent workshop on cetacean bycatch in gill nets and traps (IWC 1991) concluded that seven stocks are experiencing unsustainable levels of mortality, including northeastern US harbor porpoises (*Phocoena phocoena*), vaquitas in Mexico's Sea of Cortez, striped dolphins (*Stenella coeruleoalba*) in the Mediterranean Sea, Indian Ocean coastal humpbacked dolphins (*Sousa plumbea*), and beiji (*Lipotes vexillifer*) in China's Chang Jiang River. Available data suggested the potential for unsustainable incidental mortality in 46 other stocks.

Regrettably, there is little comprehensive information on the effects of incidental catch for taxa that are most vulnerable. Woodley and Lavigne (1991) reviewed the literature on bycatch of pinnipeds worldwide, and found that there were no data on incidental take of 13 of the 34 extant species, but nine others are being harmed, mainly by gill netting and trawling. Driftnets also drown large numbers of seabirds yearly—at least hundreds of thousands—and have seriously depressed local populations of species such as tufted puffins (*Fratercula cirrhata*) in the Aleutian Islands (Atkins and Heneman 1987).

Trawls are funnel-shaped nets that are towed behind vessels. Animals are swept into the net and collect at the end of the funnel. The proportion of nontarget species caught in trawl fisheries can be very high. Bricklemyer et al. (1989) estimated the worldwide bycatch of finfish in shrimp trawls to be 4.5 to 19 million metric tons (10 to 42 billion pounds)—a figure roughly equivalent to 5 to 20 percent of the world's entire seafood catch—compared to total shrimp landings of 2 million metric tons (4.4 billion pounds). Including other taxa killed in great numbers—sponges, jellyfishes, soft corals, crabs, squids, sand dollars, starfishes, and sea turtles—would provide an even clearer picture of the astounding magnitude of bycatch from the shrimping industry. Andrew and Pepperell (1992) estimated the total bycatch from the world's shrimp fisheries to be 8.2 to 16.5 million metric tons. Bycatch in other fisheries, while probably less than that in shrimp fisheries, can nonetheless be very large. Nontarget species constitute 76 percent of the yellowtail flounder (*Limanda ferruginea*) trawl catch off the northeastern USA (Murawski 1991). Effects of trawling on community structure have been documented in the Gulf of Mexico (Powers et al. 1987), Georges Bank in the Northwest Atlantic (Murawski and Idoine 1989), North Sea (Anonymous 1991), and the Northwest Shelf and Gulf of Carpentaria of Australia (Sainsbury 1988).

US shrimp trawl fisheries in the Southeast Atlantic and Gulf of Mexico are estimated to take and discard more than 10 billion fishes (Pellegrin et al. 1981) and 5,500 to 55,000 sea turtles annually (National Research Council 1990a), including critically endangered Kemp's ridleys. TEDs have recently been required in US shrimp fisheries to minimize sea turtle bycatch. Incidental take from shrimp trawling has severely reduced red snapper populations in the Gulf of Mexico (Goodyear 1990). The effects of shrimp trawling bycatch on invertebrates is unknown. The development of bycatch reduction devices that enhance the proportion of shrimp in trawl catches could substantially reduce the mortality of other species if the devices are widely adopted.

Aside from coastal gill netting, many artisanal fishing techniques are limited in scope, and their bycatch is less threatening to biodiversity. Glaring exceptions are muro ami, a technique in which weighted bags are used to smash coral reefs to scare fishes from their hiding places (Corpuz et al. 1983), dynamite fishing, biomass fishing, and poisoning to collect tropical coral reef fishes. Dynamite fishing (Figure 4-1) is widely used throughout the tropics despite being illegal in most areas (Zann 1982). The explosions kill everything near the blast, including corals. Fish collecting for the aquarium trade often involves poisoning reefs with sodium cyanide, rotenone, or sodium hypochlorite bleach, resulting in the indiscriminate killing of fishes, corals, and other invertebrates. Reefs with lower living coral cover have lower fish diversity (Soekarno 1989), so these indiscriminate techniques remove not only today's inhabitants but also destroy their habitats for the long-term future (Munro et al. 1987).

For most fisheries, data on incidental take and its effects are much more limited than data on targeted species. But it is clear that the amount of bycatch is very large in some fisheries, and that its effects are serious. It is also clear that population and ecosystem studies of fisheries' bycatch and actions to minimize it could make an important contribution to the conservation of marine biological diversity.

Alteration of Food Webs: Species that occur together are connected by a network of feeding relationships called food webs. In the sea, food webs can be strikingly complex. They often have multiple trophic levels. But these levels are usually blurred because so much marine production

Figure 4-1. Dynamite fishing. *Although their use is widely illegal, explosives are used by many fishers in poor tropical countries, devastating coral reefs on which future fish production depends. Dynamite fishing, muro ami, and poison fishing illustrate how desperate poverty undermines sustainability.*

becomes (and is consumed as) detritus, such as the fecal pellets of sur-geonfishes (Acanthuridae) and the mucoidal particles ("coral snow") given off by reef corals. Moreover, feeding stages of marine fishes and invertebrates grow through a wide range of body sizes and thus have a variety of food web roles. As a result, the number of possible interactions among species is very high. Early food web research described all the connections among web components, but certain connections are partic-ularly important in terms of energy or materials flow. Certain species, often generalized consumers from upper trophic levels (species that have few species that eat them, but eat many other species), serve as "strong interactors" or keystone species in marine food webs. If these are re-moved, the food web changes profoundly (Paine 1980).

Removal of species often leads to the loss of other species, to shifts in relative abundance, or to reduced resilience (Paine 1980; Pimm 1980). In some systems, there seem to be multiple stable-state communities in which removal of keystone species causes communities to shift from one state to another (Beddington 1984). If this is true, overexploitation can cause fundamental, effectively irreversible changes in marine ecosystems even if whatever removed the keystone species—say, fishing pressure—is relieved.

In the Northwest Atlantic continental shelf, overfishing has led to significant changes in biomass dominance among fish species (Sherman 1990). Populations of Atlantic mackerel (*Scomber scombrus*) and Atlantic herring (*Clupea harengus*) have declined, leading to an increase in less valuable sand lance (*Ammodytes* spp.) populations.

Exploitation of sea otters by aboriginal Aleuts in waters off North Pacific islands led to alternate stable-state communities (Simenstad et al. 1978). When otters were abundant, herbivorous sea urchins were rare and large algae were abundant. When otters were locally overexploited, the number of sea urchins increased and they grazed down the algae, reducing habitat for many species of fishes and invertebrates. But it is not just sea otters that affect benthic communities by preying on sea urchins. Over-fished sites in Haiti and the U.S. Virgin Islands (Hay 1984) and Hainan Island (China) (Hutchings and Wu 1987) have dense populations of urchins in the genus *Diadema*, and very little *Thalassia* seagrass or coral cover. The proliferation of urchins and the consequent elimination of algal or seagrass beds, forming "sea urchin barrens," occur in many places where predators of sea urchins—such as wrasses (Labridae) or triggerfishes (Balistidae)—have been removed (Personal communication with Paul Dayton, Scripps Institution of Oceanography, La Jolla, California, USA). Similarly, the removal of large sharks off South Africa to protect human swimmers led to increases in the numbers of small sharks they eat, hence to reductions in other fishes on which the small sharks prey (van der Elst 1979).

The interaction of environmental fluctuations and fishing can cause important food web effects. During the 1982–83 El Niño, physical changes in the upwelling ecosystem concentrated Peruvian anchovy schools, which made them easier to catch. The weakened upwelling also reduced primary production twentyfold. Together, these dramatically reduced anchovy populations. Both seabirds and marine mammals that depend on anchovies lost weight or had reduced reproductive success (Barber and Chavez 1983).

Breaking a strand in a spider web changes the relative position of other nodes in the web. In a similar way, the lesson from studies in the sea and the more numerous studies in fresh waters is unmistakable: Removal of one or a guild of species can dramatically disrupt the web of relationships among species.

Marine Mammals, Seabirds, and Sea Turtles

Fishes are not the only marine vertebrates that are overexploited. Overexploitation has decimated many marine mammals, seabirds, and sea turtles. It pushed sea otters and critically endangered Mediterranean monk seals (*Monachus monachus*) very close to the brink of extinction, and eliminated Caribbean monk seals (*M. tropicalis*) and great auks (Day 1981).

Because marine mammals, seabirds, and sea turtles are generally long-lived, slow-reproducing organisms, they are very vulnerable to overexploitation. Some that have been subjected to horrific hunting pressures have not recovered well. For example, sealers slaughtered millions of endemic Juan Fernandez fur seals (*Arctocephalus philippi*) in the Juan Fernandez islands (Chile) two centuries ago; a single ship carried a million skins to England. As of 1973, the population was no more than 700 to 800 individuals (King 1983). Much the same story could be told for blue whales, northern right whales, and some other species.

Great Whales: The quest for baleen, oil, and meat has profoundly affected great whale populations. Because demand exceeded supply, several species were pushed toward extinction. Basque, German, French,

Dutch, and British whalers hunted great whales in the eastern North Atlantic from the 1000s to the 1600s, then expanded their operations geographically and were joined by Americans in the 1700s. By 1800, whalers had nearly exterminated Atlantic populations of northern right and bowhead whales (*Balaena mysticetus*), which now cling precariously to survival. The third North Atlantic whale they hunted, the Atlantic gray whale, is now extinct (Mead and Mitchell 1984).

Declining right and bowhead whale stocks caused whalers to shift their efforts to humpbacks (*Megaptera novaeangliae*) and sperm whales (*Physeter macrocephalus*), which they quickly overexploited. The advent of exploding harpoons and steam engines in the late 19th century allowed whalers to kill more whales in four decades than they had in the previous four centuries. And these new technologies allowed them to target fast-swimming species—blue, fin (*Balaenoptera physalus*), sei (*B. borealis*), Bryde's (*B. edeni*), and minke (*B. acutorostrata*) whales—that had previously escaped exploitation. When blue whales became rare, whalers shifted to the smaller fin whales, then to still smaller sei whales, overexploiting each in turn (Figure 5-1).

Whaling nations, alarmed by the rapidly declining whale stocks, established the International Whaling Commission (IWC) in 1946 to encourage sustainable whaling, but even IWC accelerated the depletion of whale stocks. Under its New Management Procedure, catch limits remained high and whale populations continued to dwindle, which encouraged competition among the whalers for the remaining whales.

The outcry from a growing international "Save the Whales" movement led, in 1982, to the adoption of a moratorium on commercial whaling, but commercial whaling still exists. Even during the moratorium, commercial whaling has continued under the name "research whaling." The IWC Scientific Committee is developing a Revised Management Procedure that might allow commercial whaling to resume, and three nations—Iceland, Norway, and Japan—are planning to resume whaling for minkes.

Small Whales: Overexploitation is a growing problem for many small cetacean species as well. Some river dolphin species (family Platanistidae) are in imminent danger of extinction. Their habitats make them susceptible to directed hunting, incidental fishery take, and habitat destruction. This section surveys only directed take of small cetaceans.

Some indigenous peoples who have a long history of whaling continue to hunt populations that are now vulnerable. For example, Canadian native hunters have severely depleted several white whale or beluga (*Delphinapterus leucas*) stocks. Canadian and Greenland subsistence hunters take narwhals (*Monodon monoceros*) and harbor porpoises, and Faeroe Islanders take pilot whales (*Globicephala* spp.). But by far the largest catch of small toothed whale species is in Japan, which has a long whaling tradition.

Japanese hunters harpoon Baird's beaked whales (*Berardius bairdii*), pilot whales, Pacific white-sided dolphins, and Dall's porpoise (*Phocoena dalli*). Between 1960 and 1970, they killed 5,000 to 9,000 Dall's porpoises, and these numbers nearly doubled in the late 1980s (Brownel et al. 1989). Japanese fishers also drive small whales into bays and block their escape with nets. For example, between 1976 and 1982, Iki Islanders killed more than 4,000 bottlenose dolphins (*Tursiops truncatus gilli*), and hundreds of false killer whales (*Pseudorca crassidens*), Risso's dolphins (*Grampus griseus*), and Pacific white-sided dolphins (May 1990). The directed kill of small cetaceans in Japan's coastal waters totals over 20,000 annually.

Takes of other small cetaceans are not long-established traditions. In the 1970s, when El Niño and overfishing caused Peruvian anchovy fisheries to collapse, fishers shifted to gill netting dolphins, catching 10,000 in some years. Substantial numbers of small whales are killed for bait (IUCN 1988). In Venezuela, more than 7,000 dolphins annually are killed for shark bait. Crab fisheries in Chile's Strait of Magellan can use thousands of cetaceans annually as bait, including Commerson's dolphins (*Cephalorhynchus commersonii*), Chilean or black dolphins (*C. eutropia*), Peale's dolphins (*Lagenorhynchus australis*), dusky dolphins (*L. obscurus*), and Burmeister's porpoises (*Phocoena spinipinnis*).

How the above activities affect populations of small toothed whales is unclear. Sadly, overexploitation is just one factor threatening small cetacean populations; habitat destruction and pollution are occurring as well. The future for many small cetacean species is grim unless nations get better estimates of their population status and mortality from various causes, and institute or enforce effective conservation measures.

Birds: Of the roughly 8,700 species of birds worldwide, perhaps none have suffered from direct exploitation as much as seabirds and shorebirds, including pelicans, albatrosses, gulls, terns, and sandpipers. Many nest in dense colonies, making them vulnerable to large-scale egg collection or hunting. Flightless species such as penguins, and burrow-nesters such as shearwaters, can be easily caught ashore. Overexploitation has sharply decreased or eliminated many local colonies worldwide (Croxall et al. 1984).

Perhaps the most famous extinct seabird is the great auk, whose populations once numbered in the millions. The auk was persecuted for its meat, feathers, and eggs (one sea captain took 100,000 eggs in a single day), and the last of these penguin-like flightless North Atlantic birds was killed by Icelandic fishers about 1844 (Day 1981).

Islanders in virtually every part of the world have traditionally relied on seabirds, and seabird parts played an important ceremonial or medicinal role for some peoples. Because many seabird colonies are so restricted, even hunting for local consumption can eliminate entire colonies.

The ease of collecting eggs and birds often made commercial exploitation profitable. Some species of penguins were highly prized for their oil content, and exploitation by "oil gangs" eliminated colonies of king penguins (*Aptenodytes patagonicus*) on Macquarie Island. In the mid-19th century, more than 2 million rockhopper (*Eudyptes chrysocome*) and gentoo (*Pygoscelis papua*) penguins were killed for oil in the Falkland Islands. Plume hunters in the 1800s almost killed off short-tailed albatrosses (*Diomedea albatrus*) in Japan and common terns (*Sterna hirundo*) along the eastern USA. Similarly, market hunters in the late 19th and early 20th centuries shot shorebirds by the wagonload. By 1925, the Eskimo curlew (*Numenius borealis*), whose migrating flocks once darkened the skies of eastern North America, was all but extinct. Moreover, in many countries, seabirds are killed because fishers consider them competitors.

Many nations have restricted hunting and egging, and such legal measures have enabled the recovery of many seabird populations. Commercial exploitation of some species continues, although it has greatly diminished or been halted where seabird and shorebird populations have declined dramatically. However, illegal hunting and egging continue to threaten some seabird populations and prevent already depleted populations from recovering.

Sea Turtles: Sea turtles are very slow to mature, taking perhaps, in some cases, more than 50 years. High mortality from natural and anthropogenic sources has combined to put six of the seven species in danger of extinction. Once abundant in the world's tropical and temperate seas, many sea turtle populations have declined dramatically, largely from overexploitaiton.

In addition to the traditional use of sea turtles as food, they are also hunted for leatherback oil by folk healers in the British Virgin Islands to treat respiratory ailments (Cambares and Lima 1990), and by Asian fishers to seal their boats (Bhaskar 1981). To some societies, sea turtles are such potent symbols of fecundity that people eat their eggs as aphrodisiacs.

Sea turtles are also hunted for distant luxury markets. In particular, hawksbill, olive ridley (*Lepidochelys olivacea*), and green turtles have been heavily exploited for shell, skin, and meat. Demand is so high that many coastal people take turtles opportunistically. In Honduras, for example, spiny lobster fishers kill all turtles they encounter, knowing that Japanese shell dealers will buy the carapaces (Personal communication with Gustavo Cruz, Universidad Nacional Autonoma de Honduras).

Sea turtles' life cycles make them particularly vulnerable to human exploitation. Turtles spend their lives at sea except when, after decades of pre-adult growth, the females come ashore, in two- to four-year cycles, to lay eggs. Nesting females are predictable, defenseless targets and, hence, are easier to capture than turtles at sea. However, even with heavy exploitation, the effects of removing most females from a population may not be noticed for many years. By the time it is obvious that a population is crashing, it is too late to stop the decline.

The demise of the Kemp's ridley, the most endangered sea turtle, dramatically illustrates the long-term, devastating effect of overexploitation. Despite years of hunting and the collection of millions of eggs, in 1947 more than 42,000 Kemp's ridley females nested in a single day on a remote Mexican beach (Hildebrand 1982). Kemp's ridleys declined so precipitously from targeted exploitation and incidental drowning in shrimp trawls, that, by 1968, the largest nesting aggregation totaled 5,000 animals. Today, despite 25 years of rigorous nesting beach protection, as few as 362 turtles nest in a season, less than 1 percent of the species' 1947 abundance (Pritchard 1990).

Many countries have recognized the need to protect sea turtles, but sea turtle biology constrains conservation efforts. Not only do their populations grow slowly, at best, but sea turtles are migratory and often forage in the waters of one country but nest on the beaches of another, sometimes going from a safe area to one where they are intensely exploited. For example, for decades Australia has protected green turtles nesting in Queensland, while these same turtles are heavily exploited in Indonesia (Personal communication with Colin Limpus, Queensland National Parks and Wildlife Service, Queensland, Australia). Migration makes sea turtle conservation an issue requiring international cooperation, as has been the case with migratory birds.

To date, conservationists have been unable to resolve the debate over the use of sea turtles. Many argue that the well-controlled collection of eggs is the most sustainable use of sea turtles, and also provides an incentive to protect adults and their nesting beaches in developing nations such as Suriname, which allows collection of "doomed" eggs by Carib Indians where nests would wash out and be lost.

A handful of projects to rear turtles for commercial use have been started in the last 30 years, but these are expensive and fraught with difficulty. Although there is no consensus among sea turtle biologists about turtle farms, most conservationists argue that if adult turtles are to be killed, they should be used for subsistence only.

"Non-fishery" Overexploitation: Invertebrates, Aquarium Fishes, and Plants

Marine species provide an astounding variety of products beyond those fishes and invertebrates traditionally considered "fishery products," and the ways in which they are taken are as diverse as their uses. Examples include mangrove trees used as fuel or timber, sponges used for bathing, stony corals used as building material, and tropical reef fishes kept as pets. Overexploitation can have locally, even globally, serious effects on these species and their habitats.

Although commodities such as these are seldom considered *fishery* products, local fisheries departments are often involved in managing their exploitation. In countries where fisheries are overexploited and fishers need alternative livelihoods, such "miscellaneous marine products" are starting to receive long-overdue recognition.

Many of these species are taken in tropical multispecies subsistence and artisanal fisheries. Catches in such fisheries are usually underrecorded, so trends are difficult to ascertain. Many commercially important marine invertebrates are sedentary shallow-water species, which makes them vulnerable to collectors. The standard pattern is overfishing of shallow inshore populations followed by collecting in deeper waters using scuba.

Algae and Vascular Plants: Marine algae are taken in many places, with reports of overexploitation in the Philippines, India, and many Caribbean islands, but there have been few studies on their conservation status. The generally slower growth of mangrove trees puts them at greater risk. Although traditional uses of mangroves can be sustainable, use of mangrove foliage for grazing livestock has severely damaged forests in the Indus River Delta of Pakistan and northern India. Exploitation for fuelwood and charcoal is a major problem in all densely populated countries where mangroves occur, especially Southeast Asia. Logging for mangrove timber and wood chips is a major industry in India, Bangladesh, Thailand, Indonesia, and many South American nations. In Malaysia, these industries have been managed by government forestry departments for decades and seem sustainable, but where such management is absent (e.g., the Philippines and Indonesia), vast mangrove forests have disappeared (Saenger et al. 1983; Hamilton and Snedaker 1984). Furthermore, as is discussed later, large areas of mangrove forest have been destroyed to make way for mariculture operations in the Philippines, Ecuador, and other nations.

Sponges: About 15 species of sponges are taken commercially, mainly in the Mediterranean (especially Tunisia) and the Caribbean. Sponge fisheries have declined because of overfishing and devastating epidemics (Josupeit 1991). Management of sponge fisheries has long been overlooked. The development of a sponge farm in Pohnpei (Federated States of Micronesia), that requires little capital investment or training but seems commercially viable might provide one sustainable solution.

Coelenterates: Fisheries for the pink and red precious corals in the genus *Corallium* have a characteristic "boom and bust" pattern. Mediterranean and coastal Asian stocks are heavily overexploited, so remote oceanic seamounts in the Pacific are now the main source. Regional fishery management plans are now under development for the Mediterranean (Caddy and Savini 1989) and for US Pacific waters (Grigg 1984). No solution has been found for regulating exploitation in international waters.

Management of other corals has largely been overlooked. Black corals (Antipatharia) are collected widely for souvenirs and jewelry. The impact of this is still virtually unknown, but there are anecdotal reports of overexploitation, particularly in the Caribbean. Collecting is most intense in the

Philippines. Sustainable methods of exploitation have been developed in Hawaii (USA), where the fishery is closely monitored and regulated (Wood and Wells 1988). But even there, heavily fished areas have not fully recovered, although they are only lightly fished now (Personal communication with Richard Grigg, University of Hawaii, Manoa, Hawaii, USA).

As discussed in Chapter 3, the mining of reef rock and live corals for building can devastate coral reefs. In contrast, the collection of stony corals as curios is a rather small issue in global terms, but has increased significantly and can accelerate reef deterioration, particularly on reefs damaged by other human activities. Reefs in parts of the Philippines, the main supplier, have suffered, and there is concern in other countries as well. Ironically, the main importer, the USA, has strict legislation banning the collection of stony corals on its own reefs (Wood and Wells 1988).

Mollusks: Many of the approximately 5,000 species involved in the shell trade are reef species. The Philippines, Indonesia, Haiti, and Mexico export large amounts. Although few data are available on the impact of collecting, there are anecdotal reports of overexploitation. Especially vulnerable species include narrow endemics with small populations; large, visible species living exposed on reef flats, such as spider conchs (*Lambis* spp.) and egg cowries (*Ovula ovum*); species without planktonic larvae, such as volutes (Volutidae); and other popular but rare species, such as Triton's trumpet snail (Wells et al. 1983).

Mother-of-pearl is a valuable marine product from marine mollusks including pearl oysters (*Pinctada* spp.), trochus (*Trochus niloticus*), green snails (*Turbo marmoratus*), and paua, awabi, or abalone (*Haliotis* spp.). The main suppliers are Southeast Asia and Pacific island nations, and overfishing is widespread. However, in most countries where these species are exploited, they are treated as fisheries and regulated to some degree.

Giant clams are taken mainly from nearshore reefs in the Philippines and Indonesia and from more remote reefs in the Indo-West Pacific. Their adductor muscles are a prized specialty food in southeastern Asia, but the shell trade has also contributed to the current rarity of these species, particularly *Tridacna derasa* and *T. gigas* (Munro 1989).

Curios: In addition to corals and mollusks, many marine species are involved in the curio trade, including coconut crabs (*Birgus latro*), sea fans, starfish, the tests of sea urchins, seahorses (*Hippocampus* spp.), and inflated porcupine fish (Diodontidae). The impact of collecting these is poorly documented.

Aquarium Invertebrates and Fishes: Astronomical growth in the aquarium trade has increased demand for tropical marine animals and even living reef rocks (rocks with attached plants and animals). About 250,000 pieces of live coral were imported into the USA in 1991 for the aquarium industry (Derr 1992). No marine fishes or invertebrates seem to be threatened with extinction through the aquarium trade, although there may be local overexploitation of some high-value species such as butterflyfishes (Chaetodontidae). A much greater concern is chemical and physically destructive methods used to catch aquarium fishes that damage the reef as a whole.

Physical Alteration

Because organisms are adapted to some physical conditions but not others, physical conditions are very important in determining the community of species that live in each ecosystem. If, say, the structure of the seabed or the temperature of the water is altered, the composition and functioning of the biological community will change correspondingly.

Although conservationists and resource managers generally agree that the physical alteration of ecosystems is the greatest single threat to biological diversity on land, the importance of physical alteration on marine biodiversity is often overlooked. Yet, it probably ranks with overexploitation and pollution as the greatest near-term threats to life in the sea.

Human activities alter marine ecosystems in many ways. Some are intentional: The object is to alter the physical environment. In other cases, physical alteration has been overlooked, as with trawling, human visits, and noise pollution. Some activities do not even occur in the sea, but affect it because they alter the flow of important materials between land or fresh waters and the sea. Only a sampling of the many kinds of physical changes is covered here, but others that occur or could occur on various spatial scales and at varying intensities should not be dismissed as unimportant. They include: disposal of waste heat from power plants; changes in oceanic stratification from ocean thermal energy conversion plants; alteration of tidal cycles from tidal power plants; disturbance from ships running aground; and physical damage from a wide variety of fishery techniques, including muro ami and dynamite fishing.

Intentional Alteration

Although beavers, termites, coralline algae, red mangroves, and many other species alter their physical environments, none does so nearly as much as humans. By changing the structure of the seabed and characteristics of the overlying waters, we are simplifying, fragmenting, and even eliminating species' habitats, and thereby affecting the functioning of entire ecosystems.

The scale, intensity, duration, and timing of physical alteration all influence their impact. Kilometers-long seawalls, which exist along some coasts (and which will undoubtedly proliferate as sea level rises), have more impact than small ones. Thermal pollution that elevates temperature by 5°C (9°F) affects species more than a 2°C (3.6°F) increase. Permanent or chronic alterations have greater impact than acute events of equal intensity. And mussel dredging can have a greater effect on oxygen demand in some seasons than in others (Riemann and Hoffmann 1991).

Intentional physical alteration of marine ecosystems includes: logging of mangrove forests; construction of facilities such as marinas, jetties, and pipelines; dredging and filling; mining; and anchoring. In ecosystems where the physical structure is especially delicate or heavily visited, even direct contact with people is harmful. Marine ecosystems are also vulnerable to purposeful physical alteration in terrestrial watersheds, such as dams, road building, logging, and agriculture, that brings about major physical changes in estuaries and coastal waters downstream.

Destruction, Simplification, and Fragmentation: Some human activities so change the physical environment that few or none of the original species can remain and the ecosystem functions very differently, effectively destroying entire mangrove forests, salt marshes, mudflats, rocky intertidal and subtidal reefs, coral reefs, seagrass beds, or areas of the deepsea benthos. In other cases, the physical conditions in these ecosystems are altered enough to degrade them, diminishing or eliminating populations of many species and altering ecosystem functions such as productivity. This kind of ecosystem simplification is even more widespread than ecosystem destruction. The breaking up of continuous ecosystems into fragments surrounded by unsuitable habitat has received a great deal of attention from terrestrial conservation biologists since the mid-1970s, but has not been studied in the sea. The importance of marine ecosystem fragmentation is unclear, but it merits serious attention.

Logging Mangrove Forests: Logging in mangroves illustrates the multiple causes and consequences of physical alteration. Mangroves are logged for their wood and bark, and to make way for roads, homes, industries, and mariculture. The Philippines alone have lost almost 70 percent of their mangrove forests since the 1920s, over half of which were converted to mariculture ponds (Primavera 1991). Logging of mangroves eliminates not only the permanent residents of this ecosystem, but also the nursery grounds on which wild shrimp and fish stocks depend and mariculturists rely to stock their farms (Turner 1991). It eliminates mangrove detritus, a major food source in marine ecosystems, as far as a kilometer (0.62 miles) from the site. Removing mangroves also makes coasts more vulnerable to erosion by wind and wave action, a crucial consideration in low-lying storm-prone nations from Bangladesh to Guyana.

Shoreline Structures: Shoreline structures for waterfront protection, containment, and enhancement can have many negative impacts. In addition to destroying the ecosystems they replace, they markedly alter patterns of sediment and nutrient transport. When breakwaters, jetties, and groins accumulate sediment, they can prevent downcurrent shoreline renourishment and erode beaches and marshes. In addition, seawalls and bulkheads will prevent the migration of salt marshes and seagrass beds to favorable elevations as sea level rises (Leatherman 1991).

Levees built to protect land from estuarine flooding also prevent the deposition of nutrient-rich sediments that nourish coastal lowlands. This not only diminishes their productivity, but can cause subsidence of the very lands that the levees were built to protect. Rather than benefiting floodplain ecosystems, the sediments are carried into coastal waters, where they diminish light penetration, hence productivity (Chabreck 1988), and discourage species that locate their prey visually or filter-feed, thus altering the community composition, structure, and function.

Beach Renourishment: Beach renourishment, the replacement of sand lost to erosion, usually harms both the source and dumping areas. Sand mining is a major physical disturbance in the source area because it resuspends sediments, stressing ecosystems for long distances downcurrent. Furthermore, digging deep holes for sand can change patterns of wave refraction and perhaps concentrate wave energy onshore (Leatherman 1990), contributing to increased shoreline erosion. If sand that is too fine is added to a beach, it can quickly erode and cause an abnormally high suspended sediment load downcurrent, affecting neighboring ecosystems, especially vulnerable ones such as seagrass beds, kelp forests, or coral reefs. Careful timing and planning of sandy beach renourishment can diminish impacts on water quality and species (Hurme and Pullen 1988).

Channelization, Dredging, and Filling: The straight navigation channels cut through salt marshes and mangrove forests are not simply ugly; they fragment wetlands and eliminate land. After initial construction, currents, waves, and boat wakes cause some canals to continue to widen, increasing land loss. Moreover, channelization and ditching encourage saltwater incursion into fresher portions of estuaries, killing species intolerant of higher salinities. Loss of the roots of salt-intolerant plants decreases the strength of the sediments, encouraging erosion and subsidence. Before plants adapted to the new conditions can establish themselves, elevations may decrease to the extent that growth of emergent plants ceases and marshlands are eliminated (Chabreck 1988).

Dredging is a widespread physical alteration to marine ecosystems that can be ruinous. Its effects include the alteration of the bottom topography, the destruction of biota and their habitat, and massive sediment resuspension. Dredging near coral reefs harms suspension-feeders, fills in vital interstices in the reef, and covers the reef surface with sediments that deter settlement by reef species, thereby reducing fish, invertebrate, and algal diversity and income from tourism and fisheries. The resuspended sediment particles take up oxygen, which may cause localized areas of hypoxic (low oxygen) or anoxic (no oxygen) water (Nunny and Chillingworth 1986).

The damage caused by dredging is not limited to the site from which sediments are removed. The dumping of dredged material into marine environments harms the disposal site by burying benthic organisms, clogging the feeding apparatus of many species, and, in some cases, changing the average grain size of the seabed, which alters the benthic community. In addition to these physical effects, dredged materials are usually contaminated with toxic chemicals and thus spread the pollution from the dredge site to the dump site.

Mining and Oil and Gas Development: In this century, mining for minerals and drilling for oil and gas have extended from the land into estuaries (such as Lake Maracaibo, Venezuela), coastal waters (such as the North Sea), and onto the continental slope (such as in the Gulf of Mexico) as technological improvements have permitted. Physical threats to marine species and ecosystems from petroleum extraction and mining operations include: being smothered with spilled oil; fragmentation and loss of wetlands from pipeline construction; disposal of huge amounts of drilling muds and cuttings; and the effects of channelization, dredging, and filling mentioned above. Disposal of some types of mining wastes decreases the stability of the benthic community at the dump site (Nicolaidou et al. 1989). Of course, these processes also produce chemical pollution and solid wastes.

Unlike overexploitation and pollution, the diverse kinds of physical alteration in the sea are seldom, if ever, considered collectively. Other activities discussed under those headings, such as dynamite fishing and thermal pollution, could just as easily be considered physical alteration. The realization that human impacts in the sea include dramatic reshaping of the seabed and its overlying waters from both marine activities and land uses can help decision makers deal with the ubiquitous effects our species has on marine biological diversity.

Trawling

At least since 1376, when England's House of Commons petitioned the king over its concern about destruction of underwater vegetation by trawling (Graham 1955), people have suspected that trawling harms the seabed and its species. The reason for concern has grown as nets and boats have become larger and more powerful.

Trawling profoundly disturbs the seabed (Figure 4-2). Like storms and turbidity currents, it churns and resuspends sediments. But its effects are felt deeper than storm-generated waves, and it occurs more often. A shrimp trawler with nets 20 meters (65 feet) in width towing at 5 kilometers per hour (2.7 knots) scrapes 1 square kilometer (0.39 square miles) of seabed in ten hours. Shrimp fisheries in the northern Sea of Cortez (Mexico) and the northern Gulf of Mexico (USA) sweep the entire

Figure 4-2. Shrimp trawler. Trawling profoundly disturbs the seabed. It destroys the burrows of bottom-dwelling species, mangles huge numbers of nontarget species, and increases suspended sediments. Some extensive areas of seabed are trawled 3 to 10 times per year.

trawling grounds several times per year; in the Kattegat (Denmark), the frequency can exceed 10 times per year (Riemann and Hoffmann 1991).

Trawling affects benthic communities both directly and indirectly. Direct effects include damage (and often death) of target and nontarget species due to contact with the trawl and physical alteration of the seabed. Indirect effects include the resuspension of sediment particles, toxic chemicals, and nutrients, as well as the discarding of bycatch, which undoubtedly affects food webs.

The reason that trawling is destructive might not be obvious to non-experts. The seafloor is not covered by homogeneous, featureless accumulations of sediments. Rather, sandy and muddy bottoms are communities with complex structure that comes from nonliving objects (rocks, shells, worm-tubes, logs), living organisms (seaweeds, sponges, bryozoans, mollusks), and, most of all, the results of biological activities such as burrowing. Most of the structures are either too small to be readily seen or lie below the sediment-water interface. But the high structural complexity they provide markedly increases the diversity of species that the seabed accommodates.

Trawling disrupts this "layer-cake of life" (McAllister 1991a), altering not only species composition and structure but also ecosystem processes including carbon fixation, nitrogen and sulfur cycling, decomposition of detritus, and the return of nutrients to the water column. Communities dominated by benthic photosynthesizers, such as algal beds and seagrass beds, are particularly susceptible, a realization that led to the prohibition of trawling near eelgrass beds within Puget Sound (USA) (Reeves and DiDonato 1972).

Bottom trawling also affects the benthos when churned-up sediments resettle. Riemann and Hoffmann (1991) found that suspended particulates increased an average of 1,000 percent shortly after trawling. In addition to burying benthic organisms, resuspended sediments can diminish photosynthesis in the water column, and they inhibit feeding and raise the metabolic costs of clearing sediment particles in suspension-feeding organisms. As a result, frequent resuspension can change the dominant species in the benthic community from suspension-feeders to deposit-feeders. By removing accumulated sediments, trawling can also expose underlying layers of barren clay, rock, or sand with little potential for recolonization (Nunny and Chillingworth 1986) by soft-sediment dwellers. Given the prevalence of trawling and its potential to affect marine ecosystems, there has been astoundingly little research on its environmental impact.

Anchoring, Trampling, and Visitation

Some marine ecosystems and species are so sensitive to human presence that they are harmed when we visit; in essence, people can love them to death. Boat anchoring can cause long-term damage to the substrate and increase sedimentation in areas of high use. Coral reefs are particularly vulnerable, and anchor damage has become a problem around the world, especially where there are large numbers of tourists or recreational visitors. Off Grand Cayman Island in the Caribbean Sea, for example, a single cruise ship destroyed over 3,000 square meters (0.74 acres) of previously intact reef in one day (Smith 1988).

Large numbers of people visiting coastal areas can increase erosion, harm vegetation, and frighten away organisms that use the area for feeding or breeding. Coral reefs and beach dunes are easily disturbed by trampling.

Along rocky shores, the long-term physical impacts of human visitation are lower (Bally and Griffiths 1989), but hardly negligible (Povey and Keough 1991). Even visits by wildlife-watchers to rookeries can be damaging. For example, threatened Steller's sea lions panic and abandon beaches when humans visit. During these stampedes, adults trample young, substantially reducing reproductive success of colonies. Many shorebirds, waders, and seabirds are similarly sensitive to human visits.

Noise Pollution

Noise pollution is rarely mentioned as a risk to biodiversity, yet it is pervasive throughout the oceans. Seabird researchers have long known that noise from aircraft overflights can disrupt colonies, causing parents to abandon their eggs or chicks, thereby exposing them to the elements and predators such as gulls. Noise might actually be the greatest stress for some marine mammal species because:

1) Water can conduct sound waves for thousands of kilometers.
2) Many species have evolved special sensitivities to sound at frequencies like those produced by shipping and underwater construction.
3) A substantial number of species rely on acoustic signals as their primary means of communication.

Concern about undersea noise was heightened during the 1990 sound transmission experiment at Heard Island in the southern Indian Ocean, when researchers trying to measure global ocean temperatures pulsed the sea with sounds loud enough to be carried around the world. Although one study (Personal communication with Ann Bowles, Hubbs Seaworld, San Diego, California, USA) suggested that it affected marine mammals, more comprehensive studies are needed.

The sound pulse experiment dramatically pointed out the paucity of research on the effects of human-generated noise on marine organisms. However, shipping is clearly a far greater source of noise than any short-term experiment (Personal communication with Peter Tyack, Woods Hole Oceanographic Institution, Woods Hole, Massachusetts, USA). Super-tankers produce unremitting sounds at magnitudes reaching 209 decibels, as loud as the sound pulse of the Heard Island experiment. Considering the abundance of tankers in the world's shipping lanes, it is remarkable that the effects of this undersea cacophony have never been addressed adequately.

The U.S. Minerals Management Service (MMS) has sponsored studies on the effects of noise from oil exploration and drilling activities, and MMS and the U.S. Office of Naval Research now have the technology to address the issue more broadly. These studies clearly identify the potential for significant risk to marine mammals, particularly beaked whales (Ziphiidae) and sperm whales, and to large fish species.

There is not enough information to say whether anthropogenic under-sea noise is affecting populations of marine species. Still, imposing yet another stress on species already harmed by overexploitation or chemical pollution is unlikely to benefit them.

Siltation from Land-Based Activities

Any activity that exposes soil to erosion (such as logging, cutting roads into steep hillsides, construction, or agriculture without proper terracing) increases runoff of sediments from the land into streams and rivers and then into the sea. Marine ecosystems are harmed by careless land-use practices hundreds or even thousands of kilometers upstream.

Suspended sediments, particularly fine silts and clays, can smother marine organisms by clogging their respiratory and feeding organs, and reduce the light available for photosynthesis. Upon settling, these sediments can cover photosynthetic surfaces, bury benthic organisms, and render hard substrates unsuitable for colonization by rock- and coral-dwellers.

In ecosystems with characteristically low sediment loads, increased sedimentation and turbidity are especially harmful. Sedimentation is very low throughout most of the open ocean and the deep sea, but is also low in some nearshore ecosystems, which, as a result, attract communities of species that flourish in clear waters, such as kelp beds and coral reefs. These communities are harmed by anything that increases suspended sediments. The logging of most of the Olympic Peninsula in the north-western USA, an area of very high rainfall, has markedly increased sediment loads in rivers and coastal waters, coating rocky reefs with thick layers of silt and eliminating many kilometers of kelp forests (Personal communication with Chris Morganroth, Quileute Tribe, LaPush, Washington, USA).

Siltation has profound effects on coral reefs (Cortes and Risk 1985) and is one of the largest sources of coral reef degradation worldwide, especially for shallow reefs that fringe coastlines. Reefs generally have more coral species, higher coral cover, and higher growth rates the farther they are from sources of runoff or the lower the sediment concentration in overlying waters.

Coral reefs are based on the marvelous symbiosis between coral animals and the microscopic dinoflagellate algae (zooxanthellae) within their tissues. When increased turbidity decreases photosynthesis by the zooxanthellae, reef corals cannot build their skeletons as quickly. They are forced to divert energy from growth and reproduction to production of mucus that sheds choking sediment particles. This diminishes their ability to compete with other species and deter disease. Together, these effects increase coral colony mortality, while they decrease coral species diversity, the percent of living coral cover, and average colony size. Sediments also prevent the establishment of new coral colonies by inhibiting the fertilization of coral eggs, causing abnormal development or death of larvae, and by making the substrate unsuitable for larval settlement. As the reef is deprived of organisms that maintain the integrity of the reef structure, boring sponges, sipunculid worms, and other agents of bio-erosion weaken the substrate, causing the reef to collapse and increasing its vulnerability to physical disturbance. The declining structural complexity and number of hiding places in the reef inevitably decreases the number and biomass of reef fishes.

The coral reefs of the Philippines, probably the most diverse ones in the world, have been devastated by deforestation (Gomez et al. 1981). On the island of Palawan, deforestation and the construction of logging roads accelerated erosion, adding large amounts of sediments to a river that empties into a bay with several reefs. The high sediment deposition significantly reduced coral cover and diversity in the bay, with the reef closest to the river mouth losing almost half of its live corals following heavy rainfall. This, in turn, severely degrades local fisheries, and the local people will pay the economic cost for decades to come (Dixon 1989). Deforestation, agriculture, and the resulting erosion have degraded coral reefs in many other countries, including Madagascar, Indonesia, Costa Rica, Panama, and Barbados.

Altered Freshwater Input

Modification of river drainage basins by humans has led to dramatic changes in the flow of water and nutrients they bring to the sea. Reductions in freshwater flow have caused profound changes in places including San

Francisco Bay (USA) and the Murray River (Australia) (Rozengurt 1991). Before the Aswan High Dam was built, the Nile River discharged about 150 million metric tons of sediment per year into the eastern Mediterranean Sea, but the sediment discharge now is virtually nil. Similar changes caused by hydrological manipulation have occurred in the Colorado River in western North America, the Indus River in southern Asia, and major African rivers (Halim 1990).

In the southern part of the former USSR, dam construction blocked between 30 and 97 percent of the freshwater flow to the great estuaries and nursery grounds of the Azov, Black, and Caspian seas. In the Black Sea, major changes to the hydrology of the Danube, Dnepr, Dnestr, Don, and Kuban rivers (combined with pollution of some of them by sewage and industrial wastes) has depleted oxygen supplies in coastal areas and eliminated brackish water ecosystems. Historically one of the world's most productive seas, the Black Sea, now subject to multiple stresses, has suffered a severe decline in fish catch. Decreases of 90 to 98 percent of the commercially valuable species have been recorded in all of its major rivers and estuaries (Rozengurt 1991).

Reduction in sediment and nutrient supply from rivers has dramatic effects in downstream estuarine and coastal ecosystems. Deltas and their mangrove, marsh, and mudflat communities shrink, and offshore fish populations can decline sharply. Deprived of the annual pulse of sediments, the coast adjacent to the Nile is now subject to erosion, while total fisheries landings dropped by 80 percent after the Aswan High Dam was completed.

Dams sometimes release large amounts of fresh water at times when they would not occur naturally, sharply increasing currents and altering (usually decreasing) temperatures. Downstream effects of these releases, predictably, can be exceedingly harmful. It is not just the *amount* of annual freshwater input but also its *timing* that determines the productivity and diversity of downstream ecosystems.

Conversely, many estuarine organisms, such as swimming crabs in the genera *Callinectes* and *Scylla*, and ecosystems, such as mangrove forests and oyster reefs, depend on constant or frequent freshwater inputs, without which they are replaced by marine species. Because estuaries are important for biological diversity and vital to many fisheries, failure to maintain the amounts and timing of freshwater input and their suspended sediments and nutrients can cause unacceptable losses of these valuable ecosystems.

In nations worldwide, the private interests and government agencies that determine the amount and timing of freshwater flow into the sea are seldom the ones that deal with the health of estuaries, coastal waters, and their fisheries. Therefore, developing effective mechanisms to coordinate institutions that deal with fresh waters and the sea is essential if we are to maintain these vital ecological linkages.

Pollution

To those whose activities produce undesirable by-products, the sea is the ideal depository· Wastes simply seem to disappear. Even some scientists and engineers talk about the "assimilative capacity" of the sea as if their ability to detect pollutants were the measure of the pollutants' impact on living systems. Regrettably, a complex mix of vast amounts of pollutants takes a heavy toll on estuaries and coastal waters, and pollution problems are arising ever farther from land.

Pollutants discussed below include chemicals and solid wastes. Siltation from land-based activities and noise pollution are treated in the above section on physical alteration.

Chemical Pollution

Some of the wastes that enter the sea are harmful to life processes because they are toxic or radioactive. Others are nutrients that stimulate the growth of some kinds of living things at the expense of others. One thing these pollutants share is that once they are discharged, dumped, or washed into the sea, they are extremely difficult, if not impossible, to reclaim. As is true with most human insults to the environment, preventing pollution is ultimately less costly than cleaning it up or suffering its consequences.

Toxic Chemicals and Radionuclides: Toxic products and by-products of human activities enter the marine environment from the air, fresh waters, land, and sea. Some of these pollutants have diffuse origins and are called nonpoint-source pollution. Emissions into the atmosphere from vehicles and power plants settle onto the land and are washed into the sea by storms, or settle onto the sea surface. And water that runs off the land into drainage channels, streams, rivers, and estuaries brings the marine environment countless toxic chemicals that have been applied to the soil and vegetation, washed from impervious urban areas, or disposed of improperly. Others—the point sources—enter the sea at readily defined points. Wastewaters discharged into groundwater, streams, and rivers inject poison into oceans. And liquid industrial wastes, radioactive wastes, sewage sludge, and contaminated dredged materials are loaded on vessels and moved out to sea to be dumped. In addition to these nonpoint and point sources, pollutants enter the sea from accidents during their production or transportation. The sea is being forced to serve as the central cleansing organ of this planet, but the scientific evidence indicates that its capacity to do so is limited.

Toxic materials entering the marine environment are either naturally occurring substances in elevated concentrations or synthetic compounds from industrial processes. Human activities add naturally occurring chemicals to the sea, including trace metals such as lead, mercury, cadmium, and arsenic, radionuclides, and petroleum residues (hydrocarbons). In excessive concentrations, they become harmful environmental contaminants.

Trace metals enter the sea from power plant emissions, mining, and manufacturing, often as particles or adhering to sediment particles. In rivers, contaminated particles either settle to the bottom immediately or flow downstream, where they precipitate and add to the toxic load in estuarine and coastal sediments. Some are carried out to the deep ocean. Heavy metals also enter the sea via the atmosphere. Most of the lead and much of the cadmium, copper, zinc, and iron in open ocean waters entered via the atmosphere.

The most visible, headline-grabbing marine pollution is from oil spills (Box 4-1), but few people know that chronic oil pollution from the land and from shipping far exceeds major oil spills as a source of petroleum residues in the sea.

Radioactive forms of hydrogen, carbon, potassium, and uranium are found naturally in the sea, but almost all of the natural radioactivity in seawater comes from one potassium isotope (Clark 1989). Mining activities can increase the concentrations of natural radionuclides in coastal waters that receive mine drainage, and coal burning emits radioactive particles that settle onto the sea surface or are washed into the sea by streams and surface flow. Artificial radionuclides of cesium, strontium, plutonium, carbon, and hydrogen also enter the sea as a result of accidental releases from nuclear plants (as they have at Sellafield [UK], Three Mile Island [USA], and Chernobyl [former USSR]), nuclear weapons testing, and dumping of radioactive wastes.

BOX 4-1. Oil pollution

Petroleum products are major marine pollutants, with an average of 3.25 million metric tons entering the sea yearly, mainly from ships or runoff from the land (GESAMP 1990).

About half of the world's crude oil production is transported by sea. Although spills contribute only 10 percent of oil input to the sea, they can be devastating. Among the worst are those, like the *Amoco Cadiz* spill in Brittany (France) in 1978, that reach biologically diverse shorelines. But subtidal ecosystems are also vulnerable: A spill on Panama's Caribbean coast in 1986 killed up to 45 percent of subtidal corals at 9- to 12-meter depths (GESAMP 1990). Moreover, attempts to prevent damage to the shoreline can cause long-lasting harm. Dispersants used on the *Torrey Canyon* spill off West Cornwall (UK) in 1967 hindered biological recovery for at least 14 years afterward (Vandermeulen 1982).

Although infrequent, oil field blow-outs can spew enormous amounts of oil into the sea. Mexico's 1979 IXTOC blow-out lasted 10 months and leaked 400,000 metric tons into the Gulf of Mexico. As with spills from ships, many accidents are caused by human error.

Chronic low-level oil pollution, in contrast, gets far less attention than catastrophic spills and blow-outs, but contributes more than half the oil entering the sea. It comes from leaks at marine terminals, disposal of drilling muds from offshore oil operations, municipal and industrial wastes, urban runoff into rivers, and atmospheric fall-out from the incomplete combustion of oil in motor vehicles (Clark 1989). Only a tiny fraction of used engine oil is recovered worldwide; much of the rest must go down the drain on its way to the sea.

Crude oil and petroleum products are mixtures of hydrocarbons and other substances, with highly variable toxicity. When they enter the sea, some evaporate rather quickly, some dissolve, some emulsify into small droplets, and the heavy residue forms tar balls. Although oil degrades more slowly at lower temperatures, even tropical ecosystems are vulnerable. Protected beaches, estuaries, and wetlands are at high risk because oil trapped in sediments oozes out for decades thereafter.

Most of oil's damage to animals and plants results from coating, asphyxiation, and poisoning through direct contact or ingestion, but there are many less obvious effects. Seabirds, otters, and juvenile pinnipeds are at special risk because they must keep their feathers or fur clean in order to avoid heat loss. Even with low-level exposure, their grooming behavior causes them to ingest any oil clinging to them.

Effects on fishes, invertebrates, and algae vary widely. Eggs that float just below the sea surface are generally more vulnerable than demersal eggs, but species that deposit eggs in the intertidal zone are at high risk. Early stages, such as larvae, tend to be most sensitive; the well-developed senses of more mobile adult fishes and invertebrates can allow them to avoid oil.

National policies and international treaties have helped to lessen the frequency of oil spills and reduce chronic pollution. For example, the 1973 International Convention for the

Prevention of Pollution from Ships as modified by the Protocol of 1978 (MARPOL 73/78) requires segregated ballast and crude oil washing systems, oil/water separators, and oil discharge monitoring, and it prohibits all discharges from ships in specially designated areas (e.g., the Mediterranean and Red seas) (GESAMP 1990).

Still, far more can be done to reduce the damage from oil in the sea. These include improved training and certification of people who produce and transport oil, improved vessel traffic management, higher valuation of natural resources that would make mistakes more costly for the perpetrator, an international treaty requiring that ships of oil and other hazardous materials have double hulls, and strong incentives for motor oil recycling. The most important way to reduce oil pollution in the sea, however, is to improve energy efficiency and the use of alternative fuels. Reduced consumption would have many other benefits, from reducing air pollution to slowing global warming.

Some radionuclides are adsorbed onto microscopic organic particles that eventually settle to the seabed (Preston 1989), producing muds more than 50 times as radioactive as the water column. These can be ingested by organisms such as cumacean crustaceans and polychaete worms, which are eaten by fishes. Some species of phytoplankton can concentrate radio-nuclides 200,000 times greater than levels in ambient waters (National Academy of Sciences 1971).

Synthetic organic compounds, particularly those that persist in the environment, are the last major category of chemical pollutants in the sea. Roughly 500 to 1,000 new ones are introduced into the market—and therefore into the environment—every year. The fates and effects of most remain unknown, but many that have been studied carefully do not have promising records. Among the better-known ones are: PCBs, which have a number of industrial uses and enter the marine realm via water discharge and incineration; chlorinated hydrocarbon pesticides such as DDT and HCH (hexachlorohexane), which enter primarily as runoff from agriculture, silviculture, and lawn maintenance; dioxins, which are very toxic compounds commonly found in aerial emissions from incinerators and aqueous discharge from pulp and paper mills; and organotins such as TBT (tributyltin), which are used as anti-fouling agents in marine paints.

Like trace metals and radionuclides, synthetic organics tend to adhere to particles and accumulate in marine sediments, and can later be resuspended and contaminate the water column. In areas with many outfalls, such as river mouths and port areas, the sediments can be extremely contaminated in "hotspots." These sites are often dredged to keep shipping lanes and slips open. The dredged material is then deposited in coastal waters, on wetlands, or on dry land. More than one billion metric tons of sediments are dredged annually worldwide. Nearly all this material is contaminated with at least trace levels of toxics, and about 35 percent are significantly polluted. About 5 percent are so polluted that they are not allowed in marine environments, and the appropriateness of dumping the other 30 percent in marine ecosystems is hotly debated.

While a large fraction of the toxic materials entering the ocean stays within the coastal zone, the addition of many toxics into the atmosphere provides a pathway that carries them out to the open ocean as well. The very thin (roughly 1-millimeter [0.04-inch]) surface microlayer of the ocean now contains elevated concentrations of many contaminants, sometimes hundreds of times more concentrated than in the underlying water. This also occurs in coastal waters.

Toxics enter the sea mainly from rivers and the atmosphere. It is difficult to measure the global influx of contaminants along either of these pathways, so global estimates of the relative contributions from different sources are rough. The UN Group of Experts on Scientific Aspects of Marine Pollution (GESAMP 1990) estimates that 44 percent of all pollution entering the world ocean is from runoff and discharges from land (mainly through rivers), 33 percent is from atmospheric deposition, 12 percent is from maritime transportation (spills and operational discharges), 10 percent is from deliberate dumping of wastes, and 1 percent is from the offshore development of mineral resources.

Although the data in many areas are sparse, there are some unmistakable patterns in marine pollution. Estuaries and coastal waters, not surprisingly, are more polluted than open oceans. However, the open ocean is not pristine, and concentrations of toxic substances that are commonly airborne are high enough to be a cause for concern. In fact, concentrations of PCBs in some oceanic areas are comparable to those in coastal waters of industrialized countries.

In the North Atlantic and the North, Baltic, and Mediterranean seas, surface concentrations of contaminants including cadmium, lead, and PCBs can be high. Whereas known hotspots are primarily in and seaward of estuaries in coastal areas of industrialized nations, the ubiquity of some contaminants (such as organochlorines) in areas far removed from known sources suggests that their residence time in the sea is long; these substances do not disappear. Moreover, natural processes can transport them a long way. Breakdown products of DDT, for example, occur in the fat of Antarctic penguins (Lukowski 1983) at least many hundreds (more likely thousands) of kilometers from where the DDT was released into the environment. The shift in the use of chlorinated pesticides from industrialized to developing nations is likely to increase contamination in Southern Hemisphere waters, perhaps beyond concentrations in the industrialized North.

How do ionizing radiation and toxics affect marine biological diversity? Emissions of particles or energy from radionuclides alter the charge of atoms in biologically important molecules, causing immediate or delayed biological damage. Because radiation levels usually observed in the sea are below those known to harm organisms, observed effects are based on laboratory exposure. In general, more complex organisms are more sensitive than less complex ones. For example, significant mortality after 30 days occurs in fishes at doses only 1 percent of those needed to kill protozoa (Clark 1989). A more insidious potential impact on biodiversity occurs through genetic damage, which results in abnormalities in subsequent generations. Radionuclides seem to be a far smaller threat to marine species and ecosystems than some other pollutants (GESAMP 1990) that have caused far less public concern.

Toxic compounds are rarely immediately lethal in small quantities, so the main concern is usually long-term continuous or intermittent exposure. But even responses that are delayed or subtle can have significant consequences for the composition, structure, and functioning of marine ecosystems. For example, pollutants such as oil can interfere with the reception of sex pheromones that mediate mating behavior in lobsters and with the chemical signals critical to larval settlement. Obviously, anything that markedly diminishes reproduction in a species affects its abundance in the community and its relationships with other species.

Communities in contaminated ecosystems generally have reduced species diversity, as more sensitive species die out and pollution-tolerant species, relieved of competition or predation, proliferate. The effects of toxic pollution on marine biological diversity are unquestionably significant at myriad localities and might well be significant on a hemispheric scale as well, although this is difficult to prove conclusively. It is clear, however, that exposure to toxics can cause outright death, disease (including cancers), immune system impairment, reduced reproductive success, and developmental aberrations. Exposure is generally greatest for benthic species in contact with highly contaminated sediments and for surface-dwelling neuston and pleuston. But nekton are also vulnerable: Their larvae, like those of benthic species, tend to be more sensitive to pollutants than adults are; moreover, they concentrate at the sea surface, where they can encounter toxic substances in the microlayer.

Toxic materials might also affect genetic diversity within species by selecting for pollution-tolerant genotypes. Good places to investigate this would be areas such as the North Sea and Puget Sound (USA), where there is strong correlation between pollution and disease or developmental anomalies in marine animals.

Unfortunately, when local communities, businesses, or nations decide to stop polluting, species and ecosystems might not respond immediately, both because many pollutants are persistent and because food webs do not always return to their original state after stresses are removed. For those whose waters are already polluted who have the wisdom and courage to stop polluting, steadfastness is the key. For those whose waters are as yet unaffected, the key is prevention.

Nutrients: In the sea, as on land, the lack of certain nutrients is a major factor limiting primary production. Hence, it is ironic that nutrient pollution in the sea is a large and growing problem; indeed, the UN's Group of Experts on the Scientific Aspects of Marine Pollution (GESAMP 1990) identified nutrients as *the* most damaging class of pollutants in the marine realm. Too much of a good thing can be harmful in the sea, as elsewhere.

Vascular plants, seaweeds, and phytoplankton require certain elements as building blocks to synthesize organic molecules. There are six essential nutrients that are required in relatively large quantities: nitrogen, phosphorus, potassium, calcium, magnesium, and sulfur, and seven more required in smaller quantities: iron, chlorine, copper, manganese, zinc, molybdenum, and boron. A few other elements, such as silicon, are needed by some but not all photosynthesizers. The different species of marine "plants," however, need somewhat different amounts of these nutrient elements, so differing amounts favor different species.

When other factors (e.g., light or grazing) are not limiting, plant growth is limited by the availability of nutrients. The growth of different species and primary production in general will not rise above the level that can be supported by the nutrient that is *least* available relative to what plants require. In the sea, the nutrients that are most often limiting are nitrogen and phosphorus and, after those, silicon (required by diatoms, a major component of the phytoplankton), iron, and manganese. By affecting the species composition and the growth of photosynthesizers, nutrient availability affects the entire ecosystem, including grazers and decomposers that consume phytoplankton, the species that eat the grazers and decomposers, and so on.

All the essential minerals occur naturally, but some are so unevenly distributed that various marine ecosystems have very different species compositions and primary productivity. Estuaries, which receive nutrient-rich water and sediments from rivers, usually have higher levels of nutrients than coastal waters, which are usually richer in nutrients than open-ocean waters, except for upwelling areas. In general, nutrient-rich waters are more productive, have denser algal populations, and have fewer species than nutrient-poor waters. These algal "bloom" communities generally have shorter food chains and more decomposition by bacteria and fungi.

Eutrophication is the biological process initiated by excessive nutrient enrichment. The following progression (GESAMP 1990) characterizes eutrophicated waters:

1) increased primary production;
2) changes in plant species composition;
3) very dense, often toxic, blooms;
4) conditions of hypoxia (low oxygen concentrations) or anoxia (no oxygen);
5) adverse effects on fishes and invertebrates; and
6) changes in structure of benthic communities.

Not all of these phenomena are observed in every case, and the process may not always reach the final stages. Oxygen levels in the water can be depleted during algal blooms when increasing numbers of dead cells are decomposed by bacteria, affecting animals in the water column and on the seabed.

The land naturally sends much larger amounts of nutrients to the sea than vice versa. But increasing human populations and development have further increased the flow of nutrients to the sea (GESAMP 1987), so that anthropogenic inputs now equal or exceed natural inputs (Meybeck 1982).

Anthropogenic nutrients, including nitrogen and phosphorous compounds, enter the sea as runoff from nonpoint sources including managed farmlands, forests, and lawns, from sewage effluent, from atmospheric deposition, from dredging, and from the dumping of garbage and sewage from vessels at sea. These can initiate eutrophication that, in extreme cases, leads to oxygen depletion that devastates populations of animals that cannot tolerate low oxygen levels. Large recurrent anoxia or hypoxia events in the Mediterranean, North, and Baltic seas, the Persian/Arabian Gulf, the Inland Sea (Japan), the Gulf of Mexico, and New York Bight (USA) have been credited to anthropogenic nutrient pollution.

Nutrient pollution can have adverse effects long before oxygen is depleted. Reef corals, which rely on mutualistic algae within their tissues to supply most of their energy, grow and survive best in clear waters with low nutrients. They become vulnerable when added nutrients stimulate growth of phytoplankton or benthic algae. Because algal blooms cloud the water, they reduce the amount of sunlight that reaches the reef, slowing the growth of corals and making their survival impossible in deeper water. Similarly, higher nutrient levels favor benthic algae that compete with corals for space and light and can actually overgrow and smother them.

During major algal blooms, the phytoplankton community can become dominated by a single species to the near exclusion of others. These "red, green, or brown tides," so called because the algae are abundant enough to discolor the water, seem to be increasing worldwide. Nutrient pollution is probably partly to blame.

Many phytoplankton species exude toxins that inhibit the growth of competing species. But the chemical weapons of species that dominate blooms can also have toxic effects on some animals that ingest the cells and/or exudates, and other species in their food webs, from marine mammals to humans. The red-tide-causing dinoflagellate *Chatonella* has caused massive fish kills in Florida (USA) and the Inland Sea (Halverson and Martin 1980). The dinoflagellate *Gonyaulax,* the source of deadly paralytic shellfish poisoning in humans, has caused mass mortality of wedge clams (*Donax serra*) in South Africa (Villier 1979) and Atlantic herring in the Bay of Fundy (Canada) (White 1980), and toxicity in clams and mussels in India, China, Japan, Washington (USA), Argentina, Portugal, and Norway. In recent years, a brown tide of *Phaeocystis*, long a problem in Liverpool Bay (UK), has plagued estuaries and coastal waters of the northeastern USA. Although it is not toxic, *Phaeocystis* is unpalatable and so gelatinous that blooms can inhibit the movement of herrings (Bold and Wynne 1985). When it blooms, many other species decline, including commercially valuable shellfish.

Sewage is a major cause of eutrophication in the world's coastal waters, especially in estuaries, lagoons, and bays where restricted circulation slows their dilution, but does not seem to have affected oceanic waters. Sewage, which includes human, household, and, oftentimes, industrial wastes, enters the sea from discharges into rivers, from outfalls into coastal waters, and from dumping from ships. It is particularly rich in nitrogen, and is often contaminated with toxic organic chemicals, heavy metals, and pathogens, such as bacteria that cause cholera. Sewage enters the marine environment raw or treated to some extent. Treatment might be no more than a filtering out of large particles (primary treatment), it might include bacterial decomposition (secondary treatment), or, in very rare cases, some chemical contaminants might also be removed (tertiary treatment).

The effects of sewage on marine ecosystems depend on its composition, how it is disposed of, and the physical and biological characteristics of the site. One of the most illuminating examples of the effects of sewage is on the coral reefs of Kaneohe Bay, Hawaii (USA), which was polluted with sewage for over 30 years (Laws and Redalje 1982). The nutrients from a sewage outfall stimulated phytoplankton blooms, which decreased water clarity, and favored the growth of sponges, other filter-feeding organisms, and green algae that overgrew the corals. The reefs were fast disappearing by the late 1970s when sewer outfalls were diverted seaward. Because of the short residence time of water in the bay (14 days), results were rapid, and the reefs have begun to recover.

Atmospheric deposition is an important source of nitrogen and trace metals such as iron, manganese, and zinc, downwind of industrialized and densely settled areas. The primary anthropogenic source of atmospheric nitrogen compounds is the high-temperature combustion of fossil fuels. Large amounts are deposited in acid precipitation falling on the sea surface. As much as 40 percent of the nitrogen pollution in coastal waters near industrialized areas can come from atmospheric deposition.

Runoff from fertilized agriculture and silviculture is a huge source of phosphorus and nitrogen that enters the sea primarily through rivers. For example, the Mississippi River, which drains the agricultural heartland of the USA, ejects vast amounts of nutrient-laden sediment and water into the

Gulf of Mexico, where hypoxic and anoxic areas are expanding. On a smaller scale, nitrogen from mariculture operations also causes eutrophication. Salmon culture in Norwegian fjords has caused repeated devastating algal blooms.

Although eutrophication has many undesirable effects, some European and US engineers and scientists are now advocating dumping millions of tons of toxic-contaminated sewage sludge into the deep sea. Others want to combat global warming by adding large amounts of iron compounds to the Southern Ocean, stimulating vast Antarctic plankton blooms and hastening the removal of CO_2 from the atmosphere. Such schemes overlook the potential for profound disruption of food webs and changes in community composition.

Solid Waste

For centuries, the problems caused by various types of solid wastes in the marine environment were not obvious; metal and glass garbage that was dumped into the ocean sank out of sight, and paper and cloth wastes decayed. Hence, in 1975 when the U.S. National Research Council estimated that ships routinely dumped 6.3 million metric tons (14 billion pounds) of trash into the world's oceans each year, it concluded that the primary impact of this marine pollution was merely aesthetic.

With the increased use and subsequent disposal of plastics, however, solid wastes have taken on a different character in the marine environment. Plastic items are produced in great numbers: more than 34 billion plastic bottles yearly in the USA alone (O'Hara et al. 1987). Plastics, because of their strength, buoyancy, and durability, presently make up more than one-half of all anthropogenic debris items found at sea and on coastlines (Debenham and Younger 1991). In the Mediterranean Sea, it is estimated that 3.6 million objects large enough to be sighted by eye, mostly plastic debris, are afloat per day (McCoy 1988). Even along the world's remotest shorelines, debris from distant countries is washing ashore. But plastic in the sea is more than a litter problem; plastic debris is causing widespread mortality in marine species, through entanglement and by ingestion (Shomura and Yoshida 1985; Shomura and Godfrey 1990).

Entanglement: Marine mammals, seabirds, sea turtles, fishes, and crabs often become entangled in the loops and openings of plastic fishing gear, strapping bands, six-pack rings from beverage containers, and other items (Figure 4-3). Once ensnared, these animals might be unable to swim or feed, or might develop open wounds that become infected.

Seals and sea lions appear to be the most prone to entanglement. At least 15 of the world's 32 species of pinnipeds have been seen entangled in plastic debris (Fowler 1987). Debris entanglement kills thousands of northern fur seals (*Callorhinus ursinus*) in Alaska's Pribilof Islands each year. Despite regular efforts to remove debris items from its habitat, entanglement poses an increasingly severe threat to the remaining population of the critically endangered Hawaiian monk seal (*Monachus schauinslandi*).

"Ghost fishing," the continued capture of marine animals by lost or discarded traps, nets, and line, is a major entanglement problem. For example, in both the USA's inshore American lobster (*Homarus americanus*) fishery and Kuwait's hamoor (*Epinephalus tauvina*) and hamrah (*Lutjanus coccineus*) fishery, 25 to 100 percent of the traps or pots are lost yearly, and continue to trap and kill these species for several weeks to

Figure 4-3. Northern fur seal strangled by discarded fishnet. *Solid waste is more than unsightly litter. Countless marine crabs, fishes, birds, and mammals are killed by discarded fishing gear and plastic garbage each year. Many marine species consume plastic waste and suffer from intestinal obstructions. Such losses are almost entirely avoidable.*

several years. Ghost fishing losses in these two fisheries have been estimated at 7 percent, and 3.5 to 12.8 percent of landings, respectively
(Mathews et al. 1987).

Lost gill nets, whether floating or on the seabed, continue to catch
animals indiscriminately for years or decades. For example, ensnared in
one 1.6-kilometer (1-mile)-long segment of gill net retrieved from the
North Pacific were 100 salmon and 350 seabirds. Animals killed by lost
nets attract other animals, which themselves become entangled, and so on.
The growing use of nondegradable synthetic materials in fishing gear,
including plastic coating for traps and nylon netting, has seriously wors
ened the problem.

Ingestion: Marine animals are known to ingest everything from large
pieces of plastic sheeting to tiny plastic resin pellets (the feedstock of the
plastics industry). Juvenile sea turtles, which live and feed in areas where
buoyant food and debris concentrate, may be particularly prone to ingest-
ing marine debris (Carr 1987) such as cigarette filters. Leatherback sea
turtles, which feed largely on jellyfish, sometimes mistake plastic bags for
their prey, and can die from intestinal blockage. At least 69 seabird species
eat plastic debris, and some feed plastics to their young. For instance, on
Midway Island (USA), more than 1,600 kilometers (1,000 miles) from the
nearest populated island, all 300 Laysan albatross chicks (*Diomedea
immutabilis*) examined had plastic debris in their stomachs, including
plastic toys, bottle caps, balloons, condoms, and cigarette lighters (Shomura and Yoshida 1985). How this affects their populations is uncertain,
but it merits close scrutiny.

Initial efforts to reduce the marine debris problem have focused on at-
sea sources, including merchant ships, passenger cruise vessels, commercial fishing vessels, recreational boats, and offshore oil and gas industry
rigs. At the end of 1988, an international protocol took effect regulating
the centuries-old shipboard practice of tossing garbage over the rail.
Known as Annex V of the International Convention for the Prevention of
Pollution from Ships (MARPOL), this protocol prohibits the disposal of
plastics from vessels at sea and regulates at what distance from shore all
other solid-waste materials may be discarded. However, the limited resources of enforcement agencies and the sheer size of the oceans make the
effectiveness of this prohibition unclear.

It is becoming increasingly clear, however, that a large portion of marine debris originates on land. Plastic wastes are discharged into coastal waters via sewage systems and storm drains. Other wastes escape from improper solid waste management practices or coastal littering. Plastics manufacturing and processing facilities are another land-based source of plastic resin pellets in the sea. Due to the diversity of land-based sources of debris, there is no one simple way to control the problem. Ultimately, the most promising solution is combining sound, enforceable laws with both market incentives to use materials that are not environmentally harmful and education to reduce and properly dispose of the waste that we generate.

Alien Species

Alien species, also known as biological invasions, biological pollution, or exotic, nonindigenous, or introduced species, among many other terms, are organisms that have been transported by human activity, intentionally or accidentally, into regions where they have not historically occurred. They can result from human activities such as commercial fisheries, mariculture, the aquarium trade, scientific research, canals that link previously unconnected water bodies, and shipping.

Although invasions have always been part of Earth's natural history, human-mediated invasions differ dramatically in kind and scale: Large numbers of species are now transported virtually instantaneously across vast biogeographic barriers (e.g., from Europe to Australia). This appears to have no natural precedent: Natural range expansions, at most, enter adjacent biogeographic regions.

Alien species can:

1) cause clear—and at times devastating—impact in their new locations;
2) have no apparent effect; or
3) be viewed as being a "positive" addition to the ecosystem.

Evidence of the catastrophic and overwhelming changes in natural ecosystems by alien species invasions in terrestrial and freshwater ecosystems is well documented (Elton 1958; Courtenay and Stauffer 1984; Mooney and Drake 1986; Groves and Burdon 1986; Kitching 1986; Drake et al. 1989; Stone and Stone 1989). Alien species can possess competitive, predatory, parasitic, and defensive strategies with which the native biota have had no evolutionary experience. Island communities are particularly susceptible because their native species usually lack defensive or competitive capabilities to deal with the alien predators and competitors. Alien species have caused the extinction of many island plants, vertebrates, and invertebrates. On the other hand, the invaders benefit from having left behind the predators, parasites, and diseases that might have held their populations in check.

Disturbance often alters the environment in ways that allow weedy animals and plants to spread without competing with the native biota for resources. Indeed, the native biota may have been completely eliminated by human activities. But invasions are not always linked to disturbed environments.

For most invasions there have been no quantitative or experimental studies that elucidate the role of the invader in the new community. This has led some people to conclude that many alien species have had no impact on the environment. Unfortunately, a lack of evidence does not necessarily indicate a lack of impact.

Unintended Introductions

Buried in the history of global ocean exploration, colonization, and trade are the first human-mediated introductions of marine and estuarine organisms. Polynesian vessels arriving in Hawaii, Viking vessels arriving in North America, and other vessels radiating out around the world all carried fouling organisms on their hulls. These included barnacles, mollusks, hydroids, worms, and algae, as well as mobile organisms such as crabs, shrimp, snails, and fishes.

How many hundreds or thousands of such species owe their modern distributions to maritime history is unknown, and might remain so, although it is clear that the number of introductions is grossly underestimated (Personal communication with John Chapman, Hatfield Marine Science Center, Oregon State University, Newport, Oregon, USA). At the close of the 20th century, heavily fouled ships no longer move around the world on the scale they once did. Increased speed, reduced port residency, and anti-fouling paints have reduced (but not eliminated) ship-fouling invasions. But invasions continue at what may be an unprecedented rate. While species travel less frequently on the *outside* of ships, vast numbers now travel *inside* ships, in ballast (not bilge) water carried in ships' tanks to provide stability. Although water has been used for ballast since the 1880s, the number of ballast-mediated invasions appears to have grown dramatically since the 1970s. Ships are larger now, and there are more of them, so more ballast water is moved. Ships are also faster, allowing more organisms to survive in ballast water.

The millions or tens of millions of gallons of seawater carried by a ship contain huge numbers of living plankton. Canadian, US, and Australian studies have found hundreds of species of living plankton—both holoplankton and the larvae of benthic and nektonic species—in the ballast water of bulk cargo ships. Up to 50 species have been found in a single ship. All major phyla of marine organisms are found alive in ballast water. At any one moment there are hundreds of species being transported around the world by ships.

As a result, since the 1970s, scores of new invasions, apparently from ballast water transported across or among oceans, have been recorded worldwide. Some appeared in fresh waters, such as the zebra mussel (*Dreissena polymorpha*) and four other species of European invertebrates and fishes that colonized North America's Great Lakes. Five species of Asian copepods appeared on the US Pacific coast. A clam from Asia, *Potamocorbula amurensis*, invaded San Francisco Bay, where it is now one of the predominant benthic organisms (Carlton et al. 1990). The Japanese crab *Hemigrapsus sanguineus* has established itself on the US Atlantic coast. The northwest Atlantic comb jelly *Mnemiopsis leidyi* invaded the Black Sea, where hundreds of millions of tons are now reported.

And around the world in the last decade, hundreds of blooms of red-tide-causing dinoflagellates and other phytoplankton have happened in waters where the species had not previously been recorded. These sudden appearances, thought by some to be evidence of increasing coastal eutrophication, could well be invasions of alien species introduced in ballast water (Personal communication with James Carlton, Williams College–Mystic Seaport, Mystic, Connecticut, USA). Unfortunately, the severe worldwide erosion of support for research in marine invertebrate and algal systematics, natural history, and biogeography prevents us from recognizing the nature and scale of these phenomena.

The removal of barriers to dispersal between biogeographic regions allows wholesale invasion of alien species. More than 250 species have entered the Mediterranean from the Red Sea since the Suez Canal was completed in 1869 (Spanier and Galil 1991). If recurrent proposals to create a sea level canal across Panama ever come to fruition, thousands of species could invade the Pacific from the Atlantic, and vice versa (Briggs 1969).

Intentional Introductions

Human societies have usually viewed agricultural introductions as having positive economic and social impacts. Imported plants and animals, especially ones dating back centuries or millennia, are an accepted part of human culture, and few people lament the replaced or displaced native biota. Indroductions of species for fisheries is a more recent phenomenon, and are more likely than agricultural species to reproduce and spread without further human assistance.

Before the late 1800s, few marine species were moved intentionally for food, in part because of the difficulty of keeping fish and mollusks alive for weeks to months across oceans or continents, but toward the turn of the century technological advances began permitting more rapid intentional movements of fish and shellfish around the world. The commercial Atlantic oyster *Crassostrea virginica*, for example, was moved by railroad from the Atlantic to the Pacific Ocean, and by ship from the US coast to western Europe. These movements accidentally introduced many of the most common North American Atlantic estuarine invertebrates to the Pacific coast and England. Similar movements of oysters continued until the 1970s, including translocation of the Japanese oyster *C. gigas* to France, which also introduced Japanese algae and invertebrates to western and southern Europe. By the 1980s, however, oysters were usually transported, not as adults, but as larvae or newly settled spat.

Even in the 1990s, proponents of mariculture equate introductions of marine species with those introduced for agriculture, and point to the benefits from introducing cows and wheat. Opponents, however, point to the accidental importation of diseases and pathogens and the potential escape of the maricultured species into the wild. For example, along with maricultured Japanese oysters and Indo-West Pacific shrimp of the genus *Penaeus* have come various predators, parasites, and pathogens, such as the lethal infectious hypodermal and hematopoietic necrosis (IHHN) virus, which is now affecting wild populations of *Penaeus* spp. in the Sea of Cortez, particularly blue shrimp (*P. stylirostris*) (Personal communication with Alejandro Robles, Instituto Tecnológico y de Estudios Superiores, Monterrey Campus, Guaymas, Sonora, Mexico). Allowing an introduction to help economic development—in this case, mariculture development—can also bring severe economic hardships along with ecological damage.

Ecological Consequences of Marine Aliens

The environmental changes caused by marine invaders are known for only a few of the legions of introductions. Around the world, introduced algae, marsh grasses (particularly *Spartina* spp.), fishes, mollusks, crustaceans, worms, and many other invertebrates are now often dominant members of coastal communities. While it is tempting to equate abundance with environmental impact, for most introductions there have been no experimental studies that can verify whether invasion by alien species has caused impacts such as the disappearance of native species.

Examples of ecosystem- and community-level alterations (Race 1982; Carlton 1985, 1987, 1989; Nichols and Thompson 1985) include the invasion of North America by the European periwinkle *Littorina littorea* in the early 19th century, which led to profound changes in the community composition of rocky shores, mudflats, and salt marshes from Canada's Maritime Provinces to the US's Middle Atlantic States. The dramatic appearance in the late 1950s of the Asian green alga *Codium fragile tomentosoides* on the US Atlantic coast is reported to have had extensive impacts on local mollusk fisheries (Carlton and Scanlon 1985). The Asian eelgrass *Zostera japonica* colonized shores from British Columbia (Canada) to Oregon (USA) starting in the 1960s, creating vast areas of rooted vegetation where there were once mudflats, and leading to distinct changes in the associated infaunal organisms (Posey 1988). Destruction of wood in marine ecosystems, such as pilings in harbors, has been increased greatly by wood-boring shipworms (teredinid bivalves) and gribbles (limnoriid isopods), many of which were distributed by wooden ships for centuries.

More than a dozen marine fishes have been introduced intentionally or accidentally to the Hawaiian Islands, including the bluestripe snapper *Lutjanus kasmira* from the Marquesas, which has become an abundant predator in shallow waters (Randall 1987). The impact of another predator (of plankton) is much clearer: The introduced West Atlantic ctenophore *Mnemiopsis leidyi* had decimated fisheries in the Azov and Black seas, and it is now spreading through the Sea of Marmara toward the Mediterranean Sea (Personal communication with Richard Harbison, Woods Hole Oceanographic Institution, Woods Hole, Massachusetts, USA). Many of the Red Sea species, including fishes, shrimps, and jellyfish, that were introduced by the Suez Canal have depressed populations of ecologically similar Mediterranean Sea species. The Red Sea jellyfish *Rhopilema nomadica*, now phenomenally abundant in the eastern Mediterranean, has also harmed fisheries, coastal power plants, and tourism in Israel (Spanier and Galil 1991).

Among the most significant yet most overlooked global impacts of invasions are toxic marine phytoplankton blooms, which seem to have increased dramatically in the world's coastal waters in recent years. Millions of viable cysts of several toxic dinoflagellate species have been found in mud at the bottom of ballast tanks of ships entering Australian ports (Personal communication with Christopher Bolch, CSIRO Division of Fisheries, Hobart, Tasmania, Australia). Toxic plankton blooms often have severe economic and social impacts, including the closure of shellfishery operations and serious human health repercussions.

Terrestrial Aliens and Marine Victims

Marine species have suffered not only from species introduced from other marine areas, but from the land as well. Domestic cats introduced on Baltra Island in the Galápagos Islands (Ecuador) helped to eliminate the marine iguana (*Amblyrhynchos cristatus*). On Guadalupe Island (Mexico), introduced cats caused the extinction of the Guadalupe storm petrel, and they have killed hundreds of thousands of seabirds annually on Marion Island in the subantarctic Prince Edward Islands (South Africa) (Heneman 1992). Rabbits introduced to Laysan Island (USA) so devastated the vegetation that they very nearly eliminated the Laysan teal (*Anas laysanensis*). But the worst terrestrial aliens affecting seabird species are rats (*Rattus* spp.), which feed on eggs, chicks, and adults of ground-nesting species worldwide.

The Future of Marine Invasions

Invasions lead to fundamental changes in natural communities. Parasites and diseases commonly accompany the aliens, and the economic and social consequences of invasions can be momentous. Nevertheless, there is great pressure to move oysters, shrimps, fishes, and other species to new regions to create or supplement fisheries. The release of vast numbers of marine organisms by ballast water continues unabated in most regions. Invasions will continue until effective national and international measures are in place. And the idea of a sea level canal across Central America refuses to die. Without effective controls, alien species will be another growing, uncontrolled, irreversible, and costly experiment that we are performing on our planet.

Global Atmospheric Change

In many ways, the ocean and the atmosphere are two parts of one system. Although seawater and air behave very differently, the ocean and atmosphere exchange huge amounts of heat and gases. Because they are coupled, whatever affects the physics of the ocean affects the atmosphere, and vice versa. Two things that directly affect the atmosphere and also have profound potential effects on the ocean are the depletion of ozone in the stratosphere and the buildup of atmospheric greenhouse gases.

Increased UV-B from Stratospheric Ozone Depletion

The stratospheric ozone layer is the Earth's primary shield against damaging ultraviolet radiation from the Sun. Humans have manufactured, and continue to manufacture, certain compounds (e.g., chlorofluorocarbons [CFCs] and brominated compounds) that gradually migrate upward into the stratosphere and destroy ozone. Even with the ratification of the London and Copenhagen amendments to the Montreal Protocol accelerating the phaseout of many ozone-depleting substances, atmospheric concentrations of chlorine will continue to increase on a global scale because of continued emissions of chlorine-containing compounds from human activities worldwide (Watson and Albritton 1992). These activities include the use of ozone-depleting substances such as refrigerants, foam products, aerosol propellants, and solvents.

Recently, ground-based and satellite observations have demonstrated that there has been a decrease in total ozone over mid-latitude areas of the Northern Hemisphere in all seasons (Stolarski et al. 1992). This decrease is in addition to the previously established decreases over the Southern Hemisphere. Because of the long atmospheric lifetimes of CFCs and many brominated compounds (e.g., halons), scientists expect stratospheric ozone to continue to decrease in concentration well into the middle of the next century even if emissions are curtailed worldwide.

The decreased concentration of stratospheric ozone allows an increase in the transmission of biologically damaging solar UV-B radiation to the surface of the Earth. UV-B radiation can penetrate to ecologically significant depths in marine waters, typically tens of meters (Smith et al. 1992).

Scientists have observed significant increases in UV-B radiation in Antarctica that coincide with periods of intense ozone depletion. Research has identified many potentially serious effects of increased exposure to UV-B radiation on human health and the environment, including on marine organisms.

The proteins and nucleic acids in living things are chemically changed when they absorb UV-B radiation. On the basis of DNA damage, a 10 percent ozone reduction would result in a 28 percent increase in DNA-damaging UV levels (Worrest 1989).

UV-B radiation, even at present levels, has a detrimental effect on many marine organisms. It is known to cause population reductions in marine phytoplankton, zooplankton, and juvenile stages of some fishes, such as the northern anchovy (*Engraulis mordax*) (Häder et al. 1989).

Increased solar UV-B radiation would likely reduce productivity and affect the abundance of species throughout the entire marine food web, which would affect world food production (Häder et al. 1991). Because a sizable fraction of the world's animal protein for human consumption comes from the sea, a substantial decrease in productivity would diminish fishery resources at a time when demand is increasing. However, the relationships between damage to individual organisms and the productivity of fisheries are complex. Available scientific data are far from sufficient to make quantitative predictions.

Increased UV-B radiation could also lead to extinctions in the sea, although current data are insufficient for specific predictions. Loss of species could reduce resilience—the ability to recover from stress—in marine ecosystems. Scientists need to learn more about the interactions of species within ecosystems and their physical and chemical environment to understand how the effects of increased UV-B on single species affect the functioning of entire ecosystems.

Climatic Change

The presence in the atmosphere of small amounts of several naturally occurring gases, including carbon dioxide (CO_2) and methane, has a large effect on the Earth's distribution of heat. These "radiatively active" gases act as a blanket, keeping some of the heat radiated from the Earth's surface within the lower atmosphere, where temperatures rise, producing a "greenhouse effect." If these gases were to disappear, the Earth's surface would average $-18°C$ ($0°F$) or colder, the oceans would freeze solid, and life would end. That is *not* going to happen. Rather, human activities—including the burning of fossil fuels, the destruction of forests, the breeding of cattle, the establishment of rice paddies, and the release of CFCs from leaking refrigerators—are increasing at atmospheric concentrations of radiatively active gases. This will warm the Earth's surface, most of which is ocean.

The greenhouse effect will also change the pattern of heat distribution of the Earth's surface, which will alter ocean circulation patterns, precipitation patterns, and storm tracks. Warmer temperatures will bring rising sea level as well. Working Group II of the Intergovernmental Panel on Climate Change (IPCC 1992) states that "the combination of sea level and temperature rise, along with changes in precipitation and UV-B radiation, are expected to have strong impacts on marine ecosystems, including redistributions and changes in biotic production."

How much climatic change will affect marine biological diversity depends on how quickly it happens. One possibility is that global temperatures and warming-related phenomena will increase steadily in proportion to increasing atmospheric concentrations of greenhouse gases. This would be less damaging than a sudden change to a very different climatic regime. But the climate does not seem to be changing gradually: Although greenhouse gases have been steadily increasing for decades, average global temperatures have neither increased proportionately nor smoothly. The global average temperature has increased less than expected. Moreover, in the 1940s, the warming trend reversed itself, while since the 1980s, the Earth has experienced its warmest years since systematic weather monitoring began. Instead of a gradual increase, it is disturbingly possible that temperature increases will be sudden and irreversible as the Earth's climatic system shifts from one state to another. Perhaps even more likely is the possibility that the transition between pre-greenhouse and greenhouse climates will be marked by sudden switching back-and-forth, which might be manifested as increasing frequency or severity of large-scale phenomena such as the El Niño–Southern Oscillation.

Global climatic change is not just a terrestrial issue. To anyone concerned about saving, studying, and sustainably using marine biological diversity, how the climate changes, and how quickly, are questions of vital importance.

Altered Circulation and Upwelling: If the Earth were a landless sphere covered by an ocean of even depth, anticipated greenhouse warming in the atmosphere would have fairly predictable effects on the sea. They would probably be gradual; after a lag period caused by the slowness with which the ocean changes temperature, warming would parallel increases in atmospheric greenhouse gas concentrations. Warmer waters and their biotas would expand in area, while cooler waters and their biotas would contract. Only long-lived, sessile organisms such as massive reef corals might have difficulty shifting with the shifting climatic zones.

However, the fact that lands of varying extent, topography, and shape are interposed between topographically complex ocean basins greatly complicates the prediction of effects. Crucial processes such as down-welling—the sinking of surface waters—and upwelling happen only at certain places in the sea. The tightly coupled ocean-atmosphere system is very complex and only beginning to be understood. But important insight comes from the last great climatic changes, when continental glaciers most recently advanced and retreated.

Driven by the geographic and seasonal patterns of heat at the Earth's surface and in the lower atmosphere, currents distribute heat within oceans, among oceans, and between oceans and the atmosphere, carrying heat from the tropics to colder regions via a global heat "conveyor belt." But these colossal rivers of seawater—vastly larger than the largest rivers on land—are not confined by riverbanks. They are free to shift as global warming changes the Earth's heat distribution. Instead of responding linearly to increasing greenhouse gases, the ocean-atmosphere system could undergo sharp, worldwide reorganization, switching from one stable state to another (Broecker 1987). There is growing evidence suggesting that, throughout the last 2 million years, gradual changes in the Earth's orbit have caused fairly abrupt shifts between glacial and interglacial climates by tripping a "switch" that alters ocean currents (Broecker and Denton 1989). The gradual increase in greenhouse gases could trigger climatic changes that would be anything but gradual; at the end of the last glacial period, dramatic shifts in sea surface temperature of at least 5°C (9°F) occurred in parts of the North Atlantic over a period of less than 40 years (Lehman and Keigwin 1992).

How might such a sudden shift in ocean currents affect marine species and ecosystems? Because currents transfer huge amounts of heat, a major shift would markedly alter temperature regimes in places whose climate depends on heat carried by ocean currents, such as Western Europe. All biological processes would be affected. The growth, timing of reproduction, and predator-prey, host-parasite, and competitive interactions among species would change. There would be large shifts in ecosystems' composition, structure, and functions, including nutrient cycling, productivity, and carbon storage. Some fisheries would benefit, others would be harmed, even eliminated. Species that have instinctive migration patterns, or that time reproduction or metamorphosis using cues other than temperature (e.g., the duration of daylight), might be especially vulnerable. A sudden shift would wreak havoc with fisheries and mariculture operations specialized in taking particular species at a particular place.

In addition to heat, currents also transport organisms, dissolved gases, nutrients, and pollutants. Surface currents bring planktonic larvae to suitable habitats; if currents shift, locating suitable habitats might either be more or less difficult for larvae. Shallow coastal areas and subsurface banks in the path of currents would likely receive different amounts of settling larvae and nutrients.

Global climatic change will be felt in the deep sea as well. Surface water is cooled off Greenland and Antarctica and becomes more saline as salt is excluded when pack ice forms. This cold saline water downwells—sinks to the bottom—and creeps equatorward. If it decreases in area, deepsea species adapted to its distinctive conditions will be affected. Deepsea communities that depend on the rain of detritus from above are also likely to be affected by changes in primary production at the surface. Moreover, if upwelling weakens, upwelling-dependent fisheries (e.g., off Peru) would suffer, as occurs during El Niños.

Because there is no precedent in the last thousand centuries for warming to levels anticipated in the next century, there is inadequate basis for more precise prediction of the effects of shifting currents on marine organisms and ecosystems.

Warming: Because temperature affects all life processes, the most striking global patterns in the distribution of shallow-water species are the temperature zones—polar, sub-polar, temperate, and tropical—that stretch across oceans. Human-caused increases in atmospheric concentrations of "greenhouse gases" are likely to commit the planet to an increase in average temperatures of 1° to 3°C (2° to 5°F) during the next century

---

(Schneider 1989; IPCC 1990). As on land, global warming will introduce new stresses to coastal and marine ecosystems, which are already under multiple stresses.

Temperature affects species in both obvious and inconspicuous ways, thereby affecting interactions among species and ecological processes. Many tropical marine species live at temperatures close to their upper lethal limits, as do some Antarctic species. An innocuous-sounding increase of a few degrees could kill them outright.

The phenomenon of coral "bleaching" was discovered early in this century, but it was not until the 1980s that biologists began observing widespread bleaching of reef corals across the tropical Pacific Ocean and in the Caribbean Sea. Corals bleach when they lose the mutualistic zooxanthellae that live within their tissues or when the concentration of photosynthetic pigments in zooxanthellae is sharply reduced. Prolonged intervals of water temperatures 1° to 2°C above normal summer temperatures can cause bleaching, although corals can recover their pigments in cooler months. But even a few days of temperatures 4°C or more above normal maxima cause bleaching and then death in 90 to 95 percent of coral colonies (Jokiel and Coles 1990); bleached colonies that do not die can stop growing and reproducing.

Living so close to their upper thermal limits makes reef corals very vulnerable to warming. Bleaching has also been linked to stress from high visible light and UV levels, low salinity, parasitic infection, and pollutants; corals appear more vulnerable to these as temperature stress increases. The near-extinction of a hydrozoan coral in the East Pacific during the severe El Niño warming of 1982–83 (Glynn and de Weerdt 1991; Glynn and Feingold 1992) could portend things to come. The long life span (decades to centuries) and slow population turnover in many coral species suggest that coral reef ecosystems could experience dramatic changes as sea temperatures increase.

Some kinds of ecosystems could shrink markedly in a greenhouse world. Among them are sea-ice ecosystems of the Arctic and Southern oceans, including permanent and temporary pack ice and the ice-free waters (leads and polynyas) between the floating ice packs (Alexander 1992). All of these provide habitat to which polar species are adapted. Diatoms grow on the underside of pack ice, crustaceans on the underside of this upside-down benthos graze upon them, fishes eat the crustaceans, and marine mammals eat the fishes. If sea-ice habitat were to diminish markedly—there is not yet scientific consensus whether this would occur (McAllister 1991b; Miller and de Vernal 1992)—one of the world's most important marine food webs would be severely disrupted.

Less obvious, but hardly less important, are the effects on physiology, behavior, and reproduction. In most species, a temperature increase of 10°C (18°F) doubles an organism's metabolic rate, hence its food and oxygen needs. Even a smaller temperature increase can stress organisms in ecosystems where food or oxygen is limiting. At the same time, warmer water can hold less dissolved oxygen. The proportion of the sea in which oxygen concentrations are limiting could increase.

In sea turtles, sex is determined by temperature during embryonic development. A recent multiyear study of loggerhead turtles in Florida (USA) found that an estimated 87.0 to 99.9 percent of hatchling turtles are females (Mrosovsky and Provancha 1992). Still higher temperatures that cause males to disappear could, in time, doom this already threatened species. Even a moderate temperature change will affect the ecological relationships of species, and hence, entire communities.

Marine plankton generally have short generation times, and therefore have the capacity to respond quickly to natural environmental changes in comparison with other species. There is little doubt that global warming would change the distributions of marine plankton (IPCC 1990). It is possible that marine ecosystems might shift poleward, as has occurred since the last glaciation. However, since it appears that climate will change faster than in the past, it is uncertain how plankton and longer-lived species will respond.

There is evidence that change in the composition, structure, and functioning of communities sometimes occurs quite rapidly. And, in a given region, rather than one community, there can be two or more alternative "stable" communities, each existing within some range of conditions but capable of "switching" if some environmental factor passes a threshold. Thus, even gradually increased temperatures could cause a sudden switch in marine communities, with high potential for disruption of fisheries on which humans depend.

On top of climatic fluctuations such as El Niño, humankind is imposing mounting stresses on the sea's biological diversity, including greenhouse warming. Because changes in climate trail increases in greenhouse gases, the climatic changes that occur will likely be the result of gases that entered the atmosphere decades earlier. Together, natural variability and the lag between greenhouse gas loading and climatic change will make it difficult or impossible for humankind to detect climatic changes before we are committed to them for a long time.

Altered Rainfall Patterns and Storm Tracks: Global warming will influence marine species and ecosystems by affecting the amounts and patterns of rainfall and wind storms. Understanding the causes and effects of climatic change is vitally important for decision makers. Yet establishing chains of causation—for example, whether the epidemic that devastated *Diadema* sea urchins in the tropical Atlantic (Lessios et al. 1984) was attributable to human activities—has usually proven elusive and will likely remain so at current funding levels for marine science research.

Scientists do know, however, that rapid environmental change commonly impoverishes nature by favoring small-bodied, rapid-reproducers over larger-bodied, longer-lived organisms. The biological effects of a worldwide change in rainfall patterns may be far-reaching but slow to accumulate and difficult to assign to a single cause.

Warming will increase evapotranspiration, the loss of moisture to the atmosphere from surfaces such as water, soil, and plants. Thus, the climate on land will effectively become dryer unless there is a large enough increase in precipitation to compensate. It is also likely that precipitation will increase in many places, but the pattern of increased precipitation will not match the pattern of increased evapotranspiration. How rainfall patterns will change is one of the greatest scientific uncertainties concerning global warming, but it is clear that many places on land will, in effect, become dryer.

How does that affect marine life? Because climatic zones will tend to move poleward, the arid zones of Northern Hemisphere lands, which are now at low- and mid-latitudes, will probably expand when they move northward because the land area is greater at higher latitudes. This could profoundly affect many rivers, especially ones—such as the Euphrates, Indus, and Colorado—that drain semi-arid or arid regions and whose flow is already substantially committed to agriculture. Diminished river flow would affect estuaries and coastal waters. These richly productive ecosystems owe their wealth to the nutrients brought from the land by runoff and river flow, and are therefore especially vulnerable to changes in rainfall on land. Salt marshes and other estuarine wetlands would likely be reduced in area. Populations of estuarine and estuarine-dependent species, including the large fraction of commercially important fishes in many regions that live in estuaries at some time during their life cycles, would therefore decrease. So would productivity in nutrient-limited coastal waters.

Many models predict more intense low-latitude monsoonal circulation, driven by enhanced contrast in land and sea temperatures. This could increase seasonal rainfall in coastal areas of South and Southeast Asia, but not in the midcontinental regions of Asia or North America, which are predicted to experience summer droughts.

At the same time, scientists expect a general increase in precipitation in higher latitudes (Smith 1990). Rivers in higher latitudes will have increased flows, which could bring changes to their estuaries as harmful as those from decreased river flow in lower latitudes. But increased flows in these rivers also have the potential to affect marine biological diversity on a global scale.

The idea that a change in rainfall measurable in centimeters and the resulting changes in runoff might affect an ocean several kilometers deep seems strange. But there is growing reason to believe that major changes in oceanic circulation can occur in response to anything that changes the density of the sea, such as altered rainfall and runoff. For example, there is evidence that altered salinity in the North Atlantic can bring sharp climatic change to adjacent lands, presumably by affecting currents that carry heat northward from the tropics (Kerr 1992). Thus, altered rainfall patterns could trip the "switch" that could abruptly reorganize ocean currents, winds, temperature regimes, and related phenomena.

Virtually all current climate change models predict that greenhouse warming will be greater at higher latitudes than in the tropics, weakening the temperature gradient between the equator to the poles. This suggests that the wind-driven circulation of the atmosphere-ocean system will be less vigorous. It is possible that storm tracks in mid-latitudes will shift poleward, although predictions about changes in the intensity or frequency of such storms are not yet convincing. In the marine realm, any shift in storm tracks would probably have the largest effects on dune, estuarine, intertidal, and shallow subtidal communities.

In the tropics, it is almost certain that the extent of water warmer than 28°C (82°F) will expand. Some climatologists believe that the frequency and/or severity of tropical cyclones (hurricanes, typhoons) will therefore increase, since these storms draw their awesome energy from high water temperatures. Again, coastal ecosystems, including coral reefs and mangrove forests, which can take decades or more to recover from great storms, are most likely to be affected by any changes. Of course, the most violent typhoons and hurricanes, such as those that struck Guam, Kauai, and Florida (USA) in 1992, are also devastating to human communities that benefit from these ecosystems.

Increasing greenhouse gases may well cause ecological and economic disruption of a magnitude without precedent in history. The world's nations—most recently at the UN Conference on Environment and Development (UNCED) in Brazil in 1992—have failed to devise cooperative, effective, enforceable ways to slow and stop global climatic change. Each nation wants the burden of reducing greenhouse gases to be borne by someone else, and all are hostage to the handful of large nations that refuse to cooperate. No president, corporate executive, union leader, rancher, automobile buyer, or villager will accept responsibility for triggering the coming changes to the Earth's oceans and atmosphere. What will it take to get humankind to act to avert the planetwide changes we are setting in motion?

Rising Sea Level: As the Earth warms, sea levels will rise, for two reasons. One is that water, when heated, expands. The other is that there will be more melting of glaciers and pack ice.

World sea levels and temperatures have fluctuated dramatically through geological history. At the last glacial maximum (18,000 years ago), the Earth was 5°C (9°F) colder and sea levels were some 100 meters (328 feet) below today's levels (Titus 1990). As the glaciers retreated, sea levels began to rise rapidly until about 6,000 years ago, when the rate of change slowed significantly. For the last century, global sea levels have risen at a rate of about 1 to 2 millimeters per year (0.04 to 0.08 inches per year) (Warrick and Oerlemans 1990).

Scenarios of sea level rise under global warming vary widely. Estimates of sea level rise by the year 2100 range from 10 centimeters (4 inches) to nearly 3.4 meters (11 feet) (Hoffman et al. 1983; Polar Research Board 1985). The Intergovernmental Panel on Climate Change predicts that by 2030 global-mean sea level will be 8 to 29 centimeters (3 to 11 inches) higher than today, with a best estimate of 18 centimeters (7 inches); by 2100, sea levels are expected to be 31 to 110 centimeters (12 to 43 inches) higher, with a best estimate of 66 centimeters (26 inches) (Warrick and Oerlemans 1990). The IPCC predictions suggest rates three to six times higher than those experienced over the last century. By 2100, the rate could easily exceed one centimeter per year (0.4 inches per year). And some experts believe that these predictions are extremely conservative, that sea level could rise as much as 200 centimeters (78 inches) by the year 2100 (Titus 1991).

Coastal ecosystems could naturally withstand modest rates of sea level rise, but anticipated rises would almost certainly overtake coastal ecosystems retreating landward. Moreover, inundation is only one effect of global warming. As sea levels rise, coastal erosion and flooding will increase, and coastlines will recede unless stabilized by seawalls or beach renourishment. Saltwater intrusion into groundwater, rivers, estuaries, and bays will increase. Changes in rainfall patterns will alter rates of deltaic sedimentation. Currents and upwelling patterns are likely to shift geographically and change in intensity. All of these "sea changes" will affect biodiversity in the world's coastal zones.

Coastal wetlands are likely to suffer the most visible impacts of the rising sea. When sea level rises slowly, the combination of inorganic and organic sedimentation can allow wetlands to rise apace. But geological records suggest that little wetland formation is possible at rates of sea level rise exceeding 10 millimeters (0.4 inches) per year. For example, in the USA's Mississippi River Delta, the reduced input of sediments caused by dams and river channelization, and land subsidence caused by oil exploration, have caused subsidence that is equivalent to an annual sea level rise of 1.2 to 1.3 centimeters (0.5 inches). At this rate—comparable to predicted rates of sea level rise—the Mississippi Delta is now losing more than 129 square kilometers (50 square miles) of coastal wetlands each year.

With its extensive low-lying delta, Bangladesh would lose 12 to 28 percent of its total land area over the next century at projected rates of sea level rise. Similar fates will befall salt marshes and mangrove forests worldwide, and at the upper limit of estimates of sea level rise, even coral reef growth may not keep pace, especially if their growth is hampered by turbid water from eroding coasts.

Loss of these ecosystems would have enormous ecological and economic consequences. Coastal wetlands act as sediment traps that stabilize coastlines, protect against hurricanes and storm surges, serve as nurseries for commercially exploited crustaceans and fishes, and are habitats for other biological resources including waterfowl and fur-bearing animals.

The impact of sea level rise is exacerbated by coastal development.

Already, 6 out of 10 people live within 60 kilometers (37 miles) of coastal waters, and two-thirds of the world's cities with populations exceeding 2.5 million people are near tidal estuaries (IUCN et al. 1991). Within three decades, the coastal population is projected to double, which will further squeeze coastal ecosystems between the rising sea on one side and coastal buildings and roads on the other.

As the coastal fringe disappears, so will its species. In the USA, some 80 species already known to be threatened with extinction are found *only* in the narrow 3-meter (10-foot) band above sea level (Reid and Trexler 1991). These, and many other species worldwide, will disappear if their opportunities for landward migration are blocked.

Sea level rise could devastate some island nations. Maldives, a nation of 1,190 small islands rising barely 2 meters (6.5 feet) above the Indian Ocean, could be entirely submerged during the next century, destroying all of the natural habitat of five endemic plant species in the process. Similarly, the highest point of the nation of Tuvalu—a chain of nine Pacific atolls—is only 4.5 meters (14.8 feet). Even short of submergence, rising seas will disrupt the ecology of such islands as the freshwater lens beneath the islands shrinks and soils become salinized. In addition, rising sea level makes low-lying lands more vulnerable to storm surges.

Interacting Threats

Because the Earth's surface is constantly changing, after millions or tens of millions of years marine species and communities eventually disappear. But they are now at unprecedented risk from anthropogenic changes, for two reasons:

1) *Many of these changes are novel, either qualitatively or quantitatively.* The polychlorinated biphenyls (PCBs) and other synthetic chemicals we add to the sea are completely new to marine organisms, which, therefore, have not evolved any means of coping with these chemicals. Global warming is expected to occur faster than great climatic changes in recent millions of years, overwhelming the abilities of many species to respond. The gape of a modern commercial trawl is many times that of the largest whale, allowing fishers to swallow entire schools of some species. In time, evolution might allow species to adapt to these novel changes, but it would take tens, hundreds, or thousands of times longer than we are giving them.

2) *Anthropogenic changes usually occur in groups, not separately.* If they occurred separately, species might find "refuges" within which the threats were diminished or absent. For example, if fishers set their nets only on large schools of sardines but some sardines have a genetic predisposition for occurring in small schools, these might survive to pass on their genes.

More often, species and ecosystems can get caught in a squeeze of multiple additive or even synergistic threats, which pose far greater risk than any individual threat. For example, as Ono et al. (1983) explained, a fish endemic to the northern Sea of Cortez called the totoaba (*Totoaba macdonaldi*), the largest species in the family Sciaenidae, is caught in a crush of directed take, incidental take, and physical alteration. Sciaenids are called drums or croakers because of the sounds they make by vibrating their muscular swim bladders. The Chinese prize totoaba swim bladders for soup stock, and, in the 1920s, Mexican fishers discovered that swim bladders could fetch remarkable prices, up to US$15 apiece. At first, the fishers left the rest of the totoaba to rot, but a market for the meat later developed in the USA. Totoaba were astoundingly abundant and easy to catch as they aggregated to spawn at the mouth of the Colorado River. The directed take peaked in 1942 at 2,300 tons, then began declining sharply.

The second cause of the totoaba's decline is the northern Sea of Cortez's intensive shrimp fishery. Along with huge numbers of other nontarget species, the shrimp fishery catches young totoaba and discards them as bycatch.

Finally, totoaba are also disappearing because of physical ecosystem alteration. Most of the Colorado River flows through the USA; only its estuary is in Mexico. Starting in midcentury, the USA dammed the Colorado for irrigation. It siphons off so much water that the mouth of the Colorado no longer provides the estuarine conditions that attracted huge totoaba spawning aggregations.

The totoaba is now under protection in Mexico and is on the US endangered species list, so there has been no legal directed take since 1975. Whether the totoaba can survive the combination of poaching, incidental take in shrimp trawls, and the degradation of its spawning habitat is questionable.

Ecosystems are, if anything, even more subject to multiple threats. As Davis (1982) and Nichols et al. (1986) explained, in 1848, when gold was discovered in California (USA), San Francisco Bay supported a rich estuarine community. But gold fever had severe consequences for the bay. Miners washed vast amounts of mud through sluices into rivers, then into the bay, reducing the depth of the bay system by 25 to 100 centimeters (10 to 39 inches). Gold also stimulated agricultural, residential, and industrial development along the bay, much of it on filled-in wetlands. From 2,200 square kilometers (850 square miles) of tidal marsh in 1850, only 6 percent of undiked marsh is left.

The productivity of California's Central Valley depends on massive irrigation from the Sacramento and San Joaquin river systems, which empty into San Francisco Bay. Agriculture has reduced the freshwater inflow to the bay by more than 60 percent. Loss of water and entrainment in water pumps has severely damaged spawning runs of the bay system's anadromous fishes.

The bay receives water containing large amounts of agricultural chemicals, industrial wastes, and sewage. Together, pollution and the reduction of freshwater inflow have jeopardized species such as the endangered Delta smelt (*Hypomesus transpacificus*).

When San Francisco was connected to the US Atlantic coast by rail, trains carried hundreds of carloads of Atlantic oysters to mature on the bay's mudflats and, with them, alien stowaways in the oyster clumps. More recently, as a major harbor, San Francisco Bay has received shipments of goods from many locations worldwide, and an unintended cargo: astronomical numbers of live organisms in ships' ballast tanks. Together, oyster introduction and ships' ballast water have introduced about 100 new invertebrate species to the bay. Now nearly all the common invertebrates in the bay's shallows are aliens from the eastern USA, Japan, and other locations.

San Francisco Bay, in many ways typical of the world's urban estuaries, has suffered from a lethal combination of physical alteration, pollution, and the introduction of alien species. One result is that its once-rich commercial fisheries for slow-reproducing, high-value species, such as Pacific salmon (*Oncorhynchus* spp.), striped bass (*Morone saxatilis*), and Dungeness crab (*Cancer magister*), are gone. The commercial fisheries that remain concentrate on small, fast-reproducing, low-value, planktivorous fishes characteristic of disturbed habitats.

The decline of the totoaba and of San Francisco Bay shows how living marine systems can succumb to multiple human activities. More often, the cause of declining marine species or ecosystems cannot be determined. Large numbers of seal or dolphin corpses washing up on the shores of the heavily polluted North Sea or Gulf of Mexico might be attributed to viral or bacterial diseases. How they have been affected by multiple anthropogenic stresses, including ones that could both impair these animals' immune systems and introduce vast numbers of pathogens into their habitats, is as yet unproven. Because marine ecosystems worldwide are subjected to multiple threats, it should not be surprising when major biological changes occur without obvious cause.

Multiple insults to the integrity of marine systems provide an opportunity for people to deflect scrutiny from their contributions to the problem; when, say, a fish species disappears, fishers, polluters, and those who physically alter ecosystems point fingers at one another. Unfortunately, as pressures on the sea increase, decision makers will seldom have the luxury of alleviating one problem, thereby restoring a species or ecosystem to its original state. The repeated failure of ad hoc solutions to the multiple, interacting threats indicates the importance of comprehensive, integrated ways of managing marine species and ecosystems.

Root Causes: The Driving Forces

The proximate threats to marine biological diversity—overexploitation, physical ecosystem alteration, pollution, introduction of alien species, and global atmospheric change—are symptoms of more fundamental forces that are driving environmental degradation around the world. Biological impoverishment is the inevitable consequence of the ways in which our species has used and misused the environment during our rise to dominance. The impoverishment has extended from the land and fresh waters into the sea.

There are five basic reasons why life in the sea is at risk:

1) There are too many people.
2) We consume too much.
3) Our institutions degrade, rather than conserve biodiversity.
4) We do not have the knowledge we need.
5) We do not value nature enough.

Overpopulation

One of the long-established principles of ecology is that large animals are less abundant than smaller ones. There are fewer buffalo than rabbits, which are less abundant than mice. Because larger animals have greater resource needs, an ecosystem can support fewer of them.

But there is one glaring exception to this size/abundance relationship: *Homo sapiens*. No animal our size or larger is anywhere near as abundant. For example, our closest relatives, the great apes, are roughly our size. But the four great ape species, together, number fewer than 350,000 individuals. In contrast, there are 5.5 billion humans, that is, *15,000 of us* for each chimpanzee (*Pan troglodytes*), pygmy chimpanzee (*P. paniscus*), gorilla (*Gorilla gorilla*), and orang-utan (*Pongo pygmaeus*). Moreover, our numbers are increasing with astounding speed, doubling every 40 years or so. In many countries, the population is doubling every 30, 25, or even 20 years.

In most countries with high fertility rates, about half the population is younger than 16. The large number of people who will soon be reaching their reproductive years creates a tremendous demographic momentum. As a result, global population will continue to grow for at least the next half-century and probably longer, barring catastrophe. Another billion people are likely to be added to the world population in the next decade alone. As the world struggles to provide for an ever-increasing number of people, other living things—from the largest whales to the infinitesimal ultraplankton—are being forced aside, eliminating products and services just when they are increasingly important.

Excessive Consumption and Inequitable Distribution of Resources

As our population has increased and new technologies have developed, humanity has appropriated an ever-increasing share of the Earth's resources. People consume, divert, or destroy an estimated 39 percent of the land's primary production, the fundamental source of the energy for terrestrial life (Vitousek et al. 1986). This trend is clearly unsustainable, and the oceans cannot make up for excessive drain on the land's resources. The world's biotic systems simply cannot accommodate an ever-growing claim on primary productivity to meet further growth in consumption.

Much of what is consumed worldwide is traded internationally. International trade and debt practices foster appalling inequities among and within nations. By 1988, developing countries were transferring a net of US$32.5 billion to industrialized countries, excluding other implicit resource transfers not involving direct financial flows. The rich nations are getting richer at the expense of the poor ones. Developing nations are being forced to liquidate their capital—especially their biotic capital, such as rainforests and coral reefs—to pay their debts. In countries worldwide, most of the resources are controlled by a tiny fraction of the population. Quick profits from excessive logging or overfishing flow to the few, while local communities are left with the consequences. Inequities within nations drive the destruction of biological resources no less than those among nations.

Industrialized countries bear the principal responsibility for overconsumption, and there is overwhelming reason for them to move swiftly toward a more sustainable way of life. For example, the USA has about 300 times as many automobiles per capita as India. But Indians, indeed, people all over the world who aspire to material wealth, must realize that the world's atmosphere, lands, fresh waters, and oceans cannot possibly support all of humankind living at the industrialized nations' excessive level of resource consumption. Dramatic changes in resource use are essential if we are to avoid irreversible loss of products and services from our planet's biological diversity.

Institutions That Diminish Biological Diversity

The interconnectedness of ecosystems and economies calls for a cross-sectoral approach to biodiversity conservation, particularly in the sea. Yet many institutions—regulatory agencies, funding agencies, research institutions, and others—operate along rigidly sectoral lines. Governmental, intergovernmental, and corporate planning tend to be overcentralized, which discourages local participation by people closest to the resources and excludes citizens' groups or non-governmental organizations (NGOs). Agencies charged with nature conservation tend to be weak. Few have the personnel or financial resources needed to support even minimal programs. Many countries lack an adequate system of environmental laws that encourage the protection of nature and the sustainable use of resources, and even where adequate laws exist, enforcement is often lacking.

Largely because of legal and institutional constraints, marine biodiversity conservation is typically piecemeal and based on the traditional terrestrial model: a protected area here, a regime for managing an endangered species there. Such efforts seldom fulfill the habitat requirements of marine species that range widely and are threatened by pollution, sedimentation, and other perils originating on land.

Insufficient Understanding of Nature

Even for the best-studied marine areas—nearly all of which are in industrialized nations with substantial facilities and numbers of scientists—scientists do not have adequate knowledge of marine ecosystems and their components. Knowledge is particularly scarce about interactions among species and ecosystems, and how the sea functions on time scales of decades or more. But most developing nations have even poorer information bases; scientists and resources for research are scarcest in countries with the highest biodiversity. Ignorance is compounded by the ongoing destruction of traditional cultures and, with that, the loss of their understanding of nature.

Even where scientific knowledge exists, it does not flow efficiently to decision makers, who have, therefore, often failed to develop policies that reflect the scientific, economic, social, and ethical values of biodiversity. Decision makers in capital cities seldom communicate with the local communities who depend directly on biological resources and whose livelihood can be jeopardized by inappropriate development projects and other actions. The general public seldom knows enough about the contribution of biological diversity to people's lives to become effective advocates for policies that reduce unsustainable resource consumption.

The Failure of Economic Systems and Policies to Value the Environment

Many conversions of natural systems—such as destroying mangrove forests to create shrimp farms—are economically and biologically inefficient. They happen partly because of the urgent need for increased production of food or hard currency, regardless of whether that production can be sustained, and partly because natural ecosystems are commonly undervalued economically.

There are several reasons why economic systems undervalue biological resources. First, many biological resources are consumed directly and never enter markets. Among marine products, commercial fish catches are accounted for by markets, while much of the food gathered by coastal communities and the value of mangrove forests as nursery grounds are not. Accordingly, the economic values of commercial fisheries and other potentially exhaustive uses are overestimated while sustainable uses (and aesthetic, spiritual, and ecosystem maintenance benefits) are underestimated, creating incentives to impoverish coastal ecosystems.

Moreover, biodiversity's benefits are in large part "public goods" that no single owner can claim. Coastal wetlands protection, for example, benefits the public tangibly, but the benefits are so diffuse that no market incentives for wetland conservation ever develop. This undervaluation then justifies government policies—such as tax incentives—that further encourage wetland conversion to uses with greater "market" value.

Too many people consuming too much of the Earth's finite resources, but having inadequate institutions and understanding of the workings and economic value of biodiversity to use them sustainably, is a recipe for overexploitation, pollution, and the other threats to life. The rewards for protecting and managing biological diversity are great, but the impediments to acting effectively are formidable.

CHAPTER FIVE

Impediments to Marine Conservation

THE SEA is vitally important to humankind, yet our species treats it with carelessness, even contempt. There is now a growing movement to protect, understand, and sustainably use the sea, but there are significant obstacles that must be diminished before we can become its effective stewards.

Insufficient Scientific Information

A major reason why marine conservation lags terrestrial conservation is our ignorance of the sea's value and vulnerability to us. We know less than we need to, and the knowledge of traditional users of the sea and marine scientists is not available to all who need it.

There are several reasons for this. One is that marine ecosystems are variable and complex, defying generalization. For example, little can be extrapolated from the under-ice ecosystem of the Laptev Sea (Russia) to the Ganges River estuaries that empty into the Bay of Bengal. The lack of broadly applicable marine ecological theory prevents managers from using a simple set of standards to guide all their decisions. Most places lack even basic background information on currents, species inventories, and ecosystem dynamics that are fundamental to informed decision making.

Another reason for scientific uncertainty is that events in complex systems often cannot be tied to single causes. Still another is the lower esteem accorded applied research by members of the scientific community; most scientists would rather generate information that is intellectually stimulating than information directly pertaining to modern challenges, no matter how much it is needed.

The most important cause of insufficient information, however, is the low priority that nations have accorded marine research. Much of the world's marine scientific expertise and information resides within industrialized nations, which have invested substantial resources in studies not only within their borders, but beyond them. But even in richer nations, there is almost never enough money for scientists, technicians, or facilities for examining the health and sustainability of marine species and ecosystems. World experts in the systematics of marine organisms and their institutions have faced crippling cutbacks of funding even in the UK and USA. Ecology has not fared much better. The deficiency is greatest for the kinds of long-term population, community, and ecosystem research and monitoring that produce the largest payoffs. For example, Dayton (1989) reported tremendous population variations in an Antarctic sponge (*Homaxinella balfourensis*) on a time-scale of decades; these could not have been interpreted in a program lasting much shorter than 30 years. An occasional two-year pulse of funding that ends when economic fortunes change is no replacement for a long-term commitment to a system of research and monitoring sites and the personnel and equipment to study them (Norse 1987).

Having better information about populations, species, and ecosystems is essential, but not sufficient; decision makers also need much better information about the human causes and consequences of protecting and using living resources. On topics as diverse as how ownership patterns in traditional societies affect their management of marine resources and how industry officials decide when and when not to pollute, social scientists can make a crucial contribution to decision making.

Whether the knowledge vital to a decision concerns ecosystems or people, acquiring it is not free. Scientific uncertainty, which is widespread even in information-rich situations, is an inevitable result of insufficient investment. No matter how large the gap is between available information and what decision makers want to know, however, scientific uncertainty does not justify inaction. Only rarely can decision makers faced with scientific uncertainties generate new information as the need arises. Far more commonly, the needed information will not be available soon enough. What can decision makers do?

An answer that is gaining growing acceptance is that they must act in ways that preserve options for future decision making, when information will be more available and choices will be clearer. A decision that errs on the side of conservation is far less likely to foreclose options for future action than one that errs the other way. For example, as an alternative to permitting logging now, a mangrove forest of undocumented importance can be held in reserve and, if there is sufficient justification, turned into charcoal at a later date, a decision that preserves that option and a wide range of others. But if the forest is destroyed now, it cannot be restored for many decades, at minimum.

Because so many decisions are based on incomplete information, how decision makers rule when there is insufficient knowledge—on which side of the issue they place the burden of proof—is a central question in marine conservation. Its answer will determine whether sustainable use becomes a living, working reality or a fondly disregarded concept having no relevance to the world in which we live.

Inadequate Transfer of Information

Although there are many things that decision makers need to learn about the sea, considerable information already exists or is being generated that is not readily accessible to them. Some of this knowledge exists among traditional users of the sea, whose livelihood has long depended on their understanding of nature's patterns and processes. The other source is the natural and social scientists who study the sea.

The value of traditional knowledge is often dismissed by people in modern societies, in developing nations even more than industrialized ones, yet scientists recognize that the people with the most intimate knowledge about some aspects of marine species—such as their movements and when they breed—are coastal people who have spent a lifetime observing marine creatures while making their living from the sea. There are woefully few pathways by which decision makers can access this information and verify its validity. Because coastal people are the ones who are most affected by marine environment/development decisions, and because their participation is usually essential to their success, improved two-way communication between them and decision makers is crucial for managing marine resources.

Except for research designed for military use or on commercially important species and minerals, most marine research has scientific goals and is not designed to answer questions about the health of the sea. With suitable interpretation, however, some pure scientific research has significant implications for decision making. Unfortunately, very few scientists become decision makers, and, except when they are seeking funding, scientists seldom try to explain the implications of their work in places where decision makers seek information and in ways that decision makers find useful. A reciprocal reason for the paucity of mechanisms linking scientists and decision makers is that decision makers seldom articulate their information needs to the scientific community.

Even information directly relevant to conserving biological diversity often reaches resource managers slowly or not at all. In general, the more accessible an information source (publications in peer-reviewed scientific journals, for instance), the older and less site-specific the information tends to be. Most currently available information is in the form of books, monographs, research journal articles, and symposium volumes. Few computerized databases on marine subjects exist, and they are not widely accessible, especially internationally.

Overlapping jurisdictions and competition among agencies responsible for marine resources can also inhibit information transfer, even within nations. Most information flow is vertical, not only within countries, but within specific agencies. How to improve horizontal information transfer—how to build bridges between those who generate information and the those who need it—is one of the greatest challenges facing marine resource conservation.

Although there is a compelling argument for strengthening marine scientific research in general, a special priority should be strengthening research directly relevant to the protection and management of marine species and ecosystems. Moreover, the health of the sea depends on finding effective ways of getting research results to the people who need it.

Cultural and Biological Diversity

No matter how much biologists know about the population dynamics of queen conchs or the ecosystem dynamics of seagrass beds, it will not be possible to protect or use them sustainably unless we understand the human causes and consequences of their increasing rarity. Our success or failure is rooted in our cultures and economic activities. These can be forces for conservation and sustainable use, or they can be forces that eliminate species and ecosystems. For life in the sea, the diversity of human cultures offers both promise and risk, but the promise outweighs the risk. Over the vast majority of human history, the sea has been little affected by human activities because:

1) Our populations were small.
2) Our relatively humble technologies did not allow us to inflict damage beyond a local scale.
3) Many of our cultures developed taboos and rules concerning the ownership of resources that had the effect of preventing harm to the environment.

The world at the dawn of the industrial age had highly diverse human cultures adapted to living in highly diverse ecosystems. In the past few generations, however, a fundamental ecological shift has occurred on Earth. Economic growth, greatly expanded international trade, and improved public-health measures have spurred a rapid expansion of human population and an even more rapid increase in the consumption of resources. A diversity of cultures adapted to local environmental conditions has largely been replaced by a world culture whose defining characteristic is its extremely rapid consumption of resources. Artisanal fisheries have been replaced by industrial fisheries, community responsibility by government responsibility, economic self-sufficiency by economic interdependence, and sustainable use by kinds and intensity of use that are environmentally damaging. Although these changes affect all the threats to marine species and ecosystems, their effects on overexploitation provide an especially clear illustration.

Overexploitation is common in times of rapid cultural change, as traditional controls break down and people learn to exploit resources in new ways. Technological innovations such as motorized trawlers and dynamite for fishing tend to favor overexploitation and weaken traditional approaches to conservation. This is especially true when a technologically superior group moves into a region where technologies are less complex. Unlike the indigenous people, the technologically superior group has the option of moving on when the region's resources are exhausted and therefore has little incentive to adopt traditions of sustainable, conservative use. It can earn virtually all the cash benefits from exploiting a stock of fishes, a mangrove forest, or coral reef, but pays almost none of the environmental costs. Instead, these costs are paid by the local communities that formerly depended upon these resources for survival and had developed ways of managing them sustainably.

At the same time, modern open-access management means that the traditional groups lose any advantage from traditions of conservative use that might have been favored in times when they could exclude other groups from their territory. These traditions evolved when both costs and benefits were internalized in the decisions made by the communities. When local people are forced to bear far higher environmental costs of resource degradation, their only rational response is to join the exploiters to try to seek greater benefits as well (Gadgil 1987).

This loss of knowledge is tragic. Traditional management systems that maintained biological diversity for centuries or millennia are becoming obsolete in a few decades, replaced by systems of exploitation that bring short-term benefits for few and long-term costs for many.

The need to gain foreign-exchange earnings through international trade has promoted ever-greater pressure to exploit all biological resources. Inevitably, this reduces cultural diversity, for two main reasons. First, a significant component of cultural diversity that enables people to earn a living from the local biological environment is no longer functional. Second, traditional groups begin to imitate the culture of the dominant group, thereby losing their cultural identity (Gadgil 1987).

Are all traditional systems doomed to failure, to be replaced by government and private ownership of resources? Or do traditional community-based resource management systems still have something to contribute? A few examples suggest that traditional systems can continue to make important contributions, and remain viable alternatives to other systems of resource management.

On Ferguson Island (Papua New Guinea), local villagers became increasingly concerned over uncontrolled hunting by outsiders and the resulting scarcity of species such as crocodiles. A management committee established by the people banned all but traditional methods of hunting crocodiles and the collection of their eggs. In other parts of New

Guinea, such committees have forbidden the use of commercially manu-factured nets, lights, and poisons for fishing (Eton 1985).

In Japan, a complex system that developed over centuries provides various forms of customary village tenure and rights to fisheries in coastal marine waters. These traditions are so effective in preventing abuse of the resource that they are embodied in national legislation. Traditional approaches have been incorporated into modern Fisheries Cooperative Associations, which now hold fishing rights to virtually all coastal waters, adapting fishery regulations to regional differences in ecology, target species, fishing effort, and level of industrialization. This ensures that management strategies, processes of conflict resolution, and interpersonal or intergroup relationships will be based on local custom-ary law and codes of conduct (Ruddle 1986). Whereas Japan is one of the nations that has earned international disapproval for the activities of its distant-water fishing fleets, Japan's management of its coastal fisheries provides a model that others might do well to emulate.

In the Philippines, villages in the coastal zone have long depended on the productivity of coral reefs, but as traditional management systems have eroded, overexploitation has become rife. Efforts to rejuvenate tradi-tional approaches through establishing Marine Management Committees by local villagers have proven very successful (Savina and White 1986). Three marine reserves have been established, including a fishery breeding sanctuary and a surrounding buffer area for ecologically sustainable fish-ing. Destructive fishing methods that use dynamite (Figure 4-1), poison, and very small mesh gill nets (biomass fishing) have been halted. Species diversity has significantly increased in some families of fishes, especially favorite targets of fishers; numbers of food fish have increased 42 to 293 percent. The greater total fish yield and resulting economic benefits from this community-based management system provide a means to achieve long-term sustainability.

These three different approaches suggest how cultural diversity can help to conserve biological diversity. There are many other cases (see, for example, Johannes 1982; McNeely and Pitt 1984). Local resources can often be managed more sustainably by the local people who depend on their continued productivity and whose management approaches have passed the test of time.

Differing Benefits and Costs of Harming Marine Life: "Commons" and Private Property

Imagine the following scenarios: A fisherman goes to the small estuary from which he draws his food and income, and empties into it a large barrel of toxic pesticide. Or he fills in the wetlands that are nurseries for the fish he catches. Or he introduces a handful of oyster-drilling snails from another ocean and watches them attack the shellfish he sells. Or he catches so many fish that the species that provides most of his income virtually disappears.

These scenarios sound ludicrous for a good reason. Sane people would not knowingly do things so damaging to their interests. Yet there are many people who pour toxic chemicals into estuaries, who physically destroy productive ecosystems, who introduce dangerous alien species, and who overfish until stocks collapse. They do so because it is in their economic interest; they do what benefits them, but pass on the costs to society as a whole, a process that economists call "externalizing" costs. Of course, they also suffer the costs to some degree, but unlike the others who do so, they reap most or all of the benefits from their actions.

At the heart of many conflicts over harm to marine resources is the "tragedy of the commons" (Gordon 1954; Hardin 1968), although the English commons themselves might not have been a bad model for marine resource management (Hanna 1990). But in the sea, a common-property resource, such as a marine fish population, is typically subject to misuse because individual fishers reap the full benefits of their catch, while the costs (in terms of reduced fish production) are spread over the entire fishing community. The fishers will continue to fish until the cost of doing so exceeds the benefit. Furthermore, they have no incentive to reduce their catches in order to increase their future revenues as they would if they were sole owners of the stock, because what one fisher does not catch today will be caught tomorrow by another, and not left to multiply. Consequently, the fish stock becomes so depleted that individual fishers can barely make a living from the resource.

Even when countries attempt to manage fisheries, fishers perpetually press for increased catch quotas, even on dangerously depleted stocks. The reason is always the same: economic need. Indeed, the higher the fishers' costs, the harder they push; overcapitalized fisheries are the bane of open-access fishery management schemes. In a similar way, polluters and loggers can offer their products at lower prices by externalizing their costs, and press decision makers for exception from regulation while the general public absorbs the costs of pollution.

This is not how societies should work. But not all economic activities are in societies' best interests. A major reason why societies are not simply collections of individuals but groups that submit to tribal or governmental authority is because they need to deal with the very human tendency to maximize benefits while passing on the costs to anyone else. The taboos or laws that authorities enforce are ways to modify activities that do not create wealth in a socially acceptable way.

In many nations, however, there are large gaps in the system of legal protection, either because appropriate laws do not exist or because they are not enforced consistently. Such legal protections are lacking when the losses experienced by members of society do not motivate collective efforts that remedy the situation. For example, citizens harmed by marine polluters might:

1) feel threatened with violence or loss of jobs if they speak out;
2) have no reason to believe that government would respond to their concerns;
3) not be organized effectively enough to compel governments to acknowledge their interests; or
4) not be aware of the harm caused by pollution.

Similarly, a government might act in ways that favor polluters over their victims because:

1) The contributions that polluters make to the nation's GNP can exceed the harm they demonstrably cause.
2) Polluters provide money, personal favors, or political support to leaders in return for favorable treatment.
3) The government might not be aware of the harm caused by pollution.

Hence, governments are not always effective guardians of biological resources and their sustainable use. Government protection of the public interest is often frustrated by two fundamental asymmetries. First, people whose activities harm biological diversity commonly devote part of their profits to ensure that government decision makers look the other way. Second, those who benefit and those who are harmed are often seen as

---

constituents of different government agencies or jurisdictions. An agriculture minister might feel little obligation to heed complaints about pesticides in runoff from a fisheries minister's constituents, or a state that produces and exports pollutants might feel little obligation to compensate a neighboring state.

Given the weaknesses inherent in a regime of common property resources, privatizing resources might seem to be an appealing solution because it eliminates a disincentive for exploiting the resource sustainably. However, there can be another disincentive for sustainable use: the conflict between present and future use of a resource.

While people in desperate circumstances act even if they sharply diminish their future options by doing so, individuals, corporations, and nations whose basic needs are met constantly choose between current consumption and investment in the future. People save for their old age; corporations funnel some of their profits into research and development; nations invest in education and transportation systems. But they do not invest unless they anticipate a sufficient return on their investment. If returns are too low, it pays to liquidate capital and invest the proceeds elsewhere at a higher interest rate.

This normal economic process can be damaging for many types of renewable resources, whether privately and publicly owned. Consider what would happen if a private company were to own blue whales. Blue whale populations are very slow-growing: Females mature at seven years and produce, at most, one young every two years, not all of which survive. Biologists have estimated the growth potential of Antarctic blue whale populations at 2 to 5 percent per year, a rate lower than the interest rate that other investments could pay. As a result, it is economically rational for a private whale owner to kill all the whales and invest the profits where they would bring a higher rate of return (Clark 1973). The remarkable dividends paid by whaling companies in the heyday of Antarctic whaling (Clark and Lamberson 1982) resulted from liquidating cetacean capital, not from living sustainably off the interest. There is less incentive to liquidate capital in species that offer a higher rate of return, but many of the marine species that people consider desirable are long-lived, late-maturing, and slow-reproducing.

Few governments appear to recognize the basic consumption-investment dichotomy inherent in resource and environmental problems. Because most decision makers consider it their duty to promote economic growth, liquidating capital is always appealing because it brings short-term returns. Unfortunately, it also diminishes longer-term returns, something decision makers often ignore. This is especially pernicious because future generations cannot vote, sue, or protest when decisions affecting their fate are being made. Their only power is that which current citizens and decision makers give them. These intergenerational inequities are probably *the most* difficult problem faced by those who care about our living planet.

People who amass benefits by destroying biological diversity and shifting the cost to contemporaries or future generations are behaving rationally from a purely economic point of view. One of the most important roles of government is deciding when behavior that is economically rational for an individual or group is not in the best interest of society. Governmental intervention is warranted to prevent overexploitation, pollution, and other threats because demands to "save jobs" in, say, a polluting industry are actually demands to sacrifice our stocks of natural capital, and hence to sacrifice the welfare of future generations in favor of immediate consumption.

Biological diversity in the sea is being depleted by the failure to employ proper investment strategies that safeguard natural capital while providing a sustainable flow of interest. Conservation is more than an idealistic wish expressed by scientists and nature lovers; it is an essential condition for sustainable economic well-being.

It is only recently that much serious thought has been given to the changes in economic policy that are necessary to prevent loss of biological diversity, changes affecting taxation, trade, food production, international aid, and virtually every aspect of our lives (MacNeill et al. 1991). Such changes will happen only when economics accounts for nonmarket goods and services and the interests of all future generations, thus transforming itself into ecological economics (Costanza 1991).

Economic Valuation

In the not-so-distant past, there were "new worlds" awaiting discovery. Although humans inhabited all continents but Antarctica by 10,000 years ago, major islands such as New Zealand and Madagascar were colonized much more recently. Colonizing peoples have generally adopted a "pioneering ethic" in which resources are considered infinite and the only constraint is having enough labor, technology, and infrastructure to get the largest possible amount of product to market. Often colonizers applied this ethic even to lands, such as the Americas, Liberia, and Taiwan, that were already inhabited. In pioneer societies, natural resources are considered to have little value until they are reduced to possession and can be bought and sold in a market. Pioneers typically use species and ecosystems wastefully.

Many nations are no longer in the expansionist mode of colonizing "untamed" areas, and have enacted environmental laws that reflect evolving views about the finiteness and value of nature. The prey of a species that people eat, the structure-forming species among which valued species feed and reproduce, and the removal of pollutants by the whole ecosystem are all resources or services that are no longer considered worthless even though they are not currently marketed directly. There are also species or ecosystems that do not provide any known products or services. How can decision makers know the true value of species and ecosystems that do not enter markets?

The destruction or injury of resources carries a cost that a growing number of societies are no longer willing to ignore or absorb. While many nations have subsidized development by not requiring mitigation or payment for environmental damage, more and more are now requiring what economists call "internalization of externalities." Companies that invest inadequate sums for oil spill prevention, for example, are starting to be compelled to pay for harming the species and ecosystems that they damage when spills occur.

Governments or local communities can be compensated adequately for injuries to natural resources only if there are techniques for estimating the value of what is injured. Similarly, they can assess the economic impact of a development project only if they have a full accounting of all values that would be forgone. These needs have led to a new generation of economic tools used to establish the full value of natural resources, including restoration costs, use values, and nonuse (indirect) values. Laws that impose strict liability are beginning to provide nations with the means and the will necessary to pursue natural-resource damages claims.

As the methods for assessing costs of environmental degradation improve, commercial interests are finding it economically prudent to avoid harming marine biodiversity so that they can minimize damage payments and gain approval for new projects.

Marine Conservation and Economic/Social Development

There was once a time when the most fitting adage about resource use was the ancient Chinese proverb "Give me a fish and I eat tonight; teach me to fish and I eat for the rest of my life." The sea's bounty seemed endless and what we could take was limited only by technology. Now widespread environmental degradation and overfishing by the burgeoning human population make this adage unrealistic. Rather than just teaching people to fish, we must concentrate on ensuring that the fish will be there to catch, by teaching people to use the sea sustainably.

The term "sustainable use" has entered our lexicon because uncontrolled human activities have depleted resources one after another. But the term is used in diverse ways; some clarification is needed.

It is equally clear that some components of the biosphere are so important that they must be preserved, and that a major share of the Earth's organisms and ecosystems will be "developed" and used by people. Living systems are renewable, but humankind can maintain them as resources only by not pushing them past the point where the processes of renewal break down. Exceeding this threshold might continue to provide benefits for a while. Whalers, for example, continued to profit from killing blue whales long after populations had plunged far below their level of maximum productivity, perhaps even below the level of continuing viability (Clark and Lamberson 1982). But once that threshold is crossed, such activities cannot be sustained, and therefore, people's use of the resource cannot continue over the longer term. There are myriad, ever-increasing, unmistakable signs that we are degrading the capacity of the sea and the rest of the biosphere to support us. That is why researchers, leaders, and citizens worldwide are seeking ways for people to use resources in a manner that allows us to live and prosper sustainably.

What is sustainable use? In theory, if humans insert ourselves in an ecological system and replace, or allow natural processes to replace, all that we remove, the system will renew itself indefinitely and our use of it will be sustainable. In places such as western Amazonia, coastal northwestern North America, and northern Australia, indigenous peoples used ecosystems for long periods without apparent degradation. (How much they changed these ecosystems from their initial state is another matter.)

Crucial to such long-term use of species and ecosystems has been cultural (especially religious) beliefs and practices that prevent degradation or favor regeneration of resources.

Sadly, there are few, if any, cases of long-term sustainable use by modern industrialized societies, while examples of how they have failed are everywhere. How modern societies can live and prosper sustainably on our planet is *the* greatest challenge humankind faces.

A similar-sounding term is "sustainable growth." Without question, growth in human population or per capita resource consumption has been occurring. But because the Earth's species and ecosystems have a finite capacity to accommodate human use, and because that capacity has clearly been exceeded in many places, these kinds of growth inevitably degrade the living systems on which people depend. Hence, the term "sustainable growth" (in population or resource consumption) is an oxymoron. Whether it is possible to sustain economic growth *without* population growth or growth in resource consumption is another, extremely important question that must be answered convincingly and very quickly.

There is yet another similar-sounding term, one used more often than either sustainable use or sustainable growth: "sustainable development." It might be more prevalent because it usually goes undefined, which allows people use it to mean different things. Used as a synonym for sustainable use, sustainable development is not only a theoretical possibility but a vital imperative for humankind. Used to mean sustainable growth in population or resource consumption, sustainable development is impossible.

People have just begun to take stock of how our actions affect ecosystems and their ability to provide the resources we need to sustain ourselves and our quality of life. Although we have been particularly slow to see this in the seemingly bountiful and vast oceans, things are beginning to change.

For example, commercial fish stocks are overexploited worldwide. But nonsustainable practices are not confined to fisheries resources or to industrialized countries. Indeed, it is in developing countries that nonsustainable practices, which undermine the potential for economic and social betterment, have the most dramatic impact.

Decision makers, however, are changing their thinking about the need to conserve marine resources. Coastal nations are increasing "ownership" of adjacent waters, and have a growing stake in having healthy, productive marine species and ecosystems. Governments, industries, funding agencies, and conservation groups are seeking ways to use marine resources sustainably.

Conservation of marine ecosystems presents some daunting challenges, including multiple threats, conflicting uses, overlapping or unclear jurisdictions, and ill-defined ecosystem processes and boundaries. Despite these, the incentives for protecting the sea are great: real and quantifiable economic benefits from controlled production, coastal tourism, and managed multiple use.

The expanded field of resource economics is encouraging. New work in economic modeling and valuation has demonstrated that investments in marine and coastal conservation are well worthwhile. Such justification is very much in demand in countries that have faced the choice of economic development *or* conservation. Economic and social analyses now show that a healthy coastal system can be an invaluable asset to a nation, worthy of its protection and capable of high returns on investment.

At the same time, more nations are extending responsibility for environmental protection and marine management to local users. Many of the real and lasting success stories in sustainable use come from local efforts in which stewardship is an integral component of economic plans, particularly ones in which the benefits to local communities are visible over short time frames. An example from West Africa emphasizes this crucial point.

Guinea-Bissau's Bijagos archipelago contains more than 80 islands and islets, with ecosystems from alluvial deltas to mangrove and savanna cays. The island chain is largely pristine and only sparsely inhabited, yet overdevelopment looms large on the horizon. Plans for developing ecotourism, constructing large facilities for beach tourism, expanding industrial and artisanal fisheries, and attracting industry all threaten the health of this ecosystem by undermining the lifestyle of the Bijagos' inhabitants, whose culture is one of the most intact in Africa.

International conservation and funding agencies have succeeded in the Bijagos only when they have demonstrated the potential for economic and social viability through coastal planning and conservation. IUCN is working to empower villagers to work for long-term stewardship. Aid agencies are exploring ways that an archipelagic park can finance itself while protecting the region's species and ecosystems.

Conservation organizations and funders can help the people of the Bijagos avoid repeating the ecologically and socially ruinous effects of development. Ultimately, however, the success of all these efforts will rest with local, not foreign, commitment.

North-South Divisions

Governments, singly and as participants in regional or world organizations, can create conditions favoring the protection, study, and sustainable use of marine organisms and ecosystems, or the conditions that encourage their destruction. There are many reasons why governments and international governmental organizations (IGOs) often seem to interfere with, rather than help, efforts to conserve the sea.

The profound inequities between industrialized and developing nations are among the greatest political barriers to solving the world's environmental problems. The sea, as a global commons, is at special risk from the resulting conflict.

In the years between the 1972 UN Conference on the Human Environment in Stockholm and the 1992 UN Conference on Environment and Development in Rio de Janeiro, concerns about the health of the environment penetrated the consciousness of both industrialized and developing nations, often called "the North" and "the South," respectively. But these two diverse groups of nations tend to view the causes of environmental degradation and the remedies to it quite differently.

The industrialized nations, located mostly in the temperate zones, are comparatively rich economically but poor biologically and have mainly democratic governments and low or even negative population growth rates. They have harnessed formidable human resources and technologies to amass financial capital by liquidating their natural capital, and have sent their companies worldwide to liquidate the South's natural capital without worrying about the effects on its people and environment. Some Northern nations have found it very difficult to give up environmentally ruinous practices within their borders, and many continue to export destructive technologies and practices to the South. By the time Northern nations began awakening to environmentalism, most of the damage within their borders was already done. This has cast the Northern nations in the ambivalent role of embarrassed, repentant latecomers to environmental responsibility who, nonetheless, maintain unsustainably high levels of consumption and are unwilling to consider lowering them.

Developing nations, on the other hand, are located mainly in the tropics, tend to be poorer economically but richer biologically than the North, have scarce scientific and technological resources, and tend to have much higher population growth rates. Their natural capital tends to be more intact. Most were the victims of Northern colonialism and got their first taste of economic development from Northern companies that took their natural resources and left them no better off for it economically. Indeed, most are now deeply in debt to Northern banks that lent them money for development schemes that failed to alleviate their poverty. Many have been sites of arms races and proxy wars fueled by sophisticated Northern weaponry and primitive Northern geopolitical rivalries. Some developing nations have rapidly rising per capita gross national products, and are thrilled with their higher buying power to the point of being blinded to the consequences. Others are suffering declines from their already low per capita GNPs, and are desperate to maintain even rudimentary institutions and forestall starvation. In virtually all of them, people see televised images of a North that seems rich, amoral, and self-absorbed. Understandably, many nations of the South resent feeling helpless and object to the North's profligate consumption, yet at the same time want to emulate it.

The North's huge GNPs, deep capital pools, educated, well-fed, long-lived people, and stable political regimes allow it to take the longer view. Recognizing that people from all nations need the Earth's species and ecosystems to sustain us, the North, having undergone what the South is now experiencing, wants the South to learn from the North's mistakes and protect the environment in the course of improving its economies.

Long-frustrated aspirations prevent most Southern nations from looking beyond the short term, especially when Northern nations flout their own endangered species acts, log their last ancient forests, send distant-water fishing fleets into the South's waters, and export chemical contaminants around the world. Southern nations feel that they, too, have the right to prosper, no matter what the environmental cost.

To secure meaningful cooperation on environmental issues, the North must use its enormous scientific and technological skills to develop ways to provide for human needs that do not threaten biological diversity, and share them freely. It must help the South find affordable but effective ways to achieve democratization, population stabilization, and environmental protection. But the North needs to offer more than just advice. It must set an example by lowering its own consumption of resources. It must offer debt relief and debt restructuring, better access to markets, equitable technology transfer, lower trade barriers, and substantial direct aid for projects that use resources sustainably.

For the South, in turn, to achieve any but the briefest economic benefits from its resources, it must stabilize its population, make governmental institutions more effective, and enlist local people in solving problems. It must recognize that the North's path to prosperity cannot possibly be repeated, and find ways to improve its economic well-being that do not destroy the planet's living systems. It must develop the infrastructure and institutions that truly meet the needs of its people, now and in the future.

It is not only oceanographic, atmospheric, and ecological processes that connect the Bights of Benin and New York, the Gulfs of Tehuantepec and Lyon, the Java and Inland Seas. Nations North and South must share the Earth. To sustain the marine processes critical to the functioning of our planet, the North and South must work together for their mutual benefit. Substantially improved cooperation between them is the key to any meaningful progress in conserving the sea.

National Sovereignty

The concept of sovereignty or self-government is one of the oldest, most entrenched, and most cherished concepts that humans have established, and the dominant units of governance increasingly have been nations. In this century, as the global economy has pulled nations into a web of economic interdependence, international treaties have encroached on national sovereignty by regulating trade, human rights, and labor practices. To partake in the advantages of the global market, nations have been compelled to comply, lest they incur sanctions such as embargoes or other trade restrictions.

Since the 1972 Stockholm Conference, increasing environmental degradation has stimulated growing awareness of the need for regional and global cooperation, and a new round of treaties has been proposed to regulate nations' environmental impacts on their neighbors and the rest of the biosphere. The Earth is increasingly seen as a finite, self-contained, interacting ecological and economic system.

Nations that resist environmental agreements are finding that invoking national sovereignty as a means of continuing environmentally destructive practices at the expense of other nations is not being greeted favorably when all people depend on the climates of the same atmosphere and the fish of the same oceans. Brazil's "right" to fell rainforests within its borders, the USA's "right" to patent genes from wild organisms, China's "right" to burn its huge coal reserves, Japan's "right" to kill whales, and many other sovereign "rights" are challenged by other nations whose interests are affected by these activities.

The UN Convention on the Law of the Sea (UNCLOS III) is the current authority on what constitutes sovereignty over the marine environment. Although not in force yet, largely because of opposition from mining interests in the USA, most of UNCLOS III policies are now part of international law. UNCLOS III was the authority in creating sovereignty over coastal resources, but it also recommends that management of those resources be shared among all the parties that have a stake in them. In contrast, UNCLOS III subjects marine scientific research within EEZs to coastal nations' discretion to grant or withhold consent. Hence, in recent years, marine scientific research has sometimes been hampered when some nations have instituted lengthy permitting procedures by asserting national sovereignty over the acquisition of knowledge about the physical and biological processes in the sea.

Slowly, haltingly, but unmistakably, nations are yielding sovereignty in some areas where they cannot afford to ignore the marine environmental concerns of other nations. However, the world's need to acquire scientific knowledge is less obvious than the need to stop pollution or depletion of species, but like those needs, it will be more effectively fulfilled as nations emerge from the cover of nationalism.

Jurisdictional Gaps and Overlaps

Because governments are supposed to serve the interests of citizens, the programs they run, the laws they enact, and the treaties they sign should reflect the situations in which their involvement is warranted.

Unfortunately, that is not always the case. On local, national, and international levels, patterns of governmental jurisdiction can fail to mirror the spectrum of people's concerns. In many cases, laws to protect and manage species and ecosystems (or agency programs to implement them) do not exist; that is, there are gaps in jurisdiction. In other cases, they exist in more than one version, and the overlapping jurisdictions conflict. In both cases, people and the resources on which they depend bear the burden.

An illustrative gap appears in laws to protect and use wetlands in the USA. Under the U.S. Clean Water Act, wetlands (including seagrass beds, salt marshes, and mangrove forests) cannot be filled with dredged sediments, whether on public or private lands, unless it is specifically permitted by two federal agencies. But the provisions that cover draining of wetlands are much weaker. Thus, by digging drainage canals, thereby eliminating a wetland, an individual, company, or government agency can undergo far less stringent regulatory agency review. The illogic of such gaps would be amusing if they were not so common, and if the biological diversity they put in jeopardy were not so important or imperiled.

Although the UNCLOS III establishes 200-nautical-mile (371-kilometer) EEZs, about two-thirds of the world's marine area is not under national jurisdiction. In the Bering Sea, there is an area called the "Donut Hole" that is completely surrounded by Russia's and the USA's EEZs but falls outside their jurisdiction. Another donut hole in the middle of the Sea of Okhotsk is totally surrounded by Russia's EEZ. Under international law, these gaps are considered high seas, within which most fish species are "free for all," hence, both are heavily fished by Pacific Rim nations. Because the jurisdictional boundaries bear no relation to the ecological boundaries of the fish stocks, foreign fishers are free to deplete stocks that straddle the EEZ/high seas boundaries, undermining the coastal nations' efforts to manage these fisheries sustainably.

Jurisdictional overlaps among agencies can lead to "turf battles" that hamper prospects for protection and sustainable use of marine resources. A later section on Belize's Hol Chan Marine Reserve suggests how many interests can claim jurisdiction over just one fairly small marine area. These problems can be minimized if responsible decision-making authorities sponsor analyses of gaps and overlaps, and use these to modify existing laws or treaties or draw up new ones.

International Development Agencies

Multilateral development banks (MDBs) and nations' foreign aid programs have great potential to help or harm marine biological diversity. In the past, these institutions often ignored the effects on ecosystems—including marine ecosystems—in the design of projects. Rather, they traditionally funded projects with specific goals (e.g., to increase national energy production or to boost national agricultural exports) that paid little attention to the environmental side effects. Their record in conserving marine organisms and ecosystems has not always been good for at least four reasons:

1) *MDBs have traditionally lacked sensitivity to the environment in general.* Until recently, MBDs and foreign aid programs had not embraced the idea that a healthy environment is essential to sustainable use. The rewards of a development project are obvious and immediate. For example, the massive infrastructure projects traditionally favored by MBDs are highly visible signs of "progress". The short-term benefits they bring are especially important to a country struggling to meet pressing economic and nutritional needs. Short-term benefits, however, are often gained at the cost of long-term degradation of the environment, including the marine environment.

2) *MDBs have traditionally lacked appreciation of the sea's value.* Oceans, coastal waters and estuaries yield not only food, but ecosystem services that are vital to human existence. Yet, the value of these products and services, and what would happen if they were lost, have seldom been fully considered in evaluating and implementing projects. Even in countries with strong national policies for conservation and resource management, the marine environment is rarely a high priority.

3) *The sea is vulnerable.* Marine systems are so big that people often assume that we cannot possibly harm them, and their resources appear inexhaustible. Few people realize that the sea is biogeochemically "downstream" from the land and therefore receives the myriad products of land-based activities. And few know how systematically marine resources have been overexploited. Moreover, so little effort has been devoted to understanding marine ecosystems that their species

composition, structure, and function are poorly known. Thus, development projects have often hurt the marine environment out of ignorance.

4) *Marine ecosystems often transcend political boundaries.* MBDs traditionally fund projects in a single country. Marine ecosystems are usually large enough to transcend political boundaries, and are connected by currents and migratory patterns. The smaller the country's coastline, the greater the likelihood that impacts of a development project will extend beyond its borders. But even Russia, the world's largest nation, shares marine species and ecosystems with its neighbors in the North Atlantic, Arctic, and North Pacific oceans.

MDBs have begun to understand how destructive their projects have been and to stress the need for environmental assessment. The World Bank (1991), the largest international development agency, recently released the *Environmental Assessment Sourcebook*, which lays out the bank's new policies and procedures for studying the environmental effects of its projects. Other MDBs, along with bilateral agencies, have taken similar action. *If* these guidelines are adhered to, marine organisms and ecosystems will benefit greatly. So, of course, will the billions of people who depend on their health and productivity.

In addition, MDBs and nations' foreign aid programs can promote projects that have the specific goal of protecting marine ecosystems. For example, they have formed the Global Environment Facility (GEF) (see Chapter 8), an experimental program to fund projects with environmental goals. Properly administered, it has the potential to bring far-reaching improvements to environmental protection projects in developing nations.

Military Use and Immunity from Public Scrutiny

Despite increasing global awareness about the health of coastal and marine ecosystems, there is little available information about the effects of military activities on the sea. Military bases and weapons-producing facilities along coasts and in drainage basins generate large amounts of wastes—including some that are very hazardous—that can enter the sea accidentally or by design. The military has dumped everything from shipboard garbage to barrels of nerve gas and unexploded bombs into the sea, and the USA (Davis and VanDyke 1990) and Russia have viewed the sea as the logical place to sink decommissioned nuclear submarines or high-level nuclear wastes from the military. The former USSR used the Arctic island of Novaya Zemlya and the USA, UK, and France have used Pacific atolls for testing nuclear weapons. Many countries use islands as naval bombing and shelling targets. Wars have littered some marine areas with ordnance and sunken vessels that can leak fuel for decades. And 1991 saw the first use of a new military tactic: the intentional release of some 6 million barrels of crude oil into the Persian/Arabian Gulf by Iraq.

It is difficult to gauge the magnitude of harm from these and other activities that have not yet come to light. Most have been exempt from environmental regulatory controls by law or common practice. Even nations with generally good environmental records tend to look the other way when actions are carried out under the cloak of "national security."

There are also less obvious effects of military activities, both negative and positive. One of the negative ones is the introduction of alien species. Wars create novel shipping corridors that are distinct from historical trade routes, or impose upon older routes much higher levels of transport activity. Not surprisingly, a large number of marine organisms are thought to have been newly introduced during world wars. The Australian barnacle *Elminius modestus* appeared in England during World War II (Elton 1958). Two species of Philippine jellyfishes (Doty 1961) were carried to Hawaii (USA) during World War II. The Korean-Japanese shrimp *Palaemon macrodactylus* appeared in San Francisco Bay (USA) shortly after the Korean War (Newman 1963), and many more western and southwestern Pacific invertebrates appeared in California (USA) harbors during the Vietnam War (Carlton 1979).

"Normal" military activity can transport species as well. Noting the arrival of the rapacious blue crab (*Callinectes sapidus*) near Yokohama Naval Base (Japan), Sakai (1976) suggested that it might have come in the ballast tanks of submarines returning from the east coast of the USA.

On the other hand, military bases and target ranges are commonly closed to the public and therefore suffer far less disturbance than readily accessible areas. Areas up to thousands of square kilometers where submarine detection devices or explosives sit on the seabed are usually closed to bottom trawling, thus providing refuge for demersal fishes and benthic communities. As a result, intertidal and subtidal areas of military reservations can be far richer biologically than adjacent areas that endure commercial, subsistence, and sport collecting. Seabirds on many islands worldwide owe their nesting success to the bombs and shells that occasionally fall in their midst; the explosions, deadly as they and their chemical residues are, are far less disturbing than human visits.

Furthermore, commercial whaling in the Antarctic essentially ceased when whaling ships were diverted to military use during World War II, giving the great whales a brief respite before the killing resumed (Figure 5-1).

In addition, many of the most expensive research projects, including oceanographic studies of currents, the mapping of undersea topography, and studies in the Arctic and Antarctic, would never have happened without the military's need for information during the Cold War. It is unfortunate that nations could justify these projects only under a military definition of national security, rather than for their contribution to management of marine resources, but the fact remains that research carried out or funded by the military has generated information of considerable importance to our understanding of life in the sea.

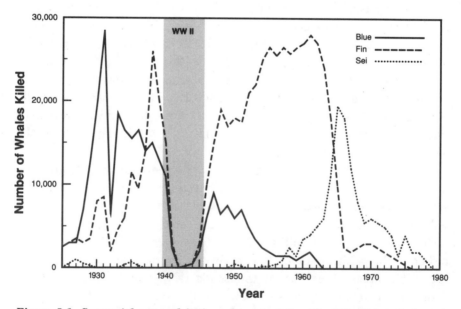

Figure 5-1. Sequential overexploitation of great whales. The kill of blue whales was already plummeting when World War II interrupted Antarctic whaling operations. After blue whales were nearly extinct, pelagic whalers turned to smaller fin whales, overexploited them, and then focused their efforts on the still smaller sei whales. Adapted from Clark and Lamberson (1982).

Increasing concern about marine pollution has increased pressure for military accountability (Davis 1990). Ideally, this could lead to diminished harm from military activities, while retaining the very real benefit of providing refuge from human predation.

Fragmented Decision Making

The threats to marine biological diversity are escalating: They are multiplying, becoming more complex, increasing in intensity, expanding geographically, and interacting synergistically, rather than additively. The net result is that losses are accelerating rapidly. The prospects for saving and sustainably using the sea depend on humankind's effectiveness in anticipating and responding to these changes.

Our rapidly increasing impact on the sea calls for concomitant changes in decision making. Although there are existing and emerging processes and institutions designed to deal with some of these issues, their potency is not commensurate with the increasing threats. Research and management institutions that deal with the threats are hard-pressed even to maintain their budgets and staffs; decision makers and managers are being called upon to do more with fewer resources. Problems that have been mounting for decades are met with short-term "fixes" rather than with long-term, fundamental changes that address driving forces. Whether chronic problems are addressed depends more on temporary economic up- or downturns than on the magnitude of the problems themselves. This is no less true in the seemingly rich nations of the North than in the poorer South.

Similarly, threats of expanding geographic scope are not being met by proportional institutional changes. Nations are still the dominant governing bodies, and most conservation efforts work vertically within them rather than horizontally among them. An official in a national marine resources agency is far more likely to interact with others in that agency and other national and local agencies, industry officials, and conservationists from that nation than with counterparts in other nations, even though officials in other nations might be struggling with the same problems, and might have found some useful solutions.

Fragmented decision making is pervasive at a time when expanding problems demand expanding solutions. It is probably the least recognized major impediment to effective marine conservation.

The Burden of Proof

Virtually every threat to life in the sea is attributable to our use of the "wait-and-see principle," which allows overexploitation, ecosystem destruction, pollution, introduction of alien species, or global atmospheric change so long as someone gains economically and the environmental consequences are uncertain. The fact that other people—often, many more—are harmed economically and in other ways, in both the short and long terms, is largely ignored.

In the ethical system common to pioneers in a new land and to today's world culture, the burden of proof that an activity does not harm the environment does not reside with individuals or organizations engaging in the activity. Rather, it is regulatory agencies and non-governmental conservation organizations—if they exist—that are required to demonstrate that an activity is harmful, even disastrous, before controls are instituted. Under this system, conservation measures are *reactive*, and are initiated only when there is irrefutable evidence that damage is occurring.

Although there are exceptions here and there, the wait-and-see principle is embedded in all the threats to marine life. Use of renewable marine resources, for example, nearly always follows a predictable pattern:

1) A new or underused resource is discovered.
2) Use of the resource begins with little or no knowledge of its size, productivity, or its relationships with other resources, and how much is used is limited only by available technology, investment capital, and market demand.
3) Technologies, capital, and markets develop faster than knowledge of the possible effects of use on either the target resources or the ecosystems of which they are a part.
4) The resource is overused and depleted, and the industry is over-capitalized and suffers from increasing costs and decreasing revenues.
5) Measures are initiated to rebuild the resource (or look for other, more immediately profitable resources) but, because of the large capital investment, the measures are structured more to protect investment than to rebuild the resource, and the resource is maintained far below its optimum sustainable yield level.

This pattern is clearly illustrated by the development of commercial whaling, particularly since World War II. In the Antarctic, for example (Figure 5-1), commercial whaling focused initially on the largest species, the blue whale, but switched to progressively smaller, less commercially valuable species—fin, sei, and then minke whales—as each stock of the larger species, in turn, was pushed toward extinction (Clark and Lamberson 1982). More recently, the wait-and-see principle governed in the evolution of large-scale, high-seas driftnet fisheries that take huge numbers of both targeted and nontarget species, yet developed with little knowledge of their impact on either.

Even the most basic ecological principles are often ignored until an overwhelming case can be made that human activities are affecting resources. For example, the phenomenon of overgrazing of marine algae by sea urchins has been observed in many places throughout the world, including Haiti and the U.S. Virgin Islands (Hay 1984), Hainan Island (China) (Hutchings and Wu 1987), Chile, California (USA), and the maritime provinces of Canada. The common thread in these situations is that fishers or hunters have removed from the ecosystem the predators of sea urchins—be they crabs, fishes, or marine otters—thereby allowing sea urchin populations to explode (Personal communication with Paul Dayton, Scripps Institution of Oceanography, La Jolla, California, USA). On the other hand, it is equally clear that removing grazers, whether sea urchins or herbivorous fishes, can trigger explosive growth of algae in coral reefs and other ecosystems. But these kinds of findings are virtually never used in setting fishery quotas. The economic activity proceeds without investigating possible consequences and produces effects that might actually be studied, but the results are not incorporated into succeeding situations. Adhering to the wait-and-see principle condemns us to repeat our mistakes.

The burden of proof typically falls on the shoulders of the public rather than on those who profit from resource use because:

1) Our modern social, economic, and political systems accept, with little or no questioning, that economic activities are inherently desirable and, often, that a healthy economy is incompatible with a healthy environment;

2) So long as resource users can invest their profits in other ventures when their current ones fail, they reap maximum rewards by taking as much

as possible as fast as possible to prevail over the competition (Clark 1973; Wise 1991), instead of sustaining their resource;

3) There is more uncertainty about the costs of using a resource than the profits coming from it; further, the benefits tend to be easy to quantify, and they go to an easily identified group of people, while the costs tend to be more difficult to quantify and are spread among a broad population.

Nonetheless, the costs that the public pays when it bears the burden of proof are very high. These include having pathogens and poisons in our food, losing species we need as resources, and being forced to choose between costly cleanup operations and the irreversible degradation of our life-support systems. Because regulators and non-governmental organizations concerned about the environment do not profit from the economic activities they deal with, they are inherently disadvantaged; that they must bear the burden of proof makes them doubly so. Placing the burden of proof on the public and on the guardians of its interest is incompatible with sustainable use.The alternative is to place the burden of proof on those who would alter the sea. Under the "do-no-harm" or "precautionary principle," polluters, developers, and other users have to demonstrate that their activities are *not* harmful to species and ecosystems before engaging in them. Although the precautionary principle has been evoked most often when pollution is at issue, it can be applied to all threats to marine biological diversity (Earll 1992). With this approach, conservation is *proactive*, not *reactive*.

Under the precautionary principle, the basic approach to renewable resource use would:

1) ensure that use of newly discovered or lightly used resources increases no faster than the acquisition of knowledge necessary to assess the effects of use on them and on other components of the ecosystems of which they are a part;

2) limit access to resources and structure developing industries and fisheries in ways that ensure that information necessary for informed decision making is available before there are irreversible losses; and

3) avoid systems in which competition encourages individual users to use renewable resources past their optimum sustainable levels.

Incorporating the precautionary principle into laws, regulations, and programs requires new analytical tools. Monitoring protocols that can predict threats to the sea and can determine the causes of degradation are urgently needed (National Research Council 1990b). Threats analysis, or risk assessment, is a promising analytical tool, but it appears to be very sensitive to assumptions, and hence, is easy to manipulate to rationalize preconceived results. Another approach is to use jury-like deliberations based on the best available evidence in order to prioritize threats, assign responsibility, and determine remedies. Reversing the burden of proof might seem to be a radical step, but there is no feasible alternative if we are to use marine species and ecosystems sustainably.

CHAPTER SIX

The Goal and
the Strategy

E V E N before the advent of humankind, our planet was ever-changing. Living things responded by evolving new means of dealing with their environment or moving to more suitable conditions. Those that could not disappeared. The disappearance of populations, species, and ecosystems is hardly new, but the *rate* at which it is currently happening is vastly greater than at any time in the last 65 million years (Figure 2-2). Humans are causing this change. We are also suffering the consequences, and will do so increasingly unless we summon the will to slow and then stop the destruction of biological diversity.

A vital imperative for our species is to save, study, and sustainably use biological diversity (WRI et al. 1992). But to do so effectively, we must determine the most appropriate goal for our biological diversity strategy, and just what this strategy should be.

The Goal: Maintaining the Integrity of Life

In the process of articulating a goal, it helps to eliminate what the goal is *not*. If we are concerned about saving, studying, and sustainably using our planet's living systems, the goal clearly is not to maximize the number of genes, species, or ecosystems. If it were, it would make sense to discharge radionuclides in the sea to increase genetic mutations, introduce species to waters where they had not occurred, and dump plastic garbage onto deepsea sediment plains to create hard substrate. Obviously, these are not sound conservation goals.

Nor is the goal simply to conserve those genes, species, or ecosystems that people "use." As we learn more about biology, it becomes ever-clearer that a far larger number of these are important to humankind than those we consciously use, and that many that are not used now will prove to be useful in the future. Moreover, there is compelling argument for conserving living things for their own sake whether or not they have utilitarian importance.

A goal that seems more reasonable is to ensure that our species does not extinguish the genes that confer adaptability and traits that we can use, cause the extinction of other species, or eliminate any kind of ecosystem. This is an extension of the approach in endangered species laws of many nations. In the USA, this approach has not always worked because the responsible federal agencies have often waited until populations are dangerously low before taking action (Orians 1993). If, however, it were successfully implemented, the list of genes, species, and ecosystems would not diminish. If scientists determined that all sea palm (*Postelsia palmaeformis*) alleles occurred in populations on six rocky headlands between California (USA) and British Columbia (Canada), those six seaweed populations would be preserved. The same would be done if scientists estimated that a minimum viable population of Mediterranean monk seals was 533 breeding individuals, or if that one New Zealand South Island fjord ecosystem was representative of all of them.

Unfortunately, this approach is analogous to ignoring a person's health until he or she is in a hospital intensive-care unit with doctors constantly monitoring vital signs and administering medications. It has three major weaknesses:

1) *It is risky.* Trying to determine how little of something can be preserved in perpetuity is risky because scientists do not (and probably never will) know all that is necessary to ensure the continuing viability of all genes, species, and ecosystems. Even the best minds can fail to anticipate some exigency, especially when scientists and decision makers are pressured not to protect anything until it is absolutely necessary.

2) *It is expensive.* The one-by-one approach can require a large investment in highly trained personnel and collection of field data. It becomes prohibitively expensive as the number of endangered "patients" increases, as will inevitably occur because humankind is not addressing the root causes (see Chapter 4) of biotic impoverishment.

3) *It eliminates most of biological diversity's goods and services.* Even if successful, this approach gives us a world of biological resources that exist but are never far from disappearing. But as any physician knows,

good health is not simply avoiding death. Many benefits from biological diversity come not only from the *kinds* of living things, but from their *quantities*. Even if one viable sampling of every carbon dioxide–removing ecosystem survived, these could not tie up enough CO_2 to help appreciably in the fight against global warming. Similarly, a population of emperor angelfish (*Pomacanthus imperator*) large enough to avoid extinction would not suffice for the legions of scuba divers who want to see them and support local economies by doing so. Significant consumptive use of species would be all but impossible.

Because the destruction of our environment has proceeded so far, preventing things from vanishing is essential but not sufficient. Rather than adopting a goal of stopping the loss of endangered biological resources, the goal of biodiversity conservation should be to ensure that they do not become endangered, that is, the goal should be to *maintain the integrity of life*. If people value an allele in an ascidian species that codes for exceptionally high production of a pharmaceutical, we must maintain the ecological processes that select for sea squirts with that allele. If we value the meat of queen conchs, we must ensure that their populations can replace what we remove. If we value stable shorelines, we must preserve enough grassy dune, marsh, mangrove, and seagrass bed ecosystems to ensure that shorelines will not wash away.

Maintaining the integrity of living systems means keeping not only what Leopold (1949) called the *parts* (e.g., species), but also the *processes* that generate and maintain the parts, the ecological connections among living things. Rather than putting our biological patient in intensive care, we must maintain its health. In a world certain to change, individuals and nations can best maintain our options for surviving and prospering by maintaining the integrity of our diverse biota.

Conservation biologists are now grappling with approaches to maintaining biological integrity (e.g., Karr 1991; Walker 1992). There are many possible definitions of integrity, and determining appropriate ones and applying them to conservation when people are struggling to improve or maintain living standards will not be easy. It will require people to apply our collective energies to revise old strategies, devise new ones, and have the knowledge, wisdom, and courage to implement them in the face of never-ending pressures to make exceptions.

Since our species first evolved, we have concentrated our efforts on ensuring the survival and well-being of our selves, our families, our tribes, and, more recently, our nations. Now we must broaden our focus once more to embrace the goal of ensuring the survival and well-being of our living planet. Deciding how we will maintain the biological parts and processes of our home, how we will save, study, and sustainably use life on Earth, is a challenge far beyond anything humankind has faced. But alternatives that fail to maintain our biotic systems will inevitably lead to diminishing living standards, to widespread misery, and, finally, to death on a scale beyond anything that has befallen our species . . . all within the life span of a single massive coral head. People and institutions can avoid this by keeping the goal of maintaining the integrity of life constantly in mind as we formulate and implement our strategies to conserve biological diversity.

The Strategy: Building Conservation into Decision Making

Most of the world's people benefit from marine foods and other products. Everyone depends on ecological services from the sea. And everyone (even in landlocked nations) affects the integrity of the sea. Therefore, everyone shares responsibility for conserving marine biodiversity. Nonetheless, that responsibility belongs most of all to people in governments, international governmental organizations, industries, and citizens' groups whose decisions directly affect the sea, the intended audience of *The Strategy*.

The Strategy is intended to complement the *Global Biodiversity Strategy* (GBS) (WRI et al. 1992) by providing basic information for decision makers. The GBS identifies three overarching imperatives to conserve life on Earth: saving, studying, and using biological diversity sustainably and equitably. Fulfilling these is fundamental to building a sustainable society. But in all the public forums and discussions that preceded the drafting of *The Strategy*, one central strategy emerged as the *sine qua non* of all such efforts, the one thing without which they will not succeed: *building conservation into decision making*.

Conservation of living things and life processes is not a luxury, something to be considered when other activities permit. Protecting our vital interests requires people to make choices for sustainability in all our actions at all levels of human society. An official in an international lending institution pondering a grant for a long-term marine monitoring program, the leader of a Pacific island nation trying to decide whether to allow its reefs to be mined for building material, a corporate executive pondering whether to buy equipment to reduce toxic effluents, a trawler captain deciding what to do with a hopelessly torn net, and a hungry consumer contemplating which fish species to buy are *all* decision makers whose choices, collectively, will determine whether our species can protect and use marine life sustainably. Less obviously affecting life in the sea are the legislator voting on funding for family planning, the NGO executive building an agenda, the engineer choosing a chemical process to be used by an inland factory, the teacher assembling a curriculum, the engine repair shop owner with 30 liters of waste oil to dispose of, and the voter marking a ballot in a local election. These decision makers also hold in their hands the health of the sea and, hence, the answer to the question "Can our species survive and prosper?"

Sustainably using marine species and ecosystems will not happen by chance. Rather, the prospects for sustainability depend on the cumulative decisions of people choosing among available alternatives. Of course, for this to happen, a sustainable alternative must be one of the options. But this has not always been the case.

All too often, because of ignorance, short-term economic interests, or ideology, decision makers are not presented with, or do not consider, alternatives that would allow sustainability. More often, they are pressured or inclined to choose actions that are capital-intensive and aimed at narrowly focused short-term benefits rather than long-term benefits to all life, human and nonhuman alike. To make conservation integral to decision making requires decision makers to have sustainable alternatives and incentives to choose them.

The rewards for decision making are diverse, but the decisions people make are always ones they think or feel are "best." Often it is the decision that advances their immediate self-interest. Sometimes it involves a measure of self-sacrifice, or a trade-off between short-term and long-term benefits. Decision makers not only need sustainable alternatives to choose from; they need rewards for opting for sustainability.

Building conservation into decision making is best achieved when governments, IGOs, and companies institutionalize sustainability in their laws and actions by including sustainable alternatives in decision-making

---

processes and providing incentives for decision makers to choose sustainability over other options. Experience with governments has shown that conservation is most likely to be incorporated into decision making when a broad range of citizens:

1) have access to all necessary information concerning societal decisions;
2) can question and provide comments to decision makers about the choices; and
3) can, through the courts or other legal means, ensure compliance with laws and regulations.

IGOs and companies can adopt analogous processes toward this end.

Finding sustainable alternatives to the overexploitation of marine species, the pollution and physical alteration of marine ecosystems, the introduction of alien species, and the alteration of our atmosphere requires us to address the root causes of biotic impoverishment (see Chapter 4) in decision making. To achieve a sustainable future:

1) There must be no more people than the Earth can support without losing our biological goods and services.
2) We must consume no more than the Earth can yield sustainably; that is, we must live off the interest and not deplete the capital.
3) Our institutions must be reconfigured to sustain, rather than degrade, our planet's living systems.
4) We need to learn more about natural and human systems so that we can harmonize them.
5) We need to value biological diversity as the source of our wealth and sustenance.

Because of the dramatic increase in threats to marine biological diversity and integrity, there is commensurate need for new, more effective mechanisms to counteract fragmented decision making. In a world that is moving ever-faster, rapid, more effective communication networks that can catalyze needed action are essential to achieving a sustainable future.

One of the most immediate and important consequences of *The Strategy* will be the establishment of a new International Marine Conservation Network to be inaugurated in stages under the auspices of IUCN through its member organizations, initially the Center for Marine Conservation and World Wildlife Fund. Its purpose is to put the health of the sea on the agendas of the world's decision makers. The Network will establish vehicles for rapid communication among marine conservation experts and officials in governments, IGOs, and NGOs. It will review *The Strategy's* recommendations (see Chapter 9), establish priorities among them, and will devise action plans and campaigns to achieve their objectives. Not only will it further North-North and North-South discussion; more important, the Network will place special emphasis on improving South-South interaction. The process of developing collaborative efforts and promoting their progress will catalyze new ideas and, very likely, new leadership in marine conservation.

Through vehicles including a newsletter, an electronic mail network, and IUCN triennial meetings, the Network will build coalitions within and among nations to conduct campaigns on matters of mutual concern. It will provide up-to-date information on topics including new scientific findings, conservation efforts, treaties, and sources of financial support. Its progress will be reviewed at IUCN triennial meetings, and its structure and functioning will change in response to input from participants.

No doubt, as campaigns develop to address, say, making fisheries sustainable or creating alternative sewage treatment systems, separate forums will develop to deal with them. The Network's success will be measured, in part, by the number of efforts that begin to function independently. As it succeeds, the Network should become a powerful catalyst for building conservation into decision making.

CHAPTER SEVEN

Tools for Conserving Marine Biological Diversity

CONSERVING something as vast, complex, and poorly understood as life in the sea, especially given existing economic and political obstacles, is a formidable challenge. But there are tools that decision makers can use to spur protection and sustainable use of the sea. These include political advocacy, expanding the knowledge base, various kinds of planning, regulating the threats to marine species and ecosystems, using economic tools for conservation, establishing protected areas, and active manipulation of populations and areas. Decision makers in governments, industries, conservation organizations, and funding institutions who can use these tools will find that they can minimize the loss of marine species and ecosystems, and the loss of benefits to people.

Political Advocacy

Political advocacy for the protection, study, and sustainable use of marine biological diversity takes many forms around world. In democracies, knowledgeable individuals and non-governmental organizations are often active partners with legislators and other government officials in conservation programs. Advocates may be involved in forming policies, passing legislation, creating regulations, and implementing conservation programs.

In some countries with authoritarian governments, such involvement would be unthinkable. No matter what the political system, however, effective citizen advocacy is greatly enhanced by having good working relationships with decision makers. For example, the designation of the Silver Bank Humpback Whale Sanctuary off the Dominican Republic in 1986 owed much to the strong working relationships that NGOs had established with government officials responsible to the President. Similarly, current efforts to establish a biosphere reserve in the Dominican Republic's Samana Bay, the largest estuary in the insular Caribbean, depend largely upon involving citizens and local government authorities in the process of developing a proposal for consideration by the national government.

Two elements are essential for effective political advocacy almost anywhere. The first is the availability of information about private and government activities and decisions. Where human rights—such as freedom of access to information about government decisions and freedom of speech—are compromised, advocacy for marine conservation will be far less effective. The thawing of the Cold War and the increased recognition of citizens' rights in most of the nations in the former Soviet Union have enabled Russian marine conservationists to talk openly about the need to address the environmental damage in areas such as the Kara Sea east of Novaya Zemlya, where the Soviet Navy had dumped vast amounts of high-level radioactive waste.

The second essential element is knowledge about the resources and possible threats to them. Individuals and groups who are considered credible sources of information are not only more likely to be heard by legislators or government officials; their concerns are also more likely to be acted upon. Headline-grabbing techniques that might be effective in raising public attention to problems also tend to diminish credibility in the eyes of decision makers, and might not appreciably help conservation efforts. Of course, neither citizens' access to information nor having credible concerns will improve conservation if industry or governmental officials are not accountable to the public.

Because awareness about the threats to marine life are so new to most people in governments and NGOs, marine conservation advocates often can make major gains simply by educating officials and the public. At the same time, marine conservationists should recognize that their well-developed sense of concern may not be quickly understood, accepted, and adopted by others. Patience, creativity, and an ability to link marine conservation concerns with those of potential allies can bring substantial achievement.

Expanding the Knowledge Base

The marine realm is at risk because the people who affect it do not understand its values and vulnerabilities. The general public needs enough knowledge to ensure that the cumulative effects of human activities will not overwhelm the capacity of marine systems to perpetuate themselves. Natural and social scientists need to undertake research and monitoring to gain deeper and more comprehensive understanding of patterns and processes affecting marine conservation. Ultimately, the health of the sea depends on decision makers' knowledge of ways to ensure sustainable use and protection of marine biological diversity.

Education

Conserving marine biological diversity is a difficult subject because it requires people to appreciate resources that are not easily seen (most are underwater) and that are little known (fishes, invertebrates, and seaweeds, which constitute the vast majority of marine species, are not as easy for people to relate to as tetrapod vertebrates), as well as comprehend ideas that are not often discussed (ecology and evolution). This makes effective education an especially important part of any marine conservation strategy. Three segments of society must be informed before substantial change can occur: the general public, the media, and decision makers in industry, government, NGOs, and funding agencies. Education, whether formal or informal, that allows people to make informed decisions, is essential for conserving life in the sea. Perhaps the clearest and most powerful statement of this is from the Senegalese ecologist Baba Dioum, who said at the IUCN General Assembly in 1968: "In the end we will conserve only what we love; we will love only what we understand; and we will understand only what we are taught."

The General Public

Whether the worldwide movement to protect the environment succeeds or fails depends on the active involvement of informed citizens. So valid is the adage "Knowledge is power" that virtually everyone trying to do something against the interests of the general public withholds information to deny power to citizens. Processes that increase understanding among citizens in general, and among opinion shapers (e.g., educators, the clergy, entertainers, sports figures), are therefore crucial to citizen empowerment. That is why education is indispensable as a defense against environmental degradation in the sea (Hounshell and Madrazo 1990) and elsewhere.

Citizens can learn about the marine environment in school, from the media, or from NGOs. In traditional classrooms, teachers can create informed citizens by teaching themes about the processes of ecology and evolution, rather than emphasizing the memorization of many isolated facts (National Research Council 1990c). If people see the world as a complex jigsaw puzzle with every piece contributing to the integrity of the whole, they will be motivated and able to act more effectively.

Movements to save tropical forests that used education to mobilize citizen action have produced significant progress in countries as diverse as India, Indonesia, Australia, Costa Rica, Brazil, Germany, and Kenya. Education was crucial to efforts to protect the great whales. Citizens who understand and care about marine species and ecosystems cannot easily be misled.

The Electronic and Print Media

A crucial group of opinion shapers deserving special mention is people in the media. If journalists do not perceive marine biodiversity as important, they will not give appropriate coverage to stories about its conservation. Reporters and editors who see the ocean as a large aquarium containing a few interesting creatures will present it as such, and marine conservation will be perceived as an amusing luxury, but not as important as, say, unemployment or sports. Education of journalists is thus an essential part of any conservationist's job. One television newscaster can reach more citizens than any mass mailing.

Decision Makers

Effective governance depends on decision makers' having accurate, up-to-date information. Moreover, in democratic and even in most authoritarian nations, industry leaders and government officials ultimately respond to public pressure, so it behooves them to anticipate the wishes of the citizenry. Leaders who understand how protecting the sea is essential to its sustainable use can make wise decisions that improve the well-being of the people, and are likely to find useful allies in the environmental movement. Education measures targeted specifically at decision makers can be very cost-effective means of ensuring conservation of marine biological diversity.

Science

Natural and social scientists have the pivotal role of conducting and interpreting studies on which the public and decision makers depend. Without this knowledge, societies can make sound decisions only by relying on traditional knowledge, on intuition, or on chance. Key scientific tools include: inventory and research, monitoring, information transfer among scientists and to decision makers and the public, and training.

Inventory and Research

In traditional societies, individuals passed on their beliefs about the natural world to their kin. Peoples whose understanding most closely matched the workings of nature could persist through good times and bad. The rest tended to disappear along with their beliefs. Today, traditional knowledge, including traditional knowledge about the sea, has largely been devalued and discarded. Marine scientific research—the organized, objective study of the sea, its biota, and the people who affect and are affected by them—has become the *sine qua non* of sustainable use and protection of the sea. And although many fishers still rely to a considerable degree on traditional understanding, personal observation, and intuition, people who govern activities in the marine realm—whether national leaders or hands-on managers—increasingly depend on the findings of scientific researchers, as do a growing number of decision makers in NGOs.

It is obvious that people whose decisions affect marine life and their uses need to understand what species and ecosystems occur in their coastal waters, where an ocean current comes from, what a fish species feeds on, how the breeding biology of a clam affects its distribution, and how pollutants affect species. These are the kinds of knowledge that come from natural sciences such as taxonomy, physical oceanography, fisheries biology, ecology, and ecotoxicology. Lacking the body of knowledge held by many people in traditional societies, decision makers clearly need the best available scientific information in these and related fields.

Oddly enough, it is less obvious to some decision makers how conservation and management of the sea are affected by understanding ways that individual people make choices, how groups deal with ownership and use of resources, or how nations structure themselves in response to foreign criticism. But social sciences, including social psychology, anthropology, and political science, provide crucial insights about the behavior of our own species, which is both the beneficiary of and the threat to marine biological diversity. Research in both the natural and social sciences is integral to decision making about conserving life in today's complex world (Stern et al. 1992).

Sciences evolve, continually changing in response to new developments, and the population of researchers grows or shrinks in response. Sciences can stagnate, becoming occupied largely with minor refinements of old paradigms, or they can undergo dramatic bursts of creativity. A long-established science that is essential to inventorying marine areas is taxonomy; renewing this vital field will have both short-term and long-term payoffs. Three other scientific disciplines that could contribute a great deal to the future of marine biodiversity—marine conservation biology, marine landscape ecology, and marine restoration ecology—do not really exist yet as coherent bodies of theory and knowledge. But their synthetic, interdisciplinary approaches hold promise for facing the challenges of conserving marine species and ecosystems in a world increasingly stressed by human activities.

The health of a science, which determines how effectively researchers can provide needed information to decision makers, depends on the training of researchers, the facilities and funding available to them, and their status in society. Research in some of the marine sciences, such as physical oceanography, relies largely on expensive facilities such as satellites and oceangoing ships, and tends to be carried out by large teams of researchers and support workers. In others, such as taxonomy, research is largely carried out by single individuals, and requires much smaller expenditures for equipment, storage of collections, and support staff, although these are needed throughout the scientist's career. Both approaches are valid and provide essential information; indeed, they are complementary. But recent years have seen far more resources devoted to large scientific projects, at the expense of "small science." Large projects can provide huge amounts of data, some of it of great value to decision making, but most scientific innovation comes from individual researchers or small teams.

Neither natural nor social sciences are untouched by geopolitical rivalries, but scientific studies can bring together researchers who would not be allowed to cooperate in other areas. In the last few years, biologists from Cuba and the USA have cooperated on surveying the rich diversity of Cuban endemic species; scientists from several Gulf Arab nations, Iran, and the USA have worked together on the impacts of the massive 1991 oil spills in the Persian/Arabian Gulf. In addition to providing important information to decision makers, these researchers are pioneers in bridging geopolitical gaps that have impeded progress in marine conservation and many other areas.

Monitoring

Conserving marine biological diversity requires a wide range of scientific activities, from fundamental biological inventories and ecological research to observations to gauge the progress of management plans. Marine environmental monitoring—the continuing observation of conditions over time—is a crucial tool in this continuum, for it can provide important information that managers need so that they can make timely, sound decisions. Monitoring can show whether variations observed in marine ecosystems are natural phenomena or anthropogenic impacts. It can reveal trends that affect the integrity of ecosystems and the prospects for their sustainable use, and can provide early warning of impending problems. And it can show managers the geographic pattern of processes of concern, from wetlands loss to shrimp recruitment.

Although some large-scale marine environmental monitoring programs exist, none have been designed explicitly to observe marine biodiversity. In the USA, for example, neither the National Oceanic and Atmospheric Administration (NOAA)'s National Status and Trends Program, which has measured toxic contaminants in fishes, bivalves, and sediments since 1983, nor the Environmental Protection Agency's more recent Environmental Monitoring and Assessment Program (EMAP) focuses on marine biodiversity *per se*. Clearly, monitoring that takes "snapshots" can be useful, but it is not sufficient.

If the objective is to measure changes in marine biodiversity, the focus must be on dynamic processes, including changes in the genetics of marine populations and interactions within and among marine ecosystems. Only an understanding of how marine biota distribute themselves in time and space, and how species respond to change, interact, and affect ecosystem processes, will allow managers to monitor and interpret changes in these patterns and choose among options to affect them. Until we have this level of understanding and monitoring in place to provide timely information to decision makers, management decisions should err on the side of conservation.

Anthropogenic change is superimposed on natural changes that occur on all scales of space and time. The scientific community's uncertainty about whether the exceptionally warm years of the last decade were natural occurrences or results of anthropogenic increases in atmospheric greenhouse gases is largely due to insufficient duration and geographic coverage of monitoring stations. Monitoring over a wide geographic scale is vitally important because the oceans are all connected; what affects one area might well affect the rest. Monitoring over the long term is vitally important because it provides a moving picture with the power to explain what we observe rather than a snapshot that provides too little context to explain anything. Programs that link existing, local monitoring programs into a network that provides a broader geographic picture are a promising development for those who need to understand how local conditions relate to the "big picture."

Information Transfer

Information from scientific inventory, research, and monitoring allows people to make informed choices among options for management. Often, needed information does not exist. No less often, it does exist but is scattered and not available to the people who need it most. Hence, better management and distribution of existing information are among the most cost-effective ways to improve marine conservation.

The systematic compilation of existing information accomplishes two goals. First, it provides a readily accessible source for that knowledge. Second, it provides a mechanism to identify actual, rather than perceived, gaps in the knowledge base, which can then be filled through new inventory, research, or monitoring. Eliminating duplication can free funds for truly needed studies.

The network of natural heritage programs and conservation data centers that The Nature Conservancy has helped to establish throughout the Western Hemisphere illustrates this approach. These centers serve as biodiversity information clearinghouses that provide a bridge between the research community and government agencies and conservation organizations charged with resource management decisions. The centers compile and analyze existing information, identify gaps in available information and help to direct research toward filling these needs, and disseminate information to an array of users. To date, these centers have focused exclusively on terrestrial and wetland ecosystems. As a result of increasing awareness of the importance of marine biodiversity and the need for solid scientific information on which to base marine conservation decisions, efforts are under way to broaden the scope of several of these data centers to include marine environments.

The CARICOMP program is another regional network for information exchange. This collaborative effort of Caribbean marine laboratories will monitor ecosystem change at selected sites and provide mechanisms for exchange of data. Such regional approaches provide decision makers with information on areas larger than local or national programs alone could provide. They can alert scientists to phenomena that affect an entire biogeographic region, such as the epidemic that devastated *Diadema* sea urchins throughout the Caribbean in the early 1980s (Lessios et al. 1984).

Training

The ability to conserve biological diversity depends on people who have gained special expertise, usually in advanced university programs. Because the number of experts is far fewer than is needed worldwide, and especially in developing countries, training is crucial. Yet taxonomy, the description and classification of diverse living things, has long been in desperate trouble (National Research Council 1980; Norse 1987). Worldwide, there are probably no more than 1,500 trained professional systematists who are competent to deal with tropical organisms (Wilson 1992), and the great majority of these few work on nonmarine species. Although taxonomy is one of the most important fields relevant to marine conservation, the training of new researchers has all but ceased, and many of the most expert taxonomists are nearing retirement (Figure 7-1), which bodes ill for efforts to inventory the diversity of marine life. The situation is scarcely better for marine biogeographers and ecologists, whose knowledge is essential to help decision makers resolve most kinds of marine environment/development conflicts. There is also a special need to train people in emerging disciplines that can provide crucial information to decision makers, including marine conservation biology, marine landscape ecology, marine restoration ecology, and ecological economics.

Figure 7-1. Taxonomist, a vanishing species. *If taxonomists—the scientists who describe and classify species—were a species, they would be classified as endangered. There are only about 1,500 worldwide, many are near retirement, and very few new ones are entering the population. Taxonomists are crucial for discerning patterns of species diversity.*

Having a pool of trained people not only provides decision makers with information and judgment; it also creates effective advocates for conservation, who are essential wherever people are unconvinced about the value of biological diversity. In some nations, a single well-trained, well-respected individual in academia, government, or a conservation organization who understands and is concerned about living things can be the difference between environmentally ruinous and environmentally sound economic activities.

In many developing nations, graduate training in marine biology or related fields at a good research university is a luxury that few can afford. But there are other ways to build needed expertise. Costa Rica's National Biodiversity Institute or InBio (Gámez 1991) has developed a biodiversity inventory program that selects intelligent, highly motivated people, some of whom lack even a high-school education, for training as parataxonomists by Costa Rican and visiting experts. Parataxonomists are to biological diversity what China's acclaimed "barefoot doctors" are to medicine. Even without advanced university training or sophisticated equipment, they can function independently and effectively. They can identify most organisms they collect, and know which taxonomic experts to contact to identify the remainder. Costa Rica's innovative way of training experts could be emulated in many nations worldwide, including those in industrialized nations where taxonomists are too few to assist in existing and planned inventory efforts.

Planning

Experience in marine conservation efforts (and most other endeavors) worldwide has shown that conscious planning has important advantages over reacting to change as it occurs, or afterward. Four particularly potent kinds of planning are environmental impact assessment, action plans, large marine ecosystem management, and integrated area management.

Environmental Impact Assessment

Environmental impact assessment (EIA) is a process used to evaluate the environmental consequences of proposed programs and individual projects. Through EIA, potential environmental problems that affect people, wildlife, and entire ecosystems can be identified and addressed.

The use of EIA began with the passage of the National Environmental Policy Act (NEPA) in the USA in 1969. More than two decades of experience in implementing NEPA has demonstrated that it has a significant impact on government decision making. During this period, EIA processes have been adopted by many other countries and multilateral institutions. Among the useful guidebooks on EIA are ones issued or sponsored by the UN Environment Programme (UNEP) (Ahmad and Sammy 1985), the Asian Development Bank (Carpenter and Maragos 1989), the World Bank (World Bank Environment Department 1991), and the US agency that has overseen NEPA (Council on Environmental Quality 1993).

Although details of the process differ in each country, certain concepts have become basic to EIA. They include:

1) establishing the scope of the assessment in order to determine issues of true significance;
2) identifying and predicting the project's impacts;
3) analyzing alternative methods of achieving the goal of the project;
4) developing measures to mitigate the project's impacts;
5) providing for participation by the public and government agencies throughout the process;
6) documenting and considering the environmental analysis by the decision maker; and
7) implementing measures to monitor the project's impacts and the effectiveness of mitigation methods.

The EIA process can lead to dramatic improvements in a project's design and planning, from both environmental and economic viewpoints. But to be effective, an EIA process must be used *as early as possible in the planning of the project*. If EIA is commenced after a decision has been tentatively reached, it will have little impact on the project. However, it is important to ensure that the EIA process can be reopened if new findings fundamentally alter the assessment of impacts.

The EIA process should take a holistic, interdisciplinary approach that considers impacts from both natural and anthropogenic causes, and how the proposed action affects the environment, economy, society, and their interactions. A method of ensuring, whether through citizen suits or the administrative process, that the EIA process is used in all appropriate situations is also of vital importance.

One important element that has not been required in the EIA process is follow-up analysis of EIAs to determine what impacts actually occurred. The process can become a far more effective tool if such retrospective studies are used to improved prediction in future EIAs.

The extraordinary effectiveness of the EIA process does not come from a requirement that the least environmentally harmful alternative be chosen. Rather, candid, comprehensive analysis of the impacts of alternatives, and participation by the spectrum of affected people, result in better decisions.

Action Plans

Action plans for conserving marine biological diversity can be powerful tools for both catalyzing and coordinating marine conservation actions. Identifying those actions that are most important in a particular region or for a particular group of species promotes the most effective and coordinated use of conservation resources. By presenting priorities for action in terms of actual projects, and by calling attention to the most critical problems, action plans can help attract funding and other support to conservation efforts.

Action plans should represent the consensus of a broad-based group of scientists, resource managers, and conservationists on the most important priorities for action. The plans should be prepared in close collaboration with organizations and individuals who are most likely to implement the proposed actions; action plans that are "handed down from above" cannot be effective in the long run. Moreover, plans should present enough background information to demonstrate that they include the most important actions.

The Species Survival Commission (SSC) of IUCN is a network of 4,000 scientists and conservationists from around the world that has played a leading role in developing action plans for conserving biological diversity in terrestrial, freshwater, and marine realms. SSC includes about 100 specialist groups, mostly arranged along taxonomic and regional lines. Marine-oriented specialist groups include those for cetaceans, sea cows, pinnipeds, sea turtles, coral reef fishes, and sharks.

The Cetacean Action Plan (Perrin 1989) is an example of how such a plan can guide and promote conservation action. Focusing particularly on the problems facing small cetaceans, the plan outlines 45 priority projects and lists another 39 important issues for cetacean conservation. Within five years, IUCN's Cetacean Specialist Group expects to see nearly all of the 45 priority projects completed or under way, thus significantly increasing conservation actions on behalf of this group of species. As testimony to the need for such plans, demand for the plan was so high that an updated reprint was required within a few months of publication.

Similarly, UNEP's Action Plan for the Protection and Development of the Mediterranean Basin led to the formation of a master plan known as the Mediterranean Pollution Monitoring and Research Programme (Med Pol), a cooperative effort by some 84 institutions in 15 countries. It initiated baseline studies, monitored estuaries, offshore areas, and atmospheric transport of pollutants, and surveyed pollutant sources to help in the drafting of the Protocol for the Protection of the Mediterranean Sea against Pollution from Land-Based Sources, which entered in force in 1983 and was ratified by eight nations and the European Community (EC) (Jeftig 1987).

Large Marine Ecosystem Management

Nations have usually reacted to threats to marine resources with an *ad hoc* approach, treating the visible symptoms of mismanagement while the underlying causes are ignored. But a concept developed in the last decade, the large marine ecosystem (LME) approach, offers nations a holistic, proactive way to deal with the threats to the sea.

Sherman and Gold (1990) define LMEs as large regions within EEZs, generally over 200,000 square kilometers (77,000 square miles), that have unique bathymetry, hydrography, and productivity, and within which biological populations have adapted reproductive, growth, and feeding strategies. The greatest strength of the LME approach is that by encompassing a functional ecological unit, it includes whole stocks, their prey, predators, and the physical factors that affect their survival. Managing this unit rather than managing fish stocks one by one encourages scientists to undertake comprehensive ecological studies, which, in turn, can provide managers with a more comprehensive understanding of ways to manage species and ecosystem

As a kind of ecosystem management, the LME approach could be strengthened further in several ways:

1) Although managing larger areas is often essential to encompass functional ecological units, not all important ecosystems are large. Under the above LME definition, Georges Bank and many other isolated banks, seamounts, atoll lagoons, and bays would not qualify. Ecosystem management can increase the effectiveness and, often, decrease the costs of managing ecological units on any scale.

2) Another problem involves borders. LMEs rarely fit neatly within a single nation's waters. The Caribbean Sea system, for example, is shared by 38 nations, the North Sea by 20. Similarly, the actual ecosystems that LMEs seek to manage do not always fit within EEZs. The fact that ecosystem borders seldom coincide with political borders can compromise the implementation of comprehensive management regimes.

3) The focus of LMEs as currently depicted is on biomass yields. Whereas the production of fishery resources is unquestionably a vitally important function of marine ecosystems, the sea is not simply a fish production

factory. There are many other components and services of marine biological diversity that merit greater focus.

4) Finally, although much, if not all, of the sea or its EEZs could be studied and managed as LMEs, many such areas are not. Ecosystem management in the sea will become more effective when more areas come under LME treatment.

The trend toward comprehensive ecosystem management started early in the century with the creation of the International Council for the Exploration of the Sea (ICES), which advocated the establishment of international multidisciplinary studies as a means of improving fisheries management (Sherman 1990). The 1972 UN Conference on the Human Environment stressed the inextricable connection between environmental quality and resource management (Smith 1984). In recent years, international treaties have increasingly recognized the need to maintain the integrity of ecosystems and the responsibility of nations not only for their own waters, but for their effects on those of their neighbors. The 1982 UNCLOS III requires nations to minimize pollution and manage fisheries within their own EEZs, and makes them responsible for activities that affect adjacent EEZs. The Convention on the Conservation of Antarctic Marine Living Resources (CCAMLR) requires a comprehensive ecosystem approach to the management of Antarctic resources.

Only 23 of the sea's functional ecosystems are currently under study as LMEs (Sherman and Gold 1990). In some of them, such as the Kuroshio, Humboldt, and Benguela current LMEs, large-scale natural environmental changes such as circulation, temperature, and productivity drive biomass yields. Overfishing is the primary factor determining biomass yields in the Gulf of Thailand, the Yellow Sea, and the Northeast US Continental Shelf LMEs, while pollution is now the driving force in the Baltic Sea. In other cases, the link between human activities is present but less obvious. For example, the Antarctic ecosystem depend mainly on the status of krill stocks, which in turn depend on large-scale changes in water movement, but also on human impacts such as stratospheric ozone depletion (SC-CAMLR-VII 1988).

Many of the current problems in marine conservation result from the fragmented, *ad hoc*, reactive approach to problem solving that prevails worldwide. As human influence on the environment increases, it becomes increasingly clear that the comprehensive, proactive research and management that are the aim of the LME approach offer hope for the protection and sustainable use of the sea in ways that the *ad hoc* approach cannot.

Integrated Area Management

Environmental quality in marine and coastal areas has been reduced so much that the damage in many areas is readily observable. Governments' traditional response—to the extent that they have responded—has been to treat each economic or social sector's activities separately. Thus, there might be one law and set of regulations that deal with commercial fishing, another for pollution from factories, and still another for conversion of mangrove forests to build tourist hotels, all administered by different agencies. This sector-by-sector approach produces piecemeal decisions that are often at cross-purposes, as each agency applies its particular tools to the problem or protects its constituency by allowing it to externalize the costs of its activities. In the absence of some authority whose express purpose is the protection, study, and sustainable use of marine or coastal areas, nations will continue to lose vitally important marine and coastal resources, and options for rational, ecologically sustainable use of resources will further diminish.

An alternative to doing nothing or dealing with problems *ad hoc* is integrated area management (IAM), an approach that applies a coherent set of resource management policies across sectors to particular coastal or marine areas so that management is coordinated. An area under IAM usually has multiple purposes. It might, for example, provide sites for:

1) scientific research;
2) monitoring as a baseline against which to measure ecological changes caused by human activity;
3) protection of sites (such as nursery grounds) that are essential to life stages of commercially or recreationally fished species;
4) commercial activities such as housing, tourism, or fishing; and
5) public appreciation and enjoyment of relatively undisturbed marine environments.

One of the keys to IAM is ensuring that all uses and activities in (or affecting) the area in question are coordinated according to an agreed set of policies. In practice, this requires a coordinating mechanism that applies policies to the actions of individuals and institutions, both public and private, that can affect the management area and its resources. It can be an inter-ministerial council or commission, a process operated by an existing national agency, a local government operating under mandate of a national government, or a special governmental "authority" with representatives from federal, state, and local governments, user groups, traditional and indigenous peoples, and community interests.

Typically, IAM is undertaken as a part of a national or local coastal zone management (CZM) program or a national system of marine protected areas, or any substantial-sized area. The managing authority encourages the various interest groups to agree on what kinds of protection or use should prevail in particular parts of the management area. Preparation of an environmental impact assessment for all activities or projects affecting the area in question is often an additional integrating mechanism.

Because it is difficult for organizations and individuals to attempt simultaneously to achieve two or more goals—particularly economic development and ecological protection—the authority should not be responsible for detailed management of individual sectoral activities, such as tourism or fisheries. Specialist agencies should continue to manage such activities. The authority is best suited to the following functions:

1) developing, in association with interested groups and the community in general, a strategic plan for the coastal zone;
2) overseeing coastal development to ensure that it is ecologically sustainable;
3) designing and managing baseline inventory studies and comprehensive monitoring programs to define the state of the marine and coastal environments and the trends in environmental parameters;
4) designing and managing contracted, multidisciplinary ecological research programs aimed at solving environmental problems; and
5) designing and implementing comprehensive community involvement and education programs designed to achieve community support for policies, programs, and actions that will lead to ecologically sustainable development. Particular emphasis should be placed on educating young people.

Thus, while existing agencies carry out specific management programs, the authority concentrates on policy, strategy, planning, design, and supervision of research programs, and coordination. The enabling legislation should override conflicting provisions of existing legislation.

IAM allows ecologically sustainable use, study, and protection of coastal and marine areas. Without it, the energies of people and governments will continue to be dissipated in intersectoral conflicts, incompatible activities, and research that fails to fulfill human needs.

Regulating Threats

To help resolve conflicts over the use of marine resources among different sectors of society, many governments make broad use of regulation. Limits on the species, tonnage, and timing of the fish catch, rules about the way materials are transported at sea, standards for discharge of pollutants entering the sea, specifications for dock and marina construction, management policies on public lands in coastal areas, and limitations on flood insurance in low-lying coastal lands are all kinds of regulation that can be used for conserving marine species and ecosystems.

Regulation is used internationally as well as within nations. The fundamental principle that nations must conduct their activities without harming the environment of other nations was first articulated in 1941 by the International Court of Justice and was codified by the United Nations (1972) in the Stockholm Declaration on the Human Environment. Ten years later, the United Nations (1982) added conservation of biological diversity to the concept of environmental protection when it adopted the World Charter of Nature, emphasizing that natural resources must be used "in a manner which ensures the preservation of the species and ecosystems for the benefit of present and future generations."

Supplementing these general international regulatory agreements are specific agreements on matters such as reasonable and equitable use of shared natural resources, a nation's responsibility for causing injury to another's environment, the obligation to cooperate in preventing transboundary pollution, and the duty to compensate for transboundary harm caused by authorized activities (Magraw 1990), as well as limits on pollution from ships and protection of the ozone layer.

Regulation is most likely to fail when it is carried out piecemeal. Regulatory agencies that coordinate activities such as dredging, pollutant discharge, vessel traffic, and the taking of marine species minimize the likelihood that they will interfere with one another and thus diminish their effectiveness in conserving marine biodiversity. Coordinated regulation, however, is not the rule within nations or internationally. For example, in the USA, the Office of Technology Assessment (1987) estimates that at least 28 federal laws relate specifically to maintaining diversity, and scores more to preventing harm to ecosystems. These authorities are enforced by a host of different agencies for which coordination is, regrettably, a low priority.

Moreover, regulation, whether on an international, national, or local level, is no more effective than its implementation. Many sound regulations fail to achieve their objective because the public has not been educated about the need for regulation, or the regulations are not enforced consistently and diligently. The following are examples of regulatory authority available to local and national governments and to the world community.

Taking and Trade Controls

Controls on taking can be used to limit the directed or incidental kill of a species or stop it altogether. CITES allows the imposition of bans against the export of listed species to any signatory nations in order to diminish the economic incentives for continued taking. The InterAmerican Tropical Tuna Commission (IATTC) monitors dolphin mortality associated with the setting of tuna purse seine nets deliberately around dolphins, and provides this information to nations seeking to lessen bycatch of dolphins. Chapter 8 provides other examples of catch control treaties.

Controls on the Introduction of Alien Species

Controls on imports on nonnative marine species can range from regulations on what can and cannot be introduced intentionally, such as regulations on importation of species used in mariculture, to regulations designed to prevent accidental introductions. The enormous economic damage from species such as zebra mussels stimulated the USA to limit discharge of ballast water from vessels transporting organisms into US waters. In 1990 the U.S. Congress passed the Non-Indigenous Species Act, calling for a study of US shipping practices to learn how alien species are introduced and how to prevent the introduction of those species, and various US states are considering regulations requiring ships to dump ballast water far at sea, not in bays or estuaries.

Pollution Controls

International consensus affirms nations' obligation to protect and preserve the marine environment. As an example, UNCLOS III directs nations to take measures to prevent, reduce, and control pollution of the marine environment from any source, and to use the best practicable means available to do so, in accordance with their capabilities (United Nations 1983). Examples of direct controls on ocean dumping include the London Dumping Convention and the Oslo Convention, which prohibit the dumping of wastes unless a permit is obtained from the appropriate authority. The 1974 Helsinki Convention on the Baltic Sea covers several pollution sources and contains provisions on land-based sources of pollution, ships' dumping, and an agreement to limit pollution from exploration and exploitation of seabed resources.

A comprehensive set of regional agreements negotiated under UNEP's Regional Seas Programme focuses on areas that are significantly vulnerable to degradation, providing a framework for specific regional protocols on a variety of issues. For example, the 1983 Caribbean Convention includes multilateral agreements to reduce and control pollution from ships, dumping, coastal disposal, and discharges from internal waters in order to ensure that shared waters are protected.

Shipping Safety and Transport Controls

Some of the earliest international endeavors to protect the sea arose in the context of shipping safety. The International Maritime Organization (IMO) brings together government officials, transportation experts, shippers, and insurers on a regular basis to formulate agreements on shipping controls. These controls range from vessel safety measures, such as requirements for double hulls on tankers, to rules for how ships are to dispose of wastes at sea. An example of this type of regulation is the MARPOL Convention, which governs the dumping of garbage, sewage, oil, chemical, and hazardous wastes (United Nations 1973). Analyses of nations' shipping safety (Townsend 1990) can be potent motivating forces for improving vessel design and operations.

In the wake of the disastrous oil spill from the *Exxon Valdez*, the U.S. Congress updated its oil pollution law to raise the limitations on liability for damages resulting from oil spills. This aims to create a financial incentive powerful enough to inspire prevention, rather than payment, for spills as another "cost of doing business."

Trade and Tax Authorities

A body of law previously missing from the toolkit of environmental regulation, but gaining attention as a way to control activities that threaten biological diversity, are laws on trade and taxation. This kind of regulation can work effectively by using market forces to encourage environmentally sound behavior by industries and nations. The oil pollution example above is one mechanism to encourage prevention by making the polluter pay. Less direct, but perhaps with more long-term potential, is linking the cost of doing business in an environmentally sound manner with trade and tax controls.

One way of doing this is to create incentives for "clean" products, including the transfer of technology to developing nations as a means to harmonize at a higher standard of environmental protection (Stavins 1990). Although modifications would be needed, the General Agreement on Tariffs and Trade (GATT) is another potentially powerful tool for reducing threats to marine biological diversity. Unfortunately, at present, GATT does not allow for consideration of the effects of trade on the environment. Depletion of common resources or the higher costs of producing environmentally sound products may not be used by one nation to confer differential trade treatment to the products of another nation.

More troubling, however, are negotiations during the most recent round of talks that would extend GATT to cover "technical barriers" to trade. These include not only national laws and regulations that control how goods are produced, but also laws and rules that set product safety standards, workplace safety standards, consumer protection labeling, and environmental quality standards. If these aspects of the GATT treaty are extended, it could mean the international standard for environmental protection will be the lowest common denominator: Any more stringent national environmental policies would be struck down as barriers to trade. Instead of "harmonizing" environmental protection standards *downward*, GATT should be a vehicle to *improve* standards worldwide, and to assist developing countries in building environmentally sound economies.

The opportunity to consider new approaches and to blend the existing mix of authorities into a global legal context for conserving biological diversity could arise as a result of impetus provided by UNCED. As stated by Professor Louis Sohn (1992), a preeminent scholar on international environmental law:

> The task will not be easy . . . and many difficult compromises and ingenious solutions will be necessary. But the reward will be great— to discharge nobly our responsibility to the next generation—to leave them not a devastated planet but a flourishing green earth.

Economic Tools

Although institutions make some decisions that are not economically based, economics is the dominant force in international bodies', governments', and businesses' decision making. For example, economics ministries hold far more sway than environmental ministries in virtually all nations. Sadly, the workings of existing economic systems tend to devalue and degrade the environment, including the sea. To maintain the diversity and integrity of the sea, humankind must devise economic systems that favor environmental protection and sustainable use. Three economic tools that can help in this process are establishing usage rights, assigning value to non-market resources, and using market forces.

Establishing Usage Rights

Although there are many ways to protect coastal and marine ecosystems, their viability often hinges on fundamental questions of ownership and jurisdiction. In Western legal doctrine, marine resources have been treated differently from their terrestrial counterparts. Classical Roman law held that use of the sea was "common to all," and that marine fishes could be brought under ownership by any person through the act of possession (Johnston 1965). From the 17th century onward, the notion of "freedom of the seas" shaped naval politics in Western European nations and in their colonies.

In the last two decades, there have been profound changes in jurisdiction over the sea. The concept of extending jurisdiction farther into the ocean gained momentum in the early 1970s and was formalized by UN-CLOS III. The Law of the Sea, as it stands now, recognizes territorial seas of 12 nautical miles (22 kilometers) and EEZs of 200 nautical miles (371 kilometers), yet the extent of territorial seas remains controversial because some nations claim territorial seas of up to 200 nautical miles. Similarly, EEZs also range up to 400 nautical miles (741 kilometers). Currently, EEZs account for about one-third of the area of the world's oceans (Kidron and Segal 1987). Unless jurisdictional boundaries stabilize at 200 miles, it is possible that in the foreseeable future, all the oceans' resources could be placed under the jurisdiction of coastal nations (Personal communication with William Burke, School of Law, University of Washington, Seattle, Washington, USA). This, presumably, would end the current problem of unregulated high-seas fisheries.

Although division of the sea into EEZs might seem to be a useful way to end the tragedy of the commons, the actual record is not impressive. Quite often, the creation of EEZs has meant that access to waters formerly open to anyone in the world is now restricted *only* to anyone in the coastal nation. In many cases, distant-water fishing fleets still plunder marine resources, but now pay a fee to do so. This revenue is welcomed by governments of developing countries, but is rarely reinvested to protect living marine resources. Although, in the short run, coastal nations may lack the capital to replace foreign fleets with their own fleets, it is only a question of time until they do. Indeed, it might be more difficult for nations to control the activities of their domestic fleets than those of foreign fleets that operate under tighter scrutiny.

Since Hardin (1968) alerted us to some harmful effects of common property, there have been many schemes to lessen damage from unrestricted use of free, common resources. Nations attempt to use limited entry, gear restrictions, seasonal closures, and simple licensing procedures to protect resources, but, regrettably, without much success. Many of these procedures are easily circumvented because they do not deal with the inherent flaw of the commons: the economic disincentive for conservation (Wieland 1992). In some cases, a remedy can be some kind of "owning" the resources. The terms ownership, property rights, right-based resource utilization, harvest privilege, harvest rights, among others, are often used interchangeably. Technically, the first two imply a stationary resource and the possession of a title clearly identifying the proprietor. The last three are more correctly applied to more fleeting resources, such as a portion of a migratory fish stock. These can differ in degree as well as substance. This discussion employs the general term "usage rights" to mean all of them, whether the extent of title covers a right or a privilege.

For areas in which resources are exploited, one management scheme that might help to prevent overexploitation is the individual fishing quota (IFQ) system. In an IFQ system, the fisher has exclusive rights, indefinitely or for a certain period, to a portion of a fish stock. When and how the allotment is taken is up to the fisher. The IFQ system has been widely used in New Zealand's inshore fisheries. IFQs are also used in Australia's southern bluefin tuna fishery, in Canada's Pacific halibut (*Hippoglossus stenolepis*) fishery, and in the Atlantic wreckfish (*Polyprion americanus*), ocean quahog clam (*Arctica islandica*), and surf clam (*Spisula solidissima*) fisheries in the USA. Because these efforts are relatively new, their effects on marine biological divesity are still being evaluated, but they hold promise as a fishery management tool.

At its best, an IFQ system can offer benefits to fishers (less dangerous working conditions, higher price for better-quality products, better use of time), to seafood consumers (fresh fish year-long instead of during short openings, better-quality products), and to the environment (less damage to the stocks and to nontarget species to the extent that a more leisurely pace allows the fisheries to be more directed). But as close as they get, IFQs are *not* private property. Because fishers have the right to use some portion (but no particular portion) of the stock, even this system can be circumvented if the fishers high-grade, that is, discard any fish that will bring less than the highest price, or if they sell outside of the observed market, or otherwise circumvent quotas.

There is precedent for the success of usage rights in some traditional societies that placed the resources and livelihood of a community under the authority of individual owners, the village, or a spiritual power that prevented the overexploitation of lagoons, estuaries, bays, or other semi-enclosed bodies. Panayotou (1982) describes a lagoon fishery in Ivory Coast in which the tribes that operate the fishery did not permit neighboring fishers with purse seines to enter, and did not switch to mobile gear themselves. Consequently, local fishers report no surplus of labor, substantial numbers of larger fish, and higher income than those in other lagoons where mobile gear was introduced and no territorial rights were claimed. Sadly, these systems are now being eroded by market forces and the encroachment of larger-scale operations (Troadec and Christy 1990).

Artisanal fisheries, as important as they are to the welfare of people in coastal communities, can also be destructive, especially when they operate in breeding or nursery grounds or other sensitive areas. FAO and other agencies are investigating ways to restore traditional control over the artisanal fisheries. In Ghana, for example, fisherchiefs have made a comeback (Lawson and Robinson 1983). In Southeast Asia and the Caribbean, territorial user-right fisheries ("TURFs") are being experimented with in connection with fish aggregation devices (Christy 1982).

The creation of usage rights where they are at odds with customary ownership and resource rights can cause conflict about self-determination and redistribution of wealth. Avoiding such conflicts requires the involvement of the local population, using a "bottom up" approach rather than the usual "top down" approach used by multilateral aid programs.

Establishing usage rights can contribute to conservation objectives, but is one tool among many, not a panacea, and it does not work for all species. If the discount rate is high, it is economically "efficient" to eliminate a species with a slow rate of growth and reproduction (Clark 1973). While establishing a sense of property could make a significant contribution to sustainability of some living marine resources, it does not, by itself, ensure sound conservation.

Assigning Value to Non-market Resources

Because marine species and ecosystems, especially those that do not enter markets, traditionally have been undervalued, valuation—the attribution of monetary value to them—is becoming a powerful tool in marine conservation and an active research topic in ecological economics. It is increasingly used by agencies that deal with oil spill litigation and wetland mitigation when they need to assign a monetary value to, say, a nesting colony of crab plovers (*Dromas ardeola*) or the seagrass beds in an estuary.

The first step of valuation is a biological inventory of species in a community. Among them might be many species important to tourism and to subsistence, commercial, and recreational fisheries, and those that are important components of ecosystems whose services benefit humans. Some values are transacted on a market, such as fish caught by commercial operations; others affect markets only indirectly, such as the species that are eaten by food fishes. In addition, humans value the existence of certain species or ecosystems independent of our use of them, and are sometimes willing to pay substantial amounts to protect them.

Various measurement techniques have been developed and refined in the last 20 years to calculate commercial, ecological, recreational, and existence values. Market values are both readily apparent and relatively easy to estimate and understand, although the true value of such resources, for the purpose of cost/benefit analyses, needs to include the multiplier effect and the externalized costs to the environment. Recreational values—especially in areas where tourism is important—although often overlooked, can be very important economically. For example, half of all damages from water pollution can be due to impacts on recreation (swimming, boating, fishing, viewing) (Freeman 1982).

Some ecological values can be estimated by knowing how they relate to commercial values. Ecological economists have begun to examine questions such as "What is the value of a coastal mangrove forest to a region's commercial fisheries?" Values can be established for recreational experiences using methods such as the *travel cost model*. It is based on the insight that the cost somebody must pay to visit a recreational site functions as the price of access to that site. Based on this premise, an economist can use travel costs to estimate how much an individual values visiting a site, even though there is no formal market for recreation experiences as there can be for commercially marketed species. Not surprisingly, how much people will pay to visit a site depends on the site's environmental quality. For example, the use of dynamite or poison fishing on coral reefs in some tropical areas that are also dependent on tourism diminishes the economic benefits for people who rely on visitors.

Alternatively, the recreational and other environmental amenity values can be estimated by using property values. *Hedonic models* can be used in any context in which one is trying to value the individual amenities that are tied together in one purchase. When buying coastal property, people are actually purchasing a variety of amenities—such as air and water quality in the neighborhood, pristine beaches, good fishing—along with a house. The hedonic method allows an economist to estimate the contribution of each of the environmental attributes to the total value.

To determine the existence value that people place on protecting endangered species, an economist can use a direct survey technique known as the *contingent valuation method* in which individuals are asked to value directly the existence of a species, such as dugongs.

Finally, how are the resources that have been valued affected by a particular activity? For example, how much does the addition of wastes from an electroplating factory diminish the value of an estuary that yields US$12 million annually in benefits? Answering such questions requires ecologists and economists to collaborate to strengthen linkages between ecological and economic models. In this case, they need to determine how the discharge affects marketed resources, other resources that affect marketed resources, and resources that humans value for other reasons.

Assessing the benefits of modifying human activities is difficult when the resources and their values are not well understood or quantified, which is a special problem in many developing nations, or when the relationships between the change in environmental quality and the human use are diffuse and distant. But understanding the values and vulnerabilities of marine resources can have uses beyond determining how much to regulate environmentally harmful activities. For example, it can be used to determine damages from accidents such as oil spills, which organizations or nations can use to claim compensation for losses. Ultimately, sustainable use of the sea depends on how much we value marine species and ecosystems in the face of the mounting threats to them.

Using Market Forces

Marine ecosystems and species often are treated as common property. The failure of a steward to manage a resource depreciates the value of everyone's property, sometimes to extinction. To provide incentives for managing a resource properly, governments can enact policies embracing some combination of penalties, such as harvest taxes, and rewards.

Regulating systems by using market incentives and disincentives can work in cases when prohibition does not. For example, a government might want a factory to control its discharge of nitrates into a coastal lagoon. The first step is to account for all externalities from the factory's activities, that is, the total cost of pollution to society, including the cost to generations to come and the cost to resources that were traditionally considered free. These can include increased health problems of people who eat shellfish from the polluted lagoon, decreased catches for local fishers, or decreased revenue from tourism. Once these costs are included in the price of the factory's cost of generating a product, the increased price will cause demand to decrease, which creates a strong economic incentive for the producer to eliminate pollution.

Pollution abatement and other ways of improving environmental quality are usually opposed on economic grounds, typically as concern about job loss or higher prices for consumers. But there are ways to make more environmentally sound practices pay. Users who can reduce their emissions or catches below the allowable level can sell the right to pollute or catch fish to others for whom lowering such costs would be too expensive. In such a system, those who can lower their environmental impact have an economic incentive to do so. This is a contrast to the regulation/technology control system most widely used, in which there are few or no economic incentives to lower environmental impacts.

Some marine ecosystems have special value for tourism or for scientific research. For example, the government of Ecuador reaps substantial economic benefits from tourism to the islands of Galápagos National Park. A government or private manager with exclusive rights to such a resource therefore has a strong economic incentive to protect it, assuming that overuse of the resource will not yield an even higher return that can be switched to other economic investments.

Economic incentives for conserving marine biological diversity can also operate internationally. Until the late 1980s, benefits from protecting biological diversity have tended to accrue to industrialized nations, while developing countries have paid the cost, by forsaking the economic benefits of activities that reduce biodiversity. But recent debt-for-nature swap arrangements, such as those that the governments of Costa Rica, Bolivia, Poland, and Madagascar have organized with the help of conservation organizations, provide economic return for nations that want to protect their species and ecosystems while reducing foreign debts that have impeded their economic progress. There is no reason that these arrangements cannot work for the marine realm as they have for tropical forests. Conserving the diversity of life in the sea calls for creative solutions that appeal to individual and national needs, as it does on land. In combination with other tools, the thoughtful use of market forces can play an important role in this process.

Protecting Areas

Marine protected areas (MPAs) are a crucial tool in almost any overall strategy for saving, studying, and sustainably using marine biological diversity. The development of MPAs and their conceptual framework have trailed their terrestrial counterparts by nearly a century (Lien and Graham 1985). Only within the past 20 years has the concept of protecting certain marine areas become widely accepted. Since then, however, this idea has taken hold and hundreds of marine protected areas have been established around the globe. As their number has increased rapidly, so too have their diversity and complexity (World Conservation Monitoring Centre 1990). MPAs vary dramatically in size, design, purpose, approach, name, and effectiveness.

As countries experiment with MPAs, scientists and resource managers are identifying ways that MPAs can best be integrated with other approaches to maintain marine biological diversity. It is clear that MPAs can be powerful means of protecting critical marine areas, increasing public awareness and understanding about the importance of life in the sea, and providing sites for research and monitoring (Salm and Clark 1984).

Answers to questions about the size, approach, degree of protection, and design of MPAs, and their relationship with other marine and coastal zone management tools, are less clear. Defined broadly, MPAs run the gamut from small, highly protected reserves to larger, multiple-use areas and biosphere reserves, which can employ LME or IAM approaches. No one approach has emerged as best in every situation; each can make a valuable contribution to maintaining biological diversity, depending on the ecological and socioeconomic factors in each area.

Key to the success of MPAs everywhere is public understanding, involvement, and support (Kelleher and Kenchington 1991). It is essential to integrate the existing needs and patterns of use and ownership of resident peoples into design and objectives of MPAs, especially fishery reserves. Large MPAs can be zoned in ways that allow various uses in some zones and more stringent protection in others. However, compromises have been made too often on fishing and other consumptive uses within MPAs, which prevent MPAs from achieving their goals (McClanahan 1990). Another serious problem is that very few MPAs, unless integrated into broader area management schemes, are able to control inputs from adjacent areas, including land areas, that affect species and ecosystems within MPAs.

There is growing evidence that true marine reserves that offer a high level of protection are effective in achieving conservation objectives. Compelling arguments for them come from successful experiences in New Zealand (Ballantine 1991) and the Philippines (Alcala 1988), and in a proposal to establish them in the southeastern USA (Bohnsack 1990). Benefits of such "no-take" areas can include insurance against recruitment failure and the resulting collapse of stocks prone to overexploitation, protection of genetic diversity in exploited species, and maintenance of ecosystem functions. However, unless the public can be made to see, understand, and participate in such benefits, efforts to establish highly protective MPAs will not succeed.

Active Manipulation: Restoration and Mitigation

If the only way to maintain biological diversity is to slow or stop its loss, then such efforts are destined to fail because, sooner or later, here or there, losses are inevitable. However, there are techniques for gaining ground: People can help to restore depleted populations and degraded ecosystems. Interest in restoration and the number of restoration projects are growing as more people understand the importance of maintaining biological diversity. Population and ecosystem restoration could one day be among the most important areas of marine conservation.

It is, however, risky to assume that populations and ecosystems can be restored at will. In some countries, populations or ecosystems have been deliberately reduced or eliminated, or such actions have been proposed, based on the notion that these losses can be "mitigated" by restoring populations or ecosystems at other locations where (presumably) they will not fall victim to development. This assumes, however, that such efforts succeed in restoring what was lost. Sadly, with the current state of knowledge, this assumption is usually unwarranted; too little is known about how living systems work to be confident that mitigation efforts will succeed. And as true as this is on land, it is even more so in the sea, where our knowledge is far less.

Restoring Areas

As nonexperts have realized the dependence of species on their habitats and the extent of chemical contamination and physical degradation of ecosystems, interest in restoring ecosystems has grown. In general, restoration has been more ecologically sophisticated and more successful on land and in fresh waters than in the sea. But where marine and estuarine ecosystems have already been damaged or destroyed, attempts to restore lost functions can be worth considering.

Actually, some efforts to manipulate areas have not been aimed toward restoration—the return to previous composition, structure, and function—

but at enhancing habitat value, usually for a particular species. For example, a wildlife agency faced with greater demand for waterfowl than nature provides might bulldoze a salt marsh to provide the depth of water that maximizes feeding opportunities for certain kinds of ducks. This also enhances the habitat for other species that need similar conditions. Unfortunately, it also eliminates habitat for species that need shallower or deeper waters, such as wading birds or diving birds. Obviously, such ecosystem manipulation requires as much understanding of what is being lost as of intended gains.

Ecosystem manipulation is a young science, and remains experimental for many ecosystem types (Thayer 1992). Scientific and practical experience is limited (Kusler and Kentula 1989). For some ecosystems, such as seagrass beds, there have been no complete restoration successes (Kirkman 1992), whereas for others, such as coral reefs, there are positive examples, although they have been costly (Maragos 1992). Salt marsh and mangrove restorations have usually achieved partial success, at best; even if structure-forming plants are reestablished, recolonization by other species and restoration of ecosystem processes have proven elusive. Moreover, follow-up has often been insufficient. Understanding of what works and what does not is fragmented and regional: Techniques for mangrove restoration that succeed on the wet coast of northern Queensland (Australia) might not work in a dryer area 200 kilometers (124 miles) to the south, let alone in Colombia or Cameroon.

Two lessons are clear. First, efforts to improve our ability to restore degraded ecosystems are critical because so much has already been lost. Second, overreliance on ecosystem manipulation to compensate for losses under permitting programs is unwise. Implementation of mitigation measures is often spotty at best, and for those measures that are implemented, success is far from certain. Many agencies correctly place priority on maintaining existing ecosystems. Thus, enhancement and restoration are appropriate only as a last resort, when proposed impacts are absolutely unavoidable.

Recovering Populations

When populations have been depleted, active manipulation is an option for recovering them. For terrestrial and freshwater organisms, there is a variety of *in situ* and *ex situ* methods for augmenting or recovering populations. In the sea, opportunities for *ex situ* conservation are far fewer;

many species cannot yet be maintained, let alone bred, in captivity. Methods available for manipulating and enhancing depleted marine populations, including captive breeding, "headstarting," and transplanting to augment existing populations or establish new ones, might appear attractive and successful, but they might also have four problems. They might:

1) cost too much;
2) be ineffective if the original threats are not removed;
3) introduce unforeseen harmful complications; and
4) prevent the correction of the real problem by creating a false sense of security.

Sea turtles have been the subject of extensive manipulation to increase populations, and illustrate problems encountered in trying to recover populations of species that have become rare because of human activities. As an illustration of problem 1, above, a US headstarting operation for the endangered Kemp's ridley sea turtle (Woody 1991) that takes turtles well beyond hatchling stage, costs approximately US$125 per nine-month turtle. Yet it is unknown whether a turtle raised in a bucket and fed *ad libitum* can feed itself, migrate, or escape predators. Even more problematic, the most important source of mortality for sea turtles in US waters is drowning in shrimp trawls (National Research Council 1990a). Hence, until trawl-induced mortality is eliminated (for example, by ensuring that shrimpers use TEDs), introducing headstarted turtles into these waters may amount to a very expensive put-and-take fishery with no gain to the sea turtle population (problem 2).

Protected hatcheries have been used worldwide to increase the success of sea turtle hatching. Because the incubation temperature of sea turtle eggs determines the sex of the hatchlings (Mrosovsky and Yntema 1980), some hatchery programs may have produced predominantly male turtles for a number of years (Mrosovsky 1983), effectively precluding future natural reproduction (problem 3). Transplanting turtle eggs to distant beaches in order to establish additional colonies subjects eggs to temperature changes and might similarly skew sex ratios (again, problem 3). Moreover, hatchery-raised turtles might not be able to find nesting beaches because they are not imprinted on them; scientists do not know whether, or if so, how sea turtles imprint on their nesting beaches (another problem 3).

Finally, augmentation of one life history stage to compensate for losses in another might look successful while doing little for population restoration (problem 4). A consequence of sea turtles' population biology is that a relatively small reduction in the mortality of large subadults (those that had already survived most natural mortality factors and would soon reproduce) is far more likely to recover southeastern US loggerhead turtle populations than large increases in egg survival and hatching (Crouse et al. 1987).

Clearly, manipulating sea turtles to increase their populations is not as easy as its proponents had hoped, despite the fact that their biology is far better known than that of most marine species. In other cases, artificially augmenting populations of commercially fished species could have deleterious effects on natural populations of that species or on other species in the ecosystem. There is growing concern that augmentation of Pacific salmon species by the USA, Canada, and Japan with hatchery stock has contributed to the endangerment of already-threatened natural salmon populations. Japanese hatcheries that have released 2 billion juvenile chum salmon (*Oncorhynchus keta*) and 24 million red sea bream (*Pagrus major*) in a year could disrupt the ecosystems flooded by these fishes (Personal communication with Makoto Omori, Tokyo University of Fisheries, Tokyo, Japan) and the genetic integrity of their populations (Omori et al. 1992). Given these concerns, the active manipulation of populations should be viewed with caution and used only as a last resort when reduction of threats and *in situ* protection methods have failed.

CHAPTER EIGHT

Existing Marine Institutions and Instruments

IN RESPONSE to the growing need to conserve marine biological diversity, governments, NGOs, and IGOs have devised programs to diminish threats, support sustainable use, and encourage research. The question is, how effective are these efforts? Among countless examples worldwide, this chapter examines a sampling of international efforts and efforts within nations.

International Organizations and Programs

Recognition in international legal regimes that marine species and ecosystems, even more than those on land, do not respect national boundaries, is largely a recent development. Some international institutions and instruments specifically focus on the sea; others deal with the terrestrial realm, but have recently begun to include marine species and ecosystems in their missions. International institutions and instruments discussed herein include treaties, assistance programs, planning and management programs, programs to increase understanding of the resources of interest, and efforts by NGOs.

Multilateral Treaties

A number of multilateral treaties that require ratifying nations to act in ways that maintain marine biodiversity have been opened for signature, and some have entered into force. For established treaties, effectiveness depends on how vigorously they are implemented. Regrettably, some treaties that could be important for marine species and ecosystems have suffered from lack of implementation. Ironically, others are generally adhered to by nations even though they have not yet entered into force. This section examines the comprehensive UNCLOS III, and a sampling of other multilateral treaties that address specific threats to marine biological diversity of areas of special concern.

Law of the Sea

UNCLOS III, opened for signature in 1982 but still lacking the 60 ratifications to bring it into force, has been described as a constitution for the oceans. It is a complex document of almost 200 pages covering 15 major topics, including conservation of the living resources of the oceans, maintenance and restoration of populations of species, and protection of the sea from pollution.

During the 20th century there have been three international conferences on the law of the sea. In 1958, the first UN Conference on the Law of the Sea (UNCLOS I) was called to prepare The Convention on the High Seas, The Convention on Fishing and Conservation of the Living Resources on the High Seas, The Convention on the Territorial Sea, and The Convention on the Continental Shelf. They were all approved by UNCLOS I, but only the fourth one ever received the necessary ratifications to come into force. In 1960, the United Nations called UNCLOS II, which again failed to reach agreement on the width of the territorial sea.

From 1973 to 1982, the largest international legislative forum ever convened hammered out UNCLOS III. At its opening for signature, 119 nations signed. There are now 159 signatories and 53 ratifications. The USA and three other countries voted against the convention, and the USA has continued to express its opposition to some elements of the seabed mining regime. However, practically all the other provisions of the convention, especially ones relating to international navigation and the rights and duties of coastal states, have become customary international law, and as such are binding on all nations.

UNCLOS III establishes a comprehensive regulatory framework for the oceans, with provisions governing jurisdiction, access to the seas, navigation, protection of the marine environment, exploitation of living resources, scientific research, seabed mining, and the settlement of disputes. It allows establishment of a territorial sea of up to 12 nautical miles (22 kilometers) over which the coastal state has sovereignty, subject to the right of innocent passage. Basically, national laws apply to this space.

There are another 12 nautical miles of contiguous zone over which a coastal nation can exercise control necessary for public safety and order.

Within the next space, the 200-nautical-mile- (371-kilometer-) wide EEZ, Articles 61 to 68 set forth principles for the exploitation of the living resources. A coastal nation,

> taking into account the best scientific evidence available to it, shall ensure through proper conservation and management measures that the maintenance of the living resources in the exclusive economic zone is not endangered by over exploitation. As appropriate, the coastal state and competent international organizations shall cooperate to this end. . . . coastal states are obligated to maintain or restore populations of associated or dependent species above levels at which their reproduction may become seriously threatened.

This enlightened requirement, however, seems to be undermined by another recommendation. Because many developing nations have limited capacity to use their own fishery resources and enforce their own national fisheries laws, fish stocks are often left open to capture by European, Asian, and other distant-water fishing fleets that seek access to EEZ resources on favorable terms. UNCLOS III advises coastal nations to "give other States access to the surplus of the allowable catch," taking into account, among other things, "the significance of the living resources of the area to the economy of the coastal state and other national interests" (Art. 62 [2]). By favoring such "full utilization" based on MSY, UNCLOS III seems to invite overexploitation.

On the high seas, the space beyond the EEZ, the rules are set forth in Articles 116 to 120. All nations have the right to fish, but at the same time have the responsibility of applying such measures as necessary for conserving the living resources, either individually or in cooperation with other nations.

Part XII of UNCLOS III contains provisions on protection of the marine environment, starting with Article 192, which says, "States have the obligation to protect and preserve the marine environment." Part XIII guarantees and promotes marine scientific research. Part XV obliges nations to settle any dispute by peaceful means and provides procedures for settlement.

The great strengths of UNCLOS III are its coverage of all marine activities, its various environmental provisions, its innovative dispute settlement provisions, and its ability as a framework convention to grow through decisions of other international or regional organizations. Its ambiguous and general character allows negotiating nations to reach consensus in areas of acute disagreement, while allowing room for individual nations and regions to address specific problems. However, the emphasis given to freedom of navigation sometimes conflicts with the UNCLOS III's environmental provisions, and many of its provisions are neither spelled out nor enforceable. Although UNCLOS III is not yet in force, it offers a set of standards and a cooperative spirit to members of the world community.

Implementation of many topics covered by UNCLOS III, such as transboundary fisheries arrangements, is proceeding at an almost imperceptible pace. An example involves the extensive fisheries off West Africa. Of more than 20 developing nations in the region, most claim an EEZ. However, the seasonal movement of fish stocks among these EEZs thwarts any national management attempts. Regional cooperation is, therefore, essential, but barely exists and is far from adequate to deal with the complex stock assessment and allocation questions that must be addressed.

The experience of other regions in implementing provisions of UNCLOS III is mixed. Regional cooperative management under the European Community has not prevented serious declines in cod (*Gadus morhua*) and haddock (*Melanogrammus aeglefinus*) stocks. The USA and Canada have developed a complicated treaty arrangement for dealing with mutual interceptions of salmon stocks originating in each other's waters, but it threatens to unravel because of dissatisfaction about allocations among fishing groups and species. Regional arrangements for tuna management do exist, but coastal states are increasingly claiming control over access to tuna stocks.

UNCLOS III is an imperfect but still useful international compromise. When fully in force, it has the potential to help in the resolution of many of the most pressing issues affecting biological diversity in the sea.

Antarctic Treaties

More than on any other continent, conservation in Antarctica is marine conservation. So hostile to life are its bitter winds and vast, frozen expanses that nearly all Antarctic life derives food directly or indirectly from the sea, and inhabits the continent's coasts and adjacent waters. Parts of the Southern Ocean that surrounds Antarctica, although as cold as seas can get, are spectacularly rich in life. Too dark to support photosynthesis for much of the year, the Southern Ocean's productivity depends on the rapid growth of phytoplankton in the four to five months when sunlight *is* available (El-Sayed 1985). The phytoplankton are eaten by zooplankton (especially huge aggregations of shrimplike krill) which then become food for squids, fishes, penguins, seals, and whales. Although Antarctica and surrounding waters have been considered the most pristine area on Earth, their blue whales have been hunted to the edge of extinction, increased ultraviolet-B radiation is causing profound concern, and the coast has even been the site of an oil spill.

The Antarctic is protected by the Antarctic Treaty of 1959 and its subsidiary conventions, known collectively as the Antarctic Treaty System (ATS). The ATS preserves the legal positions of those countries with territorial claims to the Antarctic and those that do not recognize claims, establishes a legal framework to set Antarctica aside as a continent for science, and reserves the area south of 60°S as a zone of peace. It also prohibits nuclear explosions, the disposal of radioactive waste, and all military activities not of a scientific or peaceful nature. Recognizing the need to protect Antarctica's remarkable biota, the treaty system has adopted international regimes to prohibit mineral resource activities and regulate the exploitation of Antarctic species.

Yet, recent developments, such as increased human habitation and tourism, increasing pollution, and krill fisheries, have raised serious questions about the adequacy and implementation of some components of the treaty system. In response, the ATS Consultative Parties (the original parties, plus nations that carry out substantial scientific activity on the continent) recently negotiated an environmental protection protocol to the Antarctic Treaty. In 1991, the Consultative Parties met in Madrid to sign the Protocol on Environmental Protection to the Antarctic Treaty. The protocol supplements the treaty by dealing with issues such as mineral resource activity, the conservation of plants and animals, waste disposal and management, and the prevention of marine pollution. Although the new protocol makes major progress toward comprehensive protection for Antarctic ecosystems, concerns over exploitation of mineral and living resources remain.

The Protocol contains provisions and annexes affecting Antarctica's biological diversity. It seeks to protect the abundance, productivity, and populations of Antarctic species. Perhaps its most important and controversial provisions deal with minerals. Article 7 prohibits mineral resource activities (other than scientific research), but it is qualified by Article 25, which opens the door to modification of the ban after 50 years. Many countries and environmental organizations believe that the protocol needs a permanent prohibition on mineral resource activities, and that several "loopholes" in the protocol should be closed.

Another question concerning the environmental protection protocol revolves around its enforcement. Article 19 contains compulsory dispute settlement procedures whose results are binding. The hurdle—as with international law generally—might be getting a government faced with an alleged violation to take the issue to court.

In addition to the Protocol, the ATS protects marine biological diversity through CCAMLR. CCAMLR aims to prevent overexploitation of species south of the Antarctic Convergence, and pioneered the concept of the ecosystem management in the sea. By regulating fishing, it aims to avoid irreversible damage to both target species, such as fishes and krill, and the species that depend on them.

Regrettably, implementation of CCAMLR is hindered by the protocol's requirement that decisions be made by consensus. A single country, claiming there are insufficient scientific data to make a determination of negative impact, may block all measures to regulate the exploitation of marine life. Unanimity has proved virtually impossible to achieve, so conservation measures have moved slowly.

This is especially troubling because, under CCAMLR, the burden of proof is on those who favor conservation: Member nations may take marine resources until harm to the ecosystem is shown. This requires the evaluation of available scientific evidence of direct and indirect impacts of exploitation. Often, the data needed for decision making are incomplete or are submitted late by the fishing nations, making it almost impossible to stop exploitation until a fishery has all but collapsed. The combination of the burden of proof, decision making by consensus, and the dearth of data greatly weakens CCAMLR's effectiveness. To maintain biological diversity under CCAMLR, the burden of proof would need to be reversed; that is, fishing nations would have to demonstrate that their catch will *not* harm target species and their ecosystems before any fishing could occur.

The Convention for the Conservation of Antarctic Seals protects all species of seals in Antarctic waters. It strictly limits the killing of crabeater

(*Lobodon carcinophagus*), leopard (*Hydrurga leptonyx*), and Weddell (*Leptonychotes weddellii*) seals, and prohibits the taking of Ross seals (*Ommatophoca rossii*), southern elephant seals (*Mirounga leonina*), and Antarctic fur seals (*Arctocephalus gazella*). Yet like CCAMLR, it places the burden of proof on those who would protect, rather than exploit, seal populations.

The ATS has a number of other problems, such as lack of rules to deal with increasing tourism. Antarctic tourism has grown substantially and could pose problems to parts of the continent and its wildlife. It may eventually be necessary to limit the numbers, timing, and location of tourist visits to protect the integrity of the continent's fragile ecosystems.

The threats posed by mining and exploitation of living marine resources may compel Consultative Parties to the Antarctic Treaty to provide more protection than the ATS affords under its current conventions. The weaknesses in CCAMLR suggest that the Consultative Parties need to agree on new procedures for managing stocks of Antarctic krill, fishes, birds, and other organisms to prevent their depletion, and to allow populations of endangered species, such as the great whales, to recover.

Fishery Treaties

Marine fisheries tend to be international in scope because the ranges of marine species seldom conform to nations' political boundaries. The ranges of many species also transcend boundaries between nations' EEZs and the high seas, an international commons where fishing is considered a legal right of all countries. Therefore, the fate of many exploited fish and invertebrate populations depends on international cooperation in research, management, and enforcement.

In the years after World War II, world population growth increased the demand for protein, and rapid technological advances allowed the expansion of fishing fleet ranges and capacities. It became apparent that distant-water fleets had the potential to overexploit the world's major fishing grounds, prompting the formation of organizations such as the International Commission for the Northwest Atlantic Fisheries (ICNAF) to conserve fisheries resources. As overexploitation became reality but international cooperation proved elusive, the concept of 200-nautical-mile (371-kilometer) fishing zones gained international acceptance. This gave coastal nations jurisdiction over fisheries within their coastal waters but, in many cases, did not ensure management throughout a given species' range. There was still a need for international cooperation, especially for anadromous species such as salmon, for the many marine species, such as herrings, halibut, and gadoids (cod, haddock, pollack, and their kin) that straddle nations' borders, and for highly migratory species, such as tunas, billfishes, and some sharks.

Among its many provisions that affect fisheries, UNCLOS III has articles that address management of anadromous species, transboundary stocks, and highly migratory species. These articles encourage cooperation among coastal nations with shared stocks through appropriate regional and international organizations. The following examples illustrate the achievements and failures of some international and regional fisheries organizations.

Anadromous Species: Anadromous species spend most of their lives feeding and growing in the sea, returning to rivers only to spawn. The most famous examples are salmon, which can range thousands of kilometers throughout the North Pacific or North Atlantic from their native streams. High-seas fisheries can intercept salmon before they can be caught by coastal fisheries or before they can reproduce, sometimes jeopardizing specific populations.

The International North Pacific Fisheries Commission (INPFC) and the North Atlantic Salmon Conservation Organization (NASCO) are two international organizations governing high-seas salmon fisheries. They both strictly regulate catches on the high seas, yet high-seas catches have remained a problem. Coastal nations argue that they alone expend resources maintaining salmon-spawning habitat and augmenting populations with hatchery fish while the high-seas fishers remain "free riders." There is now progress toward regulating or eliminating high-seas salmon fisheries in both regions. High-seas driftnets are being phased out in the Northcentral Pacific, while population declines of North Atlantic salmon indicate that the high-seas fishery's days are numbered.

Transboundary Stocks: The ranges of marine fish and invertebrate populations that overlap adjacent zones of national ocean jurisdiction are considered transboundary, shared, or straddling stocks. Regional cooperation in management becomes especially important in cases such as West Africa, where 10 nations' EEZs are sandwiched between Ivory Coast and Congo. Obviously, anything that affects a given fish population within one EEZ has ramifications for neighboring EEZs. Sadly, in spite of the obvious need, cooperation between nations with adjacent EEZs is rare, and is usually an *ad hoc* response that deals with one species and one crisis at a time.

For example, the USA and Canada finally agreed on an ocean boundary across Georges Bank, one of the richest fishing grounds in the Northwest Atlantic. Yet the management approaches of the two countries have been completely isolated from each another. Fishing and management have continued as though populations of haddock and cod observe the political boundary as drawn by the World Court.

In another case, the Northwest Atlantic Fisheries Organization (NAFO) was formed in 1978 after the Northwest Atlantic coastal nations extended their fisheries jurisdictions. Important fish stocks straddle the Canadian zone and the NAFO regulatory area, and are fished on both sides. Spain and Portugal have objected to proposed NAFO management measures, and the USA has not joined NAFO, although US fishers continue to fish in the regulatory area. Unfortunately, the combined catches of these straddling stocks has exceeded MSY, especially during the late 1980s. Although recent negotiations show promise, failure to cooperate has already damaged one of the world's most productive fisheries.

The International Pacific Halibut Commission (IPHC) has been more successful. A 1923 agreement between Canada and the USA evolved into the IPHC. Its purpose is to manage Northern Pacific and Bering Sea halibut stocks for MSY. IPHC seems to have worked for two basic reasons. First, there are only two members, and both recognize the value of maintaining halibut populations at high levels. Second, the areas covered include much of the halibut's range.

Highly Migratory Species: Highly migratory species, including tunas, billfishes, and oceanic sharks, are some of the most valued food and recreational fishes and are among the most difficult and controversial ones to manage. They are the extreme case of transboundary stocks because their oceanwide movements repeatedly cross between EEZs and the high seas.

Concern over continued production and viability of tuna stocks prompted nations and industries to establish international organizations to study and manage highly migratory species (Joseph 1983). An example is the International Commission for the Conservation of Atlantic Tuna (ICCAT), which entered into force in 1969. Its main aim is the conservation of tuna-like fishes and billfishes throughout the Atlantic Ocean and adjacent seas (Edeson and Pulvenis 1983). Composed of 22 member nations, ICCAT is constrained by the lack of funding for a permanent staff, data acquisition, and timely formulation of management advice. Member nations conduct most research, carry out analyses, and enforce ICCAT recommendations for their own nationals.

As with many international organizations, ICCAT members have had difficulty reaching consensus on management. ICCAT has been unable to take decisive action in the face of dangerous population declines of Northwest Atlantic bluefin tuna (Figure 8-1) and Atlantic swordfish (*Xiphias gladius*). ICCAT's scientific committee recorded the bluefin tuna's decline throughout the 1970s and 1980s, yet ICCAT management recommendations have been both inadequate and too late to recover the population. ICCAT's main contribution seems to be its role in documenting the decline of highly migratory species.

In the South Pacific, efforts to manage tunas by such regional groups as the South Pacific Fisheries Forum Agency and the Nauru Group have been quite successful. The members, most of which are small island nations with huge EEZs, had common problems: Tunas are among their few

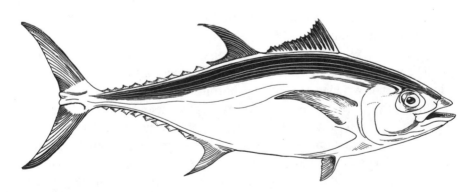

Figure 8-1. North Atlantic bluefin tuna. *Among the world's largest bony fish species, this powerful, warm-blooded, long-distance migrator's population has fallen to a few percent of pristine levels because of overfishing. Existing international organizations seem powerless to stop their decline.*

natural resources, and distant-water fleets from the USA and Japan were fishing in their EEZs without offering compensation. The Pacific nations banded together to gain an equitable share of benefits from the tuna catch through cooperative enforcement and permitting schemes. In the late 1980s, when foreign driftnetters threatened their resources, they coalesced again to oppose it. Although the conservation decisions they face are not as thorny as those faced by ICCAT, their example shows how international cooperation can benefit all the nations in the region.

International management of fisheries resources is growing from the foundations provided by UNCLOS III and earlier treaties. The ecological interconnections in the sea are the basis for international cooperation in fisheries. In the short term, individual nations might reap greater benefits through unregulated fishing, but maintaining stocks at high levels is a benefit that all nations gain through cooperation. Cooperation ensures that their short-term and long-term interests coincide.

The MARPOL Convention

Pollution from intentional discharges by ships is regulated by the International Convention for the Prevention of Pollution from Ships of 1973, which has been superseded by the 1978 Protocol (MARPOL 73/78), and was again amended in 1992. It is administered by IMO. Within IMO, the Marine Environment Protection Committee (MEPC) coordinates IMO pollution-control activities and continuously reviews the MARPOL Convention. MEPC develops international standards for reducing pollution from vessels. Members of IMO can participate fully in the MEPC even if they have not ratified MARPOL.

The MARPOL Convention, passed in the wake of the *Torrey Canyon* oil spill, addresses five types of pollution by ships: oil (Annex I), noxious liquid substances in bulk (Annex II), harmful substances in packaged forms (Annex III), sewage (Annex IV), and garbage (Annex V). Regulations have been developed for each of these, but Annex IV is not yet in force.

Annex I regulations require ships to be fitted with technical facilities (e.g., oil discharge monitoring and control systems, oily water separating equipment, or oil filtering equipment) to prevent or limit discharges. Ports must have facilities for receiving wastes. Annex I prohibits oil tankers from making discharges less than 50 nautical miles (93 kilometers) from the nearest land. The discharge of oil by ships other than oil tankers is also prohibited unless they are more than 12 nautical miles (22 kilometers) from the nearest land. Maximum quantities, rates, and levels of discharge are regulated beyond these areas.

Annex II regulations require ships to be fitted with equipment to limit the generation of wastes on chemical tankers, and require ports to provide facilities to receive such wastes. Annex II identifies four categories of chemicals (A, B, C, and D), A being the most hazardous. Category A, B, and C substances can be discharged more than 12 nautical miles from the nearest land, and in water with a depth of more than 25 meters (82 feet). For category D substances, the same distance from the nearest land applies, but there are no limits about water depth.

Annex III contains regulations for handling harmful substances carried by sea in packaged forms, or in freight containers, portable tanks, or road and rail tank wagons. It contains rules with regard to packaging, marking and labeling, documentation, stowage, and quantity limitation. These regulations are relevant primarily to the prevention of any incidental loss of packaged chemicals and the eventual recovery of such lost packages.

Annex IV regulates the discharge of sewage from ships through technical requirements and areal limitations. Sewage discharge from ships is prohibited less than 12 nautical miles from the nearest land (for sewage that is not comminuted or disinfected), or 4 nautical miles (7 kilometers) or less from the nearest land (for sewage that is comminuted or disinfected). States may impose less stringent requirements within their jurisdictions.

Annex V includes requirements for ships to control and prevent pollution from garbage (including plastics and other persistent materials). Ports must provide facilities to receive such wastes. Annex V prohibits the disposal of all plastics into the sea, the discharge of floating materials including dunnage, lining, and packing materials less than 25 nautical miles (46 kilometers) from the nearest land; discharge of food wastes and other garbage less than 12 nautical miles from the nearest land; and discharge of food wastes or other materials that have passed through a grinder or comminuter less than 3 nautical miles (6 kilometers) from the nearest land.

Oil, toxic chemicals, sewage, and plastic garbage all damage marine organisms and ecosystems. The area in which MARPOL has been most effective is in limiting discharges in nearshore (within 12 nautical miles) areas. The implementation and enforcement of MARPOL restrictions, historically spotty, are improving in regions that have established cooperative enforcement mechanisms such as the Memorandum of Understanding signed by 14 European coastal nations.

In shipping lanes on the high seas or in areas of restricted circulation, substantial concentrations of pollutants continue to accumulate. MARPOL also allows the disposal of wastes in submerged ecosystems such as kelp beds, coral reefs, and seagrass beds that are not within the protected zone "nearest land." Consequently, the minimum discharge limits, which are intended to protect such sensitive ecosystems, are not applicable.

The MARPOL Convention exemplifies the problems facing IMO. Like all IMO Conventions, MARPOL was adopted in reaction to a disaster. Thus, the drafters lacked sufficient time for reasoned and careful planning, and were unable to develop a comprehensive approach for the control and prevention of all types of vessel-generated pollution, proceeding instead with a piecemeal approach based on pollutant type (Tharpes 1989). Thus, although commendable, IMO efforts to control marine pollution tend to be characterized by stop-gap, fragmented installation measures.

Some enclosed and semi-enclosed seas with restricted circulation have been given additional protection as "Special Areas" under the various MARPOL annexes. For example, to control oil pollution, the Baltic, Mediterranean, Black, and Red seas, the Persian/Arabian Gulf, and Gulf of Aden are designated Special Areas under Annex I. A Special Area is defined as:

> a sea area, where for recognized technical reasons in relation to its oceanographical and ecological condition and to the particular character of its traffic, the adoption of special mandatory methods for the prevention of sea pollution by oil (as in Annex I), noxious liquid substances (as in Annex II), or garbage (as in Annex V) is required.

Within Special Areas, discharges are prohibited or strictly curtailed. Some members of MEPC are starting to discuss the need to strengthen discharge limitations outside Special Areas. To avoid the tedious process of renegotiating MARPOL, the standards applicable to Special Areas could be extended to cover the sea worldwide. The likelihood for greater protection would increase if more nations participated in the MEPC. To date, most decisions made through IMO have resulted from negotiations among a small number of industrialized countries.

London Dumping Convention

The London Dumping Convention (LDC) came into force in 1975, and applies to all "marine waters other than the internal waters of States." It defines dumping as any deliberate disposal at sea of wastes or other matter, or any deliberate disposal of vessels or other human-made structures. Dumping does not include wastes derived from normal vessel operations or wastes related to the development of seabed mineral resources. A 1978 amendment to Annex I provides for the incineration of certain substances at sea if the operator has a special permit and complies with the Regulations for the Control of Incineration set forth within an Addendum to Annex I.

The LDC's central provisions are found in Article IV, which uses a so-called black list/gray list method of pollutant regulation. Dumping of substances in Annex I (the black list) is prohibited. Those listed in Annex II (the gray list), including low-level radioactive wastes, may be dumped if a special permit has first been obtained. All other substances may be dumped if a general permit is first obtained. Unfortunately, no guidelines are provided for issuing either general or special permits. Although issuing states are obligated to first carefully consider factors set forth in Annex III of the LDC, permitting is left solely to the discretion of the issuing state, so long as it:

1) keeps records of the nature and quantities of all matter permitted to be dumped and the location, time, and method of dumping; and
2) monitors individually, or with other contracting parties and competent international organizations, the conditions of the seas.

The LDC does not apply to:

1) ships and aircraft entitled to sovereign immunity (e.g., warships);
2) dumping necessary for the safety of human life or vessels;
3) dumping of substances listed in Annex I by special permit in cases where emergencies pose an unacceptable risk to human health and no other solutions are feasible;
4) dumping of substances listed in Annex I if they are "rapidly rendered harmless" by physical, chemical, or biological processes in the sea, provided that they do not make edible marine organisms unpalatable or endanger human health or that of domestic animals; and

5) dumping of substances listed in Annex I if they appear as trace contaminants in wastes.

In effect, the LDC creates a minimum international standard for dumping. Supporters of ocean dumping assert that the LDC merely recognizes the sea's ability to serve as a waste treatment facility for limited quantities of waste, but their assertion is not supported in many areas where ocean dumping has been carried out. Their argument also assumes that all contracting parties to the LDC will act responsibly, observing and enforcing dumping practices from their vessels. Without vigilant compliance, the LDC does little more than legitimize ocean dumping.

International Assistance Programs

Development projects can both harm and benefit the marine environment (Boxes 8-1 and 8-2). It is obvious that projects directly based upon marine species or ecosystems can have significant impacts, but land-based development projects far from the coasts can cause serious harm as well. There are also development projects specifically designed to maintain or improve the health of the sea. The following three examples illustrate the strengths and weaknesses of international assistance programs.

Mangrove Deforestation for Mariculture, Ecuador

Mangrove forests are among the world's most productive ecosystems, playing a vital role in marine food webs along many tropical coasts. As with many other types of tropical forests, they are rapidly disappearing, but their loss has attracted less attention than it merits.

In Ecuador, development projects funded by organizations including the World Bank, the Inter-American Development Bank, and the U.S. Agency for International Development have promoted the conversion of mangrove forests into ponds for growing penaeid shrimp. The goal of these mariculture projects is to increase earnings from shrimp exports. In the short run, they have been highly profitable, but they have not benefited the rural communities that they were supposed to help (Personal communication with Patrick Dugan, IUCN Wetlands Programme, Gland, Switzerland). Further, in the long run, destruction of mangroves has proven to be counterproductive. Shrimp production in converted areas has actually declined because mangroves are the shrimps' primary nurseries. The result is a severe shortage of the wild juvenile shrimp needed to stock

---

the artificial ponds. Other possible problems are that maricultured alien shrimp species that escape can establish themselves in the sea, competing with, introducing pathogens to, and altering gene pools of closely related native shrimp.

Both Ecuador's government and its shrimp industry have recognized the destructive nature of mangrove conversion. Development agencies are also beginning to realize that mangroves are most productive when left intact. Their legacy of mangrove forest destruction, however, exemplifies development agencies' enthusiasm for producing quick gains in export earnings at the expense of long-term sustainable use.

Hydroelectric Power Plant Construction, India

India has been the largest recipient of project loans from multilateral development banks. Hydroelectric projects have represented a significant percentage of these projects. An example is the World Bank-funded

US$450 million Sardar Sarowar Dam Project in the state of Gujarat, India. This project, approved in 1985, involves the construction of 30 major dams on the Narmada River system in western India. As is the case with many MDB projects, a detailed environmental impact assessment was incomplete at the time of project approval. The potential benefits of the project—providing hydroelectric power and irrigation—were seen by the World Bank as being more important than the potential environmental damage.

As Chapter 4 explains, large dam projects can have significant direct impacts upon marine ecosystems. This project also has the potential for indirect effects because it will displace approximately 90,000 people, who will have to find land and work elsewhere. This can only increase the pressure on India's overstressed coastal areas and marine resources.

This project exemplifies longstanding problems with MDBs' environmental impact assessments. When assessments have been made, they have often been inadequate for decision making or have simply been ignored. There are still development officials who see maintaining environmental quality as an obstacle to development, as opposed to an integral part of the process. The World Bank has promised to be more careful about avoiding environmental damage and India has now refused further World Bank funding for Sardar Sarowar as a means of avoiding the bank's increasingly stringent measures for protecting the environment. Still, the bank continues to support projects whose environmental consequences, including those affecting marine species and ecosystems, that are poorly understood.

Global Environment Facility

The GEF, a cooperative venture among national governments, the World Bank, the UN Development Programme (UNDP), and UNEP, is a three-year experimental program to fund projects with environmental goals. Among its initial proposals are projects to develop a global set of high-priority areas for establishing marine protected areas, and to improve the waste disposal process in Caribbean and Mediterranean ports.

While the existence of the GEF indicates an encouraging recognition by leading international agencies that environmental protection is a vitally important kind of development, its success will depend on its commitment to this concept, its ability to identify projects that will bring the greatest long-term benefits to living systems and people who seek to use them sustainably, and the commitment of its backers to ensure that the GEF will not be used to divert attention from the need to reform the MDBs' lending programs. If these banks are serious about halting global environmental degradation, the greatest contribution they can make is funding only those development projects that are ecologically sound and sustainable.

International Governmental Organizations

IGOs are formed by member nations to deal with a particular set of issues within a framework defined by their constitutions. IGOs influence policies of member nations in a variety of ways, but usually by issuing codes of conduct ("soft law") rather than legally binding agreements. They encourage marine conservation by increasing information sharing, by offering scientific, technical, or financial help to collaborating nations, or by sponsoring conventions and accompanying protocols that are binding to their signatories.

Regional and global IGOs are a response to an ever-more complex, interconnected human society; what a nation does affects and is affected by what other nations do. Because technological developments and scientific findings have steadily eliminated the factors that isolate nations, IGOs offer a peaceful means for nations to deal with matters of mutual interest. The four organizations that are highlighted below cover very different marine fields, namely marine mammals (IWC), fisheries (FAO), shipping (IMO), and pollution and environment (UNEP), yet their respective spheres of influence overlap, and they sometimes cooperate in conventions, action plans, and conferences.

The effectiveness of IGOs is hampered by claims of national sovereignty, conflicts within regions, and the burgeoning North-South friction that permeates most IGOs' agendas. The global IGOs described here are also encumbered by their size and complexity, especially when they attempt to solve regional or local questions. Smaller regional units within IGOs can be more effective because they deal with more homogeneous interests and needs and can respond more readily to changing circumstances.

International Whaling Commission

The International Convention for the Regulation of Whaling (ICRW), which created the International Whaling Commission (IWC), entered into force in 1948 upon ratification by 14 whaling nations. The ICRW is intended to promote the conservation of whale stocks and the orderly development of the whaling industry. Its greatest challenge has been the setting of overall catch limits.

Until the 1970s, IWC could not stem the continuing depletion of whale populations because whaling nations dominated its membership. Quotas often ignored scientific evidence and enforcement was lax. Further, when quotas were reduced, members frequently objected or withdrew from IWC, thereby freeing themselves from regulation.

In the 1970s, IWC came under growing pressure to improve the scientific basis for its decision making and to adopt a more formal management approach. Concern over the plight of whales prompted many NGOs to seek observer status at IWC annual meetings. These groups quickly publicized the shortcomings of IWC policies and stimulated international public support for changes in IWC policy. More non-whaling nations also became IWC members, which increased support for whale protection (Davis 1985).

Pressure from NGOs and non-whaling IWC members resulted in a "new management procedure," which required species quotas based on MSY. In 1982, concern for the status of whale stocks led to the adoption of a moratorium on commercial whaling starting in the 1985–86 season. IWC voted to continue the moratorium in 1991. The moratorium was upheld at the 1993 meeting of the IWC.

The moratorium has not ended whaling, in part because there are two exceptions to it. One exception, for aboriginal subsistence whaling, was established largely at the urging of the USA to protect bowhead whaling by Alaskan natives even though the bowhead is an endangered species under the U.S. Endangered Species Act. The other exception is for "scientific whaling," which has led to annual kills of as many as 300 minke whales (Martin and Brennan 1989), leading some to criticize scientific whaling as a guise for commercial whaling. Furthermore, the moratorium applies only to great whales, and has prompted the commercial killing of small cetaceans, including dolphins, porpoises, and bottlenose whales, and has caused heated debates about the jurisdiction of the IWC over small cetaceans (Birnie 1985).

In the 1980s, IWC implemented a limited International Observer Scheme, but conflicts of interest problems have weakened the credibility of observers. The lack of an effective, unbiased enforcement scheme is the major weakness of the ICRW.

The greatest threat that IWC faces is opposition to the moratorium. Japan, Iceland, and Norway argue that stocks of some whale species can support commercial kills, and these nations threaten to withdraw from the IWC if the moratorium continues. In fact, Iceland has declared its intention to withdraw.

Assuming that commercial whaling resumes, more effective conservation will be imperative in order to avoid a second round of destruction of whale stocks. Despite its weaknesses, IWC is the most viable means of ensuring the conservation of great whale species, perhaps even their continued existence.

United Nations Food and Agriculture Organization

The Food and Agriculture Organization (FAO) is the largest specialized agency within the UN system, both in personnel and budget. FAO was created in 1945 to alleviate hunger and malnutrition through improvement of food production and distribution. Headquartered in Rome (Italy), its present membership is about 160 nations.

FAO's Fisheries Department focuses on the use of marine and freshwater species as food, and is the principal coordinating agency for the world's fisheries. Its objectives are to serve member states by:

1) collecting, analyzing, and disseminating information on the occurrence, production, and use of living aquatic resources;
2) encouraging national and international action in research, education, and administration; and
3) promoting the wise use and conservation of living aquatic resources.

FAO's attempts to encourage coordination among and within its 25 bilateral or multilateral organizations or "fisheries commissions" involved in the management of fish stocks have been limited by the extension of coastal state jurisdiction to 200 nautical miles (371 kilometers), claims of national sovereignty, regional issues and conflicts, and the migratory nature of many of the stocks they attempt to manage.

The 1984 FAO World Conference on Fisheries Management and Development endorsed a strategy for fisheries management and development and approved action programs reflecting various priority areas for FAO support. These include planning, management and development of fisheries and aquaculture, assistance to developing countries in increasing the benefits obtained from intra- and inter-regional trade in fish and fishery products, and promotion of the role of fisheries in alleviating undernutrition. A more recent emphasis is the development of small-scale fisheries to ensure sustained production (FAO 1985).

Some FAO activities have addressed the need to protect marine species and ecosystems, but not all of its policies are environmentally sound. Indeed, FAO activities sometimes affect the environment in contradictory ways. For example, although FAO opposes the destruction of natural ecosystems, its active support of mariculture can have ecologically harmful consequences (Caldwell 1990). While it has recently shifted its emphasis to a less technological approach emphasizing sustainability, FAO has traditionally supported large-scale, intensive fisheries, infusing large sums of money for improved vessel and gear technology such as nylon gill nets that promote overfishing and incidental take, and it has consistently placed immediate social concerns above long-term environmental considerations. By encouraging overcapitalization of the fisheries, FAO's traditional approach has led to impoverishment of resources and coastal communities who depend on them (Troadec and Christy 1990). Similarly, FAO's stock assessment studies have usually aimed at developing the large-scale fishing sector rather than artisanal fisheries.

FAO's efforts to establish an international order to conserve marine life have been hampered by continued difficulties in obtaining the cooperation of national governments. FAO is funded mainly by Northern nations, some of which have powerful distant-water fishing fleets that operate mainly in the South. As a result, FAO is often criticized by coastal nations that fear that information FAO collects or FAO management plans will be used to advance the cause of distant-water fishing nations at their expense. Thus, international antagonisms have defeated attempts to implement comprehensive management of marine living resources.

FAO's greatest strength has been in collecting and disseminating biological data and in training fishery managers, particularly in countries where this type of training would not be available for any but top-level managers. While FAO's regional agencies have been occasionally called upon to help coastal nations set catch quotas, FAO has never been given regulatory power by any member nations. Its role remains, therefore, advisory at a time when regulation and enforcement are dearly needed. Smaller multilateral organizations whose members come exclusively from the countries bordering their area of influence appear more effective at proposing and implementing management plans.

United Nations International Maritime Organization

The Convention on the Inter-Governmental Maritime Consultative Organization, officially renamed the Convention on the International Maritime Organization, created the IMO in 1958 to encourage cooperation among member states in matters concerning shipping, including the adoption of the highest practicable standards for maritime safety, the efficiency of navigation, and the prevention and control of marine pollution from ships. IMO has a membership of 135 nations and is headquartered in London (UK).

The most notable of the numerous conventions that IMO's Legal Committee has drafted or implemented are the International Convention on the Establishment of an International Fund for Compensation, the Convention on the Prevention of Marine Pollution by Dumping of Wastes and Other Matter, the 1973 International Convention for the Prevention of Pollution from Ships as modified by the Protocol of 1978 (MARPOL 73/78), and the International Convention on Oil Pollution Preparedness, Response and Co-Operation of 1990.

IMO's approach to protecting the sea is strongly influenced by the problems that many member nations say they have in implementing such measures without harming their own economies. But it has also developed a Global Programme for the Protection of the Marine Environment, which seeks to apply the "anticipate and prevent" approach (Figure 8-2) advocated by the World Commission on Environment and Development (1987). This program focuses on training, but also incorporates efforts to identify problems and risks that developing countries face. IMO has recently turned its attention to reducing air pollution from ships.

In contrast with other UN specialized agencies, IMO is controlled by private interests: representatives from the shipping, insurance, and banking systems. Northern nations dominate and the USA is a driving force on the issue of pollution control.

IMO's activities are reported in *IMO News*.

United Nations Environment Programme and the Regional Seas Programme

The UN Environment Programme (UNEP) was established by the General Assembly of the United Nations following the landmark 1972 UN Conference on the Human Environment in Stockholm. UNEP is vested with the

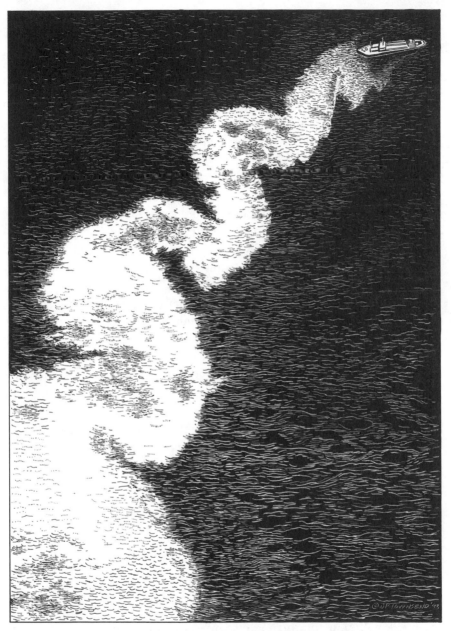

Figure 8-2. Tanker leaking oil. *Despite decades of attempts to improve tanker safety, huge oil spills in late 1992 and early 1993—from the* Aegean Sea *near Spain, the* Braer *near the UK, and the* Maersk Navigator *off India's Nicobar Islands—reenforced the need for the "anticipate and prevent" approach, including greatly strengthened tanker design and operation standards.*

overall responsibility for implementing the recommendations in the action plan drafted at the Stockholm Conference. It is headquartered in Nairobi (Kenya).

UNEP was designed to ensure that governments give adequate consideration to emerging environmental problems having global significance. Its basic functions are catalytic and include the dissemination of information, the cultivation of understanding, and collaboration with the environmental programs of other agencies and with national governments and NGOs. Although UNEP has no power to implement the programs it proposes, it relies on governments and other international organizations to carry out its programs.

UNEP is involved in numerous environmental programs, including an environmental assessment program (Earthwatch), the Global Environmental Monitoring System (GEMS), the International Referral System for Sources of Environmental Information (Infoterra), and the International Register of Potentially Toxic Chemicals (IRPTC).

UNEP has been the leader or a cosponsor of a number of major documents on biological diversity, including the *World Conservation Strategy* (IUCN et al. 1980), *Coral Reefs of the World* (UNEP and IUCN 1988/89), *Caring for the Earth* (IUCN et al. 1991), the *Global Biodiversity Strategy* (WRI et al. 1992), and *Global Biodiversity: Status of the Earth's Living Resources* (Groombridge 1992).

The most significant of UNEP's achievements in the marine environment is its Regional Seas Programme, which attempts to address common environmental problems in selected shared bodies of water by promoting cooperation on coastal and marine matters of regional common concern. By 1990, over 120 countries were participating in 10 Regional Seas Programmes: the Mediterranean, the Kuwait region of the Persian/Arabian Gulf, the Red Sea, the Eastern Africa coast, the South Asian Seas, East Asia, the South Pacific islands, the Pacific coast of South America, the wider Caribbean, and the Atlantic coast of West and Central Africa. Regional Seas Programmes are aimed at fostering:

1) an Action Plan for cooperation in coastal and marine resource development, pollution control, and research and monitoring;
2) a convention setting forth the general commitments of the countries of the region, and

3) more specific and detailed protocols to deal with particular issues and objectives. Each regional action plan has a research and training component and seeks to promote standardized monitoring and assessment.

With the Mediterranean Sea Programme, the first Regional Sea Programme, which 18 nations have joined since 1976, UNEP sponsored the Barcelona Conferences, which led to the Convention for the Protection of the Mediterranean Sea Against Pollution (Barcelona Convention) and the Mediterranean Action Plan. The Barcelona Convention is a general commitment by 11 coastal nations to cooperate in ongoing environmental programs, including monitoring and research, data exchange, waste dumping control, and pollution emergencies. The Convention is considered by some to be one of the most comprehensive regional agreements regulating sources of marine pollution (Caldwell 1990). The Mediterranean Action Plan now has four protocols, each binding to its signatories, of which the most significant deals with land-based pollution. UNEP launched a similar program in 1978 in the Persian/Arabian Gulf, resulting in the adoption by eight coastal states of the Kuwait Regional Convention for Cooperation on the Protection of the Marine Environment from Pollution.

Among existing Regional Seas cooperative arrangements, one of the most fully developed and innovative is the Caribbean Environment Programme, the Action Plan for which was adopted in 1981, with its accompanying Convention for Protection and Development of the Marine Environment of the Wider Caribbean Region (Cartagena Convention)—which was adopted in 1983 and had 19 Parties as of 1992—and its two protocols: the Protocol Concerning Cooperation in Combatting Oil Spills in the Wider Caribbean Region and the Protocol on Specially Protected Areas and Wildlife of the Wider Caribbean Region (SPAW). Negotiations began in 1992 on a third protocol dealing with land-based sources of marine pollution.

The SPAW Protocol, which had 14 Contracting Parties by late 1991, calls for the creation of a regional network of protected areas to conserve and restore ecosystems in the Wider Caribbean and to maintain the ecological processes essential to their functioning. The parties agree to protect specific components of these ecosystems, such as coral reefs, mangrove forests, and seagrass beds, in light of their particular vulnerability to physical alteration, pollution, global climatic change, and resulting sea level rise.

Regional Seas Programmes are devised to solve the particular problems of the various regions. The West and Central Africa Action Plan is especially concerned with land-based pollution and coastal erosion. The Action Plan of the Philippines, Thailand, Malaysia, Indonesia, and Singapore focuses particularly on coral reefs and fisheries and combating marine pollution.

It is difficult to assess the actual effect of UNEP and its Regional Seas Programme on the world's ecological systems. Although coverage of the Regional Seas to date is not complete, the drafting of action plans for the North-West Pacific and the Black Sea is well under way, while the South-West Atlantic Action Plan is still in the planning phase. UNEP's Regional Seas Programme is important because it balances its influence against member nations' own assessments of hazards to the environment. Moreover, it has provided a forum acceptable to the South to examine mutual problems free from the suspicion generally attached to solutions proposed by Northern countries.

The Global Plan of Action for the Conservation, Management and Utilization of Marine Mammals (MMAP) was developed between 1978 and 1983 by UNEP and FAO, in collaboration with other international governmental bodies and NGOs concerned with marine mammal issues. The objective of the Plan is to promote implementation of a policy to conserve, manage, and use marine mammals that would be widely accepted to governments and the public. MMAP is built around five concentration areas, namely, policy formulation, regulatory and protective measures, improvement of scientific knowledge, improvement of law and its application, and enhancement of public understanding.

UNEP came into being because governments face common environmental problems, are disturbed by environmental threats beyond their jurisdiction, or need to harmonize environmental policies relating to international trade (Caldwell 1990). Its shortcomings reflect the low priorities that some nations give to solving environmental issues, nations' jealously guarded sovereignty, and the complexity and magnitude of the problems UNEP is seeking to address. In light of those drawbacks, UNEP's achievements are substantial (Gray 1990).

UNEP's Regional Seas activities are reported in *The Siren*.

The World Conservation Union

The World Conservation Union (IUCN, formerly the International Union for Conservation of Nature and Natural Resources), with a membership of some 61 countries, 128 government agencies, and 416 NGOs spread across 118 countries, is unique in its size and composition, in having elements of both IGOs (previous section) and NGOs (following section), in its success in serving as a liaison between governments and NGOs, and in its influence. Established in 1948 following an international conference sponsored by the UN Education, Scientific, and Cultural Organization (UNESCO) and France, IUCN is now headquartered in Gland (Switzerland), with an environmental law center in Bonn (Germany).

IUCN has been a world leader in explaining the importance of and threats to biological diversity. In 1980, IUCN, UNEP, and World Wildlife Fund published the *World Conservation Strategy* (WCS), a global conceptual framework to integrate conservation with sustainable development. Its focus was primarily terrestrial, but it addressed marine conservation issues within the larger context of maintaining essential ecological processes and life-support systems, preserving genetic diversity, and ensuring the sustainable use of species and ecosystems upon which the world's people depend. In 1991, the same partnership joined forces again to produce *Caring for the Earth*. This work is intended to renew, update, and extend the global commitment to the basic concepts of the WCS and to guide their implementation in the years ahead by promoting a new ethic for sustainable living on Earth that integrates conservation and use of living resources. IUCN was also a major partner in producing seminal documents including *Conserving the World's Biological Diversity* (McNeely et al. 1990), *Guidelines for Establishing Marine Protected Areas* (Kelleher and Kenchington 1991), the *Global Biodiversity Strategy* (WRI et al. 1992), and *Global Biodiversity: Status of the Earth's Living Resources* (Groombridge 1992).

IUCN convened a conference in Tokyo (Japan) in 1975 that identified the need for a global network of marine protected areas (IUCN 1976), and has worked since then to establish it. At its 1982 IIIrd World Congress on National Parks and Protected Areas in Bali (Indonesia), IUCN held a workshop that resulted in a widely used guide (Salm and Clark 1984) for planners and managers of marine protected areas. At the 1992 IVth World Congress on National Parks and Protected Areas in Caracas (Venezuela),

IUCN's Commission on National Parks and Protected Areas (CNPPA) held a workshop to establish criteria for according priority and to delineate areas that merit highest priority for protection. IUCN's Ecology Commission has addressed some issues of marine ecosystem conservation, and its SSC has become a major force in conserving certain groups of marine species.

IUCN's activities are reported in the *IUCN Bulletin*.

Non-governmental Organizations

NGOs play crucial roles in conservation research, education, and advocacy, and are uniquely positioned to influence public debate and decision making about life in the sea. They analyze questions of vital importance that would otherwise be neglected, increase public awareness, and promote passage of and compliance with domestic laws and international agreements. They are guardians of the public interest, a niche that should be filled by government but often is not. NGOs have grown dramatically in number and influence worldwide, including, most recently, in Eastern Europe. Their growth has both benefited from and been instrumental in the emergence of democratic governments. Many new NGOs look to IUCN or to established NGOs for help in managing their organizations, securing funding, and carrying out their programs.

Interests that favor short-term over long-term economic benefits often characterize NGOs as "vocal special interests" that are motivated by emotions and philosophies held by only a minority of the public. These criticisms ring hollow. Increasingly, NGOs represent their constituents with professional knowledge that gives them special credibility along with passionate commitment. Polls in countries worldwide show that the public increasingly shares the views of conservationists and trusts NGOs more than industry or government.

Opponents of sustainable use and environmental protection often depict NGOs as naysayers that offer no alternatives to proposed development activities. This criticism hits closer to the mark; some NGOs relish the role of gadfly. But, sensitive to this criticism, an increasing number have responded by hiring people who specialize in finding environmentally responsible solutions with broad economic and social benefits.

Understandably, there is often tension between industries or government agencies and NGOs. Nevertheless, many people in these institutions agree that NGOs play a vital role in citizen oversight of government and industry by "keeping them honest," and that development schemes are often better environmentally *and* economically because of NGO involvement. This respect can pave the way for effective cooperation. Indeed, in some countries, NGOs are funded directly by the central government. NGOs must be aware, however, that the benefits of cooperation must not undermine their primary responsibility to provide oversight and advocacy for those who want to live sustainably.

When vital issues are ignored by decision makers, when laws are unenforced, when ignorance, greed, and corruption threaten the sustainability and future of life locally, nationally, or globally, it is usually NGOs that alert the public. They must act through established administrative, judicial or legislative processes, or, where there is no alternative, outside them.

The California Institute of Public Affairs, in cooperation with the Sierra Club and IUCN, has published a *World Directory of Environmental Organizations* (Trzyna 1989), and IUCN (1991) publishes a compendium of government agencies, their affiliates, and international NGOs that are members of IUCN. Yolen (1991) and Gordon (1993) provide useful information about US and Canadian NGOs working in marine conservation, and Partners of the Americas (1988) has published a compendium of governmental and non-government organizations concerned with natural resources in Latin America and the Caribbean.

Many NGOs worldwide do at least some work on marine issues. Some NGOs that make them a major part of their agendas include:

Center for Marine Conservation (CMC). This Washington, D.C. (USA)–based NGO focuses entirely on the sea. Strongly science-oriented, its research, education, and interactions with government emphasize fisheries, protected areas, solid-waste pollution, marine mammals, sea turtles, and other protected species in the USA and, increasingly, in other nations. CMC administers the annual International Coastal Cleanup Campaign, with nearly 150,000 volunteers cleaning up and documenting solid-waste pollution, and its president chairs IUCN's SSC Trade Specialist Group. CMC is a leader in encouraging the growth of marine conservation biology, and is cooperating with IUCN and World Wildlife Fund to encourage implementation of *The Strategy*'s recommendations.

Chesapeake Bay Foundation (CBF). Although this US–based organization focuses almost exclusively on the largest estuarine bay in the USA, the CBF is one of the world's largest marine conservation NGOs. Emphasizing the education of schoolchildren and the general public, the Foundation also conserves land and is the leading advocate working with government agencies and private organizations for restoring the bay. Its success has led it to share its expertise with other organizations. Recently, for example, it helped the Center for Marine Conservation and the Centro para la Conservación y Ecodesarrollo de la Bahía de Samana y su Entorno establish new conservation efforts for Samana Bay (Dominican Republic), the largest estuarine bay in the Caribbean.

The Cousteau Society. Headquartered in Chesapeake, Virginia (USA), this NGO is among the world's best-known conservation organizations and has achieved great success in raising public awareness about the sea and threats to it. The producer of highly acclaimed television programs and books based on Cousteau expeditions for underwater exploration around the world, Captain Jacques Yves Cousteau and his son Jean-Michel helped secure international support for the new Antarctic treaty that includes a moratorium on mining and have recently launched a campaign directed at UN acceptance of a Bill of Rights for Future Generations.

Environmental Defense Fund (EDF). Headquartered in New York City, New York (USA), EDF conducts marine conservation work that focuses primarily on marine aspects of global atmospheric change, estuaries and wetlands protection, chemical pollution, and Antarctica. Emphasizing the interactions among science, economics, and law, EDF is notable for efforts to harness market forces as a means of diminishing pollution, a prominent advocate for the U.S. Endangered Species Act, and a world leader in wildlife law. It has played a major role in the research and advocacy of treaties to slow stratospheric ozone depletion and greenhouse warming.

Greenpeace. With national organizations in 30 countries, Greenpeace is probably the world's largest environmental NGO. In addition to its efforts to stop nuclear testing and keep the sea free from nuclear weapons, Greenpeace gained world attention by trying to halt the slaughter of whales and seals. It devotes considerable resources to toxic pollution, offshore oil and gas development, marine fishes, coral reefs, Antarctica and the Arctic. Greenpeace is known worldwide for nonviolent civil disobedience, but also conducts a broad range of education, research, hands-on conservation, and advocacy programs. An example of these less-known activities is a program to help install mooring buoys in coral reefs to prevent anchor damage.

Marine Conservation Society. Based in the UK, the Marine Conservation Society conducts programs both within the UK and in other countries, mainly in the British Commonwealth. Its science-based activities include efforts to document souvenir trade in marine wildlife and to conserve coral reefs.

Ocean Voice International (OVI). Headquartered in Ottawa (Canada), OVI is a small organization that has tackled some sizable projects in tropical countries, Canada, and the USA. With IUCN, it has worked to save Shiraho Coral Reef (Japan) from a jumbo-jet airport. With the Haribon Foundation in the Philippines, it trains aquarium fish collectors at the community level to use small nets instead of poisons, has produced a global GIS equal-area grid to show species and human-impact hotspots for conservation of coral reefs and other marine areas, and has developed a coral reef conservation manual for coastal communities. It publishes a quarterly educational bulletin, *Sea Wind*.

World Wide Fund for Nature and World Wildlife Fund (WWF). Based in Switzerland and the USA, WWF and its associated national organizations are world-renowned for conducting and funding research and hands-on conservation on terrestrial, freshwater, and marine species and ecosystems worldwide. Its new marine initiative focuses on diminishing threats from pollution; developing good models of CZM, encouraging ecosystem-oriented fishery management, and protecting species of special concern. WWF-US's marine conservation scientist cochairs the IUCN's Ecology Commission, which increasingly focuses on marine issues.

Inventory, Research, and Monitoring Programs

Inventory, research, and monitoring of biological diversity are much more difficult in the sea than on land because of factors unique to the sea:

1) its geography, three-dimensionality, and inaccessibility of life;
2) the difficulties of sampling and studying marine life; and
3) the oceans' status as a global commons.

Much of the marine research relevant to conserving marine biological diversity has employed time tested methods used on land. On land, biologists can use these techniques to determine patterns of biological diversity and the geographic ranges and life cycles of species. Biologists can then map them, and use reasonably established ecological theory to determine minimum boundaries that would allow maintenance of biological diversity. Government or private conservation agencies might then acquire the land for a protected area, usually working within one country, perhaps using innovative economic mechanisms such as debt-for-nature swaps.

Unfortunately, methods used on land seldom work in the oceans because of the differences between land and sea. Without taking into account the duration and spatial scale of processes that maintain marine biological diversity, efforts to conserve marine life cannot succeed. Thus, any meaningful strategy to protect marine biological diversity must be international and long term in scope.

Inventory

There have been few attempts to determine the extent of species diversity within marine ecosystems. Recent decades have seen an alarming decrease in the number of experts in taxonomy and systematics collections of marine organisms. Moreover, sampling difficulties have slowed attempts to inventory even a significant fraction of the taxa in coral reefs, which are the most diverse shallow-water tropical ecosystems. While the deep ocean is very diverse, sampling difficulties have been formidable, and deepsea diversity is still poorly known (Grassle 1991). On the other hand, the systematics of some marine groups, such as shallow-water tropical fishes, are well studied, and there are reliable estimates of their diversity (Personal communication with John Randall, Bernice Pauahi Bishop Museum, Honolulu, Hawaii, USA; Sale 1977).

Given these difficulties, gaining sufficient understanding of the patterns of marine biological diversity soon enough to be useful for conservation will have to rely on indirect measures. One way is to assume that overall species diversity will vary with the diversity of species in well-known groups. For example, where marine fish species diversity is high, the diversity of other groups supporting that diversity should also be high. However, groups of organisms with different life histories can have very different distribution patterns, so these correlations need to be established carefully. Another indirect approach is to assume that diversity will vary with the taxa that provide much of the physical structure of the ecosystem. For example, relatively few coral species form the principal structure of most reefs. If this structural framework is disturbed or destroyed, it is reasonable to assume that the diversity of the ecosystem will be diminished. Other important tropical ecosystems, such as mangroves and seagrass beds, have only a few major structural species and might be thought of in the same way. The importance of structure-forming species to species diversity in some ecosystems suggests that satellites that can map distributions of structure-formers could be used to inventory, study, and monitor species diversity.

The need for international cooperation in marine biodiversity inventories is nowhere clearer than in the Caribbean Sea. The extensive coastal ecosystems surrounding the Caribbean's thousands of islands and cays are the most biologically diverse part of the Atlantic Ocean basin. Unfortunately, this geographic complexity also subdivides the region politically. There are 36 governments, the world's highest concentration of small nations. Some of the Basin's most significant biological processes operate on a scale that is far larger than the institutional capacity of these small countries acting separately.

The Center for Marine Conservation recently established CARIDAT, an international database of marine biodiversity that provides a computerized inventory of Caribbean marine life and is supported by a network of non-governmental organizations, natural history institutions, and government agencies. CARIDAT's primary role is to improve the scientific baseline for conserving Caribbean biodiversity by creating a regional mechanism for information flow, for collaborative inventories of marine biota, and for intra-Caribbean cooperation in collection-based research.

The most extensive body of existing data on Caribbean biological diversity is associated with reference collections that are mostly in institutions outside the Basin. This resource contains critical baseline data on patterns of biodiversity and records of past trends and ecological change. By providing access to these data, CARIDAT will jump-start the national inventory programs that are being instituted in several countries. The project empowers Caribbean-based specialists by making data available not only in the form of published reports, but also through access to the raw data that can be used to produce them. CARIDAT provides access to ecological, geographic, and taxonomic data in research collections in North, Central, and South America.

CARIDAT is intended to promote the careers of biodiversity specialists within the Basin. Inventories are conducted by multinational teams. At little additional cost, such inventories provide specialists from island countries with broadened geographic experience that will help them to recognize new taxa or other special features of their own biotas. Working groups have been organized to establish taxonomic, ecological, and geographic standards; these groups, comprised primarily of specialists inside the project area, will help create peer communities of resident Caribbean scientists and conservationists. Together with database networks and multinational inventory teams, they provide the means to overcome the barriers to collaboration. Participating institutions are provided with a geographic interface designed to produce publication-quality maps of species distributions based on museum locality records.

Data from inventories will have double value because initial inventory sites have been chosen so that each will contribute to research programs of local professionals and will emphasize areas where marine biodiversity preserves have been proposed. As the database accumulates distributional data, it will also help assure that economic development takes place in a sustainable manner. Most economic ventures must become established within a narrow window of opportunity in order to be successful, which can bring about conflict with conservationists concerned about ecosystems that are complex but usually poorly understood. A comprehensive baseline of distribution data will allow the timely evaluation of the likely effects of proposed new ventures and development policies, in addition to improved biodiversity management. By building horizontal links among scientists in different nations, such networks can remedy the pressing need for decision makers to know which areas have special biological significance.

Research and Monitoring

There are few international research programs on ecosystem composition, structure, and function, although many recent discussions and workshops, stimulated largely by the potential impact of global climatic change, have called for such programs (Pernetta and Hughes 1990; UNEP et al. 1990). While research and monitoring have traditionally been distinct activities, scientific understanding of biological diversity and its functioning depend upon long-term data sets from monitoring programs. Furthermore, ecosystem research has usually been constrained to single sites or small spatial scales that cannot encompass the range of ecosystem variability that scientists and managers need to understand. Gaining enough understanding to ensure the sound management of marine ecosystems requires long-term study at an expanded geographic scale, including comparative work at multiple sites within ecologically coherent regions as defined by patterns of endemism and their connection by ocean currents.

There are strong reasons for adopting a regional approach to research and monitoring. Currents and the long planktonic larval lives of many species are unifying forces within regions that are subjected to the same natural processes and whose species are generally in reproductive contact. Regional research using remote sensing technology and events that have affected entire regions suggest that the Caribbean Sea, for example, can be treated as an LME (Lessios et al. 1984; Ogden and Wicklund 1988; Muller-Karger et al. 1989). Research and monitoring of this scale and complexity depend upon recognition by nations in the region about the need for cooperation.

The Caribbean Coastal Marine Productivity (CARICOMP) program began in 1986 to study comparative ecosystem structure and function over the long term on a scale encompassing the range of natural ecosystem variability (Ogden and Gladfelter 1986). Based at over 20 marine laboratories and research institutions (Table 8-1) that are the centers of historical data and expertise on coastal ecosystems within their nations, CARICOMP has established standardized methods of observing ecological and physical variables in time series. A Data Management Center at the University of the West Indies in Kingston (Jamaica) will facilitate data exchange and analysis. The program is supported by funding from UNESCO, the U.S. National Science Foundation, and the John D. and Catherine T. MacArthur Foundation. A similar program, PACICOMP, linking Pacific marine laboratories, was born at the Pacific Science Congress in Honolulu, Hawaii (USA) in 1991 and is being developed.

Such large-scale, long-term ecosystem research and monitoring programs have been helped by regional conventions. For example, the political basis for cooperation in the Caribbean was established by the Cartagena Convention, to which many Caribbean countries are signatory.

New global research and monitoring programs have also been proposed to provide the information needed to manage the sea. The International Union of Biological Sciences (IUBS), with the Scientific Committee on Problems of the Environment (SCOPE) and UNESCO, has created an administrative and scientific structure to examine the ecosystem function, origin, maintenance, loss, and inventory of biological diversity (di Castri

TABLE 8-1. Laboratories in the CARICOMP Network

The institutions marked (*) attended the December 1990 Workshop on Ecosystem Monitoring Methods held December 2–7, 1990, at Discovery Bay Marine Laboratory, University of the West Indies (Jamaica). Those marked (**) have also signed a Memorandum of Understanding with the Steering Committee pledging their long-term support for the program.

** Universidad Simón Bolívar, Venezuela
** Fundación La Salle de Ciencias Naturales, Venezuela
Instituto Oceanographico, Universidad de Oriente, Venezuela
* Fundación Cientifica Los Roques, Venezuela
** Foundation CARMABI, Curaçao, Netherlands Antilles
** INVEMAR, Santa Marta, Colombia
Centro de Investigaciónes, Marinas de Punta Betin, Colombia
Universidad de Panama, Panama
* Hol Chan Marine Reserve, Belize
* Smithsonian Institution, Carrie Bow Cay, Belize
* Instituto de Recursos Naturales (IRENA), Nicaragua
Secretaria de Recursos Naturales, Honduras
** Universidad Nacional Autonoma de Mexico, Puerto Morelos, Mexico
** Centro de Investigaciónes (IPN), Merida, Mexico
* EPOMEX, Campeche, Mexico

* Universidad de la Habana, Cuba
** Instituto de Oceanologia, Cuba
** University of the West Indies, Jamaica
** Natural Resources Unit, Cayman Islands
** Universidad de Puerto Rico, Puerto Rico, USA
* University of the Virgin Islands, USA
** Florida Institute of Oceanography, Florida, USA
** Saba Marine Park, Saba, Netherlands Antilles
* Université Antilles-Guyane, Guadeloupe
** Caribbean Environmental Health Institute, St. Lucia
* Institute of Marine Affairs, Trinidad and Tobago
** Bellaires Research Institute, Barbados
Ministry of Natural Resources, Suriname
** Bermuda Biological Station, Bermuda
Caribbean Marine Research Center, Bahamas

and Younes 1990). And, the ecological role of biological diversity has been proposed for investigation within an International Network of Marine Research Stations (MARS Network) that would examine ecosystem processes on a regional scale and on a time scale of 100 years (Grassle et al. 1991).

Even more than on land, understanding the workings of life in the sea requires international, interdisciplinary efforts at larger geographic and time scales than previous efforts. Developing global and regional programs that link research institutions and use standardized methods is a crucial step in learning how to protect and sustainably use marine species and ecosystems.

National and Local Organizations and Programs

The large extent of marine ecosystems relative to the length of many nations' coasts often requires international cooperation on marine conservation questions. Yet international institutions are generally weak compared with national and local governments and other organizing forces within nations. This underscores the wisdom of the adage "Think globally; act locally." No matter how capably international institutions work to diminish threats to the sea, most effort will be expended at local or national levels. The following sections include examples that suggest what works and what does not.

Shiraho Reef, Japan

Although the condition of coral reefs is increasingly becoming a global concern, their fate is being determined at thousands of individual localities. Among the world's most endangered coral reefs are those of Japan's subtropical Ryukyu Archipelago (Veron 1992). There, along the entire 1,000-kilometer (620-mile) stretch of islands, more than 95 percent of the reefs are severely stressed or "dead" (Muzik 1985), largely as a result of intensive modern development, including agriculture and construction for industry and tourism. Until recently, one exception was a thriving reef near Shiraho village, on Ishigaki Island, southwest of Okinawa.

Although it is only five kilometers (three miles) long, Shiraho is one of the most diverse coral reefs remaining in the Ryukyus. More than 135 species of corals, including the world's largest known colonies of rare blue coral (*Heliopora coerulea*), and several hundred fish species can be found there (Moyer 1989).

In 1979, Ishigaki township proposed a new airport with runways large enough to accommodate jumbo jets for Ishigaki Island. Intended to increase tourism, the airport was to be built on land reclaimed from the reef in Shiraho Lagoon. Funds were to come from the national government. Opposition from Shiraho villagers, who were not consulted in the planning process, was immediate, and continues unabated.

The local Fisherman's Cooperative first opposed the plan, but later waived its fishing rights to the landfill area. Japan's national Finance and Transportation Ministries approved the plan in 1981 and 1982, respectively. By 1982, surveyors were guarded by riot police and peaceful demonstrators were arrested.

In 1984, the first scientific survey of the reef was published in a Japanese scientific journal, describing the unusual diversity at Shiraho and proposing its designation as a national park. Surveys by IUCN in 1987 and World Wide Fund for Nature in 1989 contradicted the Japanese Environmental Agency's assessment of the value of Shiraho reef and challenged assertions that technological solutions would protect the reef from construction damage. The surveys, and increasing local, national, and international protest, led to alterations of the airport plan, but opponents still deem these inadequate to protect the reef.

No less threatening are agricultural "improvements" begun in 1975 that will cover approximately 60 percent of Ishigaki's arable land. Logging, soil disturbance, and river channelization are greatly increasing sedimentation in Shiraho Lagoon. Doumenge et al. (1990) underscored the threat from agriculture and other development to the survival of Shiraho's coral reefs.

Shiraho has become a symbol for environmental conservation in the Ryukyus and Japan in general. Years of opposition to the airport have helped to spawn and nurture growing grass-roots environmental movements everywhere in Japan, and to draw worldwide attention to the issues. Whether Japan will ultimately decide to protect its finest coral reefs, however, is uncertain.

Federal Management of Marine Fisheries, USA

No species in the USA are so relentlessly hunted for commercial profit and recreational pleasure as marine finfishes and shellfishes. Until 1976, marine fishing was regulated only rarely by federal and state governments. But alarm over the heavy take of groundfish off New England by foreign factory trawlers and discussions in UNCLOS III led the US Congress to establish one of the most complex regimes of fisheries management in the world.

Like many other regimes that have been cobbled together by committee and political compromise, however, marine fisheries management in the USA suffers from some congenital problems. One arises from the political fiction of boundaries. Only under exceptional circumstances does the federal government exercise management authority over fishing within state waters, which stretch 3 miles (4.8 kilometers) from shore off most states. State and federal agencies jealously guard these boundaries, but fish populations and fishers do not. The result is fishery regulations that are inconsistent between state and federal waters and among states. This not only confuses fishers; it also increases the difficulty of enforcement. When Congress passed the Magnuson Fishery Conservation and Management Act of 1976, it was aware of the difficulties that these boundaries would create, but could not overcome opposition from the states to federal management within state waters. This artificial division has hampered fishery management ever since.

A second defect concerns one of the innovations of the Magnuson Act: the eight regional fishery management councils. Congress created these councils, which have responsibility for developing fishery management and are composed largely of state fishery officials and commercial and recreational fishers, in the belief that fishers would observe regulations that they had developed. The Secretary of Commerce, through the National Marine Fisheries Service (NMFS), is responsible for determining whether these plans comply with the standards in the Magnuson Act.

This arrangement has institutionalized special interests in fishery management. Such interests also have substantial influence on the Congressional committees that oversee the fishery management process because the committees are dominated by members of Congress who have strong constituencies in marine fisheries. The public interest in sound resource management takes a back seat to ingrained business and political practices.

Furthermore, fishery management councils are not only responsible for allocating fishery resources but for determining the size of the catch. Although some councils have sought the advice of knowledgeable scientists, others have spurned scientific advice. As open access has attracted more and more fishers, competition has grown and the councils have often responded by increasing the catch quota in a futile attempt to maintain everyone's traditional share. In the New England groundfish fishery, which in the late 1970s was recovering from heavy fishing by foreign factory trawlers, the council abandoned quotas altogether and allowed the groundfish fleet to double, driving groundfish stocks to record low levels.

The Magnuson Act places only scant restriction on open access to US fishery resources. For example, the introduction of more effective fishing gear is almost unfettered. Dramatic improvements in net materials, boats, motors, and navigational equipment would have sharply increased the effective fishing effort in many fisheries even if no additional fishers had entered them.

The councils and NMFS have interpreted the Magnuson Act's support for fishing efficiency narrowly, to mean whatever will take the most target fish quickly and cheaply. The result in some fisheries is massive bycatch and discarding of invertebrates, fishes, sea turtles, marine mammals, and seabirds. In the southeastern US shrimp fishery, an average of 10 kilograms of finfish are discarded for every kilogram of shrimp. In many instances, NMFS and programs such as the Sea Grant Extension Service have promoted the use of fishing gear and techniques that cause high levels of bycatch and discard.

Often at the insistence of Congress, NMFS has contributed to overfishing through vessel subsidies and marketing programs. In the late 1980s, Congress established a National Seafood Promotion Council, whose goal has been to increase US fish consumption from 6.8 kilograms (15 pounds) to 9.1 kilograms (20 pounds) by the year 2000. It is difficult to find the source of this additional biomass without assuming further overfishing of stocks.

Fragmented decision making also harms federal fishery management in the USA. As elsewhere in the world, US agencies responsible for conserving fishery resources only control fishing, not activities that degrade the fishes' habitats. For example, NMFS can only *advise* another federal agency on whether a permit should be granted to fill a coastal wetland. The loss of coastal wetlands along the southeastern US coast, which are the major nursery grounds for finfish and shellfish stocks, has been accompanied by dramatic declines in commercial finfish and shellfish. In the Pacific Northwest, NMFS has little or no control over operation of the great dams on the Columbia River, which kill about 90 percent of the salmon smolt migrating to sea, or over logging practices that degrade water quality and smother salmon nursery areas. It makes little sense to promote fishery management if the habitats essential for fish reproduction are destroyed or degraded.

Finally, fisheries conservation in the USA and elsewhere is handicapped by the limited involvement of NGOs. The USA's fishery management process is complex and intensely political. Some conservation groups have expanded their focus from popular sea turtles and whales to fishes, but their influence does not yet match that of the fishing industry. Until NGOs have a better knowledge base and commit more resources to fisheries management, the public interest in having sustainable fisheries and maintaining marine biodiversity will be poorly served.

Sea Turtle Conservation, Costa Rica

A number of innovative projects to protect sea turtles have been developed in Costa Rica, where green, olive ridley, and leatherback turtles nest in abundance. Rich in biotic resources but with limited funds to protect them, Costa Rica promotes conservation in ways that are compatible with local needs. The value of this approach is indicated by the fact that Costa Rica now has the most abundant sea turtle populations in Central America.

On the Caribbean coast, the green turtle rookery at Tortuguero is the site of the world's longest sea turtle study, a program initiated more than 30 years ago by the eminent sea turtle biologist Archie Carr. In recent years, increasing numbers of visitors have flocked to this once-remote beach, which has been a mixed blessing because tourists provided the local village with much-needed income but disrupted nesting and research. The problem was resolved by developing a guide program, in which researchers train local villagers in the basics of sea turtle biology and the "etiquette" of watching turtles without disturbing them. The guides receive a small reimbursement, which allows the local populace to benefit from this remarkable turtle resource.

At Ostional on the Pacific coast, the Costa Rican government has addressed local needs by instituting a regulated collection of olive ridley eggs. Ridleys are unique among sea turtles in that they nest in large aggregations known as *arribadas*. In the past, eggs were a popular commodity and source of income for many local villages. To conserve sea turtles, the government halted egg collection in 1983. The policy was subsequently altered at the urging of leading conservationists, to allow communities to take eggs from the first arribada (many of which would be destroyed by later nesting turtles) to be sold under a regulated system. Despite some problems, the system appears to be working overall, providing a valuable economic benefit to local people from protection of adult turtles.

Although unregulated collecting of sea turtle eggs is a punishable offense in Costa Rica, authorities have been reluctant to enforce this

---

regulation vigorously, which has spurred widespread egg poaching. When an experimental program placed Boy Scouts on a Caribbean beach during nesting season to clear debris, move nests threatened by erosion, and collect data on nesting leatherbacks, there was an unanticipated benefit: The egg collectors stayed away. This program, which has now been expanded to other Costa Rican beaches, exemplifies conservation at its best. The scouts learn about conservation first-hand, local businesses participate by funding the program, and the turtles are protected from poachers.

Costa Rica is a small, developing nation. But programs such as these have made it the world leader in finding ways for biological diversity to benefit local people on a sustainable basis.

Sian Ka'an Biosphere Reserve, Mexico/Belize

The Sian Ka'an Biosphere Reserve is located on the Caribbean coast of the Yucatán peninsula. Its objectives are varied: conserving nearshore marine resources associated with the Mexican-Belize Barrier Reef, sustaining the ways of life of local Mayan peoples, and protecting lowland dry tropical forest of the coastal zone. The biosphere reserve management structure gives the local community management authority and paves the way for a strong commitment to stewardship.

Sian Ka'an provides one of the best examples of small-scale marine conservation. A lucrative spiny lobster (*Panulirus* spp.) fishery is based in Punta Allen, in the heart of the coastal reserve. The locally managed fishing cooperative limits access to the fishery and negative impacts on the marine environment, while keeping profit margins high.

In the spiny lobster fishery, fishers are leased tracts of the nearshore waters for long-term use. Their activities are not highly regulated, but they must keep to their own leases. With a combination of traditional knowledge and recent technical training, the fishers have come to see that ecological sustainability and economic profitability over long time spans are synonymous.

The system has evolved to a point where fishers take only adult male lobsters, not egg-carrying females or immatures. To boost lobster populations, the Sian Ka'an fishers build "cassitas:" small "lobster condominiums" that act as artificial reefs in providing safe hiding spaces for the young lobsters. Every effort is made to maintain water quality and protect the habitats on which the spiny lobster population depends: mangrove forests, seagrass beds, and coral reefs.

The success of the spiny lobster fishing cooperative has opened the door to marine conservation on a wider scale. Fishers of Sian Ka'an are highly vocal about the importance of a thriving coastal ecosystem, and are vigilant environmental guardians. The Biosphere Reserve has given them a voice in local government, and they are often the first to speak out about environmental issues. They seem to be taking the lead in identifying indirect impacts on the marine environment, such as the widespread damage caused by siltation on the reefs. Good stewards they most certainly have become: a model for community-based sustainable use and conservation everywhere.

Hol Chan Marine Reserve, Belize

In 1987, Belize established its first marine protected area. The Hol Chan Marine Reserve conserves a sampling of Belize's major nearshore ecosystems: coral reefs, seagrass beds, sediment plains, and mangroves.

Before the reserve was established, the reefs near the coastal town of San Pedro had been showing signs of degradation from overfishing and physical damage from divers and anchoring. San Pedro, the nation's major tourism center, was experiencing the rapid development of hotels, marinas, and holiday homes. Concerned San Pedranos and biologists therefore felt that a reserve would conserve biological diversity, offer an example of sustainable use of the reef for tourism, demonstrate the role of marine reserves in sustainable fisheries management, and serve as a site for marine science research and education.

Hol Chan has been a great success: It has protected the reef, generated a sense of pride in the community, and produced economic benefits for those involved in tourism and fisheries, all in a relatively short period. It serves as a model for future marine protected areas; several other coastal communities have expressed the desire for similar reserves. Fish populations within the reserve have shown a remarkable increase, and preliminary results show that catches by fishers working in the areas adjacent to the reserve have also increased. In a recent socioeconomic survey, 65 percent of fishers interviewed wanted more marine reserves to be established.

Because the coastline of Belize is bracketed by the world's second longest barrier reef and most of the coastal strip is lined with productive mangrove forests, the entire coast is a resource of international significance. To declare the entire coastal zone a fully protected marine park, however, would be impractical. The challenge, therefore, is to implement an IAM plan that will ensure protection of coastal resources while permitting their sustainable use.

A major problem, however, is the many sectors with an interest in the coastal zone, and the overlapping jurisdiction of government agencies that deal with them. These sectors include fisheries, tourism, forestry, mining, petroleum exploration, navigation and ports, archaeology, land use and planning, protected areas, and education. In many instances, the activities of one sector are carried out without the knowledge of the others, leading to conflicts between the various user groups. Belize is looking to the example of the Australian Great Barrier Reef Marine Park Authority, which has successfully implemented a comprehensive scheme of zoning and regulation. A similar multiple-use system might work in Belize, in which a mosaic of different zones for particular uses (e.g., fishing, tourism development, strict protection, and scientific research) can be accommodated to minimize conflicts. This zoning scheme would be implemented in conjunction with specific criteria for coastal development and resource use.

At first, however, the emphases must be on community participation in planning and ongoing activities, and coordination and improved dialogue among the various government agencies. With limited human and financial resources, it might be more practical to concentrate on a few hotspots. Based on the lessons learned from these special management areas, such as at Hol Chan, the program can then be expanded.

The health of coastal waters is tightly linked to land use practices. This is especially apparent in Belize, a narrow country with a high rainfall in the southern portion and many short rivers that quickly affect the coastal zone. Poor agricultural practices, deforestation (including clearing of mangroves), and the heavy use of agrochemicals are deteriorating coastal water quality, which, in turn, threatens the coral reefs on which both tourism and fisheries depend. Tourism is Belize's foremost industry and has expanded rapidly; from 1980 to 1990, the number of visitors to Belize tripled. Fisheries products rank fourth among export commodities.

Well-directed management of the coastal zone (as evidenced by Hol Chan) is vital to Belize's long-term economic prosperity and ecological health and to maintaining a globally significant biological resource. Many Belizean resource managers have recognized this need, and a CZM project is now under way.

Coastal Zone Management, Oman

The sea is a vast system connected throughout by currents, nutrient transfer, and species movements. Winds, drifting larvae, migrating fishes, suspended sediments, tarballs, plastic debris, and acid rain do not respect the artificial boundaries nations impose in the sea, so efforts to protect discrete areas (especially small ones) that do not fully account for interactions with their surroundings are doomed to failure (Salm 1987). This implies something obvious, but awesome in its implications: Humankind needs an integrated global plan for managing the world ocean! The day must come when we have one.

IAM of the coastal zone is a first step, a tool that focuses on the vulnerable land-sea interface, yet is confined within practical planning limits, namely areas of national jurisdiction and institutional ability. CZM legislates control over a large multiple-use area in which damaging activities and pollution can be avoided, regulated, or contained. The entire coastal zone functions as a conservation area in which all significant ecosystems receive protection without the deployment of permanent field managers. Managers are free to focus on critical areas, reducing conflict among user groups and restoring damaged environments.

Each country has distinct management needs and constraints. In Oman, for example, staffing is a primary constraint on protected area management. Staffing needs must be minimized by decreasing the size and number of essential protected areas. Oman is able to achieve this through CZM planning that extends controls over surrounding areas and shares management responsibility there among different authorities (Salm and Dobbin 1987, 1989).

Over a seven-year period, IUCN has completed a series of CZM plans for Oman. A goal of the plans is to establish protected areas as an integral component of land-use policies and planning. The CZM plans use land-use planning to facilitate the protection of sensitive or valuable coastal and marine environments, and to control activities that affect them, such as recreation, fisheries, shoreline structures, pollution, floods, and erosion.

The CZM plans contain zone and boundary maps as well as management guidelines for proposed coastal and marine protected areas. Three action components of the plans help secure the protected areas and extend conservation beyond their boundaries:

1) instituting *planning policies* that provide a broad-brush approach to guide development in the coastal zone, and to achieve protection for a range of sensitive and scenic environments (e.g., beaches, dunes, coastal cliffs, wetlands, coral reefs) throughout the coastal zone without the need to define, legislate, and manage numerous small and scattered reserves (or many corresponding zones within a single large reserve);

2) establishing *protected areas* that allow intensive management of specific sites of particular value or sensitivity, or those needing close monitoring, restoration, or separation of conflicting activities; and

3) identifying *specific actions* for management issues in the coastal zone, and assigning these to a responsible authority. Consequently, damaging uses are controlled throughout the coastal zone.

Oman lacks problems that damage biological diversity in some other nations. Dynamite and poison fishing are illegal and do not occur there. Fisheries legislation prohibits spearfishing and collecting and exporting shells and corals without a permit, although there are problems with law enforcement. Dive clubs ensure that members respect legislation and avoid damaging marine life while diving. Land-based pollution of coastal environments is negligible, and a requirement for environmental impact assessments reduces the risk of new developments damaging coastal and marine areas (if correct procedures are followed).

Many problems do, however, arise from the use of modern fishing gear. Discarded gill nets smash corals and kill lobsters, crabs, and fishes, and, together with trawls and driftnets, drown many sea turtles, dolphins, and some large whales. This is an example of traditional use evolving, through the introduction of modern technology (synthetic fiber nets, modern boats, and outboard engines), into abuse that is difficult to resolve.

Unlike the Great Barrier Reef Marine Park (next section), when Oman's CZM project started, there was no authority charged with implementing it; plans were being made with nowhere to submit them. The Ministry of Environment now has the mandate for CZM, nature conservation, and pollution control, an ideal formula for comprehensive protection of the marine realm. However, some other agencies resist implementing plans made by another government authority, and are delaying progress in CZM.

The CZM plans have raised awareness of the need to conserve Oman's coastal and marine resources, which has led to other programs to protect the sea. Three underwater cleanups by volunteer divers each yielded two to three tons of old nets, alerting the public and government to this problem. A comprehensive study is under way to provide detailed regulations for controlling coastal erosion, and Oman is developing a code of practice for the construction of shoreline structures. Planning authorities have accepted policies including a coastal setback that gives extra protection to sea turtle nesting beaches. Some development proposals have been scrapped or relocated to appropriate sites, and many management issues (e.g., beach litter and sand mining, solid waste disposal at sea or in mangroves) have been resolved. Environmental data have been incorporated into regional land use plans, significantly influencing their design.

IAM in the coastal zone has distinct advantages. It draws protected area management into a broader integrated land- and sea-use planning process, effectively establishing the entire coastal zone as a multiple use reserve. This avoids the unpopular, often impossible task of designating a strictly protected area of huge proportions. Also, CZM shares responsibility for management among concerned authorities by building on existing infrastructure, policies, plans, and management frameworks. This reduces dependence on a large management force, and enables each authority to tackle one issue at a time. Oman's experience suggests that controlling large areas through multisectoral cooperation can be an effective tool for conserving biological diversity.

Great Barrier Reef Marine Park, Australia

The Great Barrier Reef (GBR) is the largest system of corals and associated life forms in the world. It is located on the Australasian continental shelf along Australia's northeastern coast and up to southern Papua New Guinea. Most of the reef—approximately 350,000 square kilometers (135,000 square miles)—is encompassed within the Great Barrier Reef Marine Park. The GBR hosts a diversity of islands and individual reefs, and is home to an exceptional diversity of species. It is listed as a World Heritage Site (Great Barrier Reef Marine Park Authority 1981).

Commercial fishing and tourism, recreational pursuits (including fishing, diving, and camping), traditional fishing, scientific research, and shipping all occur in and around the GBR. The Great Barrier Reef Marine Park Authority was established to balance these uses in ways that minimize conflict and provide for sustainable use of the reef.

The Authority's primary goal is: "To provide for the protection, wise use, understanding and enjoyment of the Great Barrier Reef in perpetuity through the development and care of the Great Barrier Reef Marine Park." The Authority adopted this goal because it believes that no use should threaten the essential ecological characteristics and processes of the GBR and associated areas.

Support for the park is strong, and is likely to remain so if the public is informed about the basis for management decisions and the cost and effectiveness of programs, and if it is involved in management decisions. Hence, education and information programs are central elements of the Authority's work.

Australia has a federal system of government, so there is potential for conflict between state and federal interests. The federal government maintains overriding power in the marine part of the region, while involving the state of Queensland cooperatively in all aspects of the establishment and management of the marine park. Queensland government agencies are responsible for day-to-day management of the park, subject to the Authority's zoning plans and policy decisions, and are fully responsible for most of the islands, subject only to Australia's international obligations under the World Heritage Convention.

Legislation establishing the park and the Authority was passed by the federal Parliament in 1975 with support from all major parties. It has strengths that provide crucial legislative backing to the Authority:

1) Conservation of the GBR is defined as the explicit goal of management and is not coupled with requirements for optimization of resource use.
2) In situations where there is conflicting jurisdiction with other legislation, the Great Barrier Reef Marine Park Act prevails.
3) There is potential to regulate activities outside the boundaries of the park if the federal government considers that these might cause pollution having unacceptable impact within the park.
4) The Act provides for meaningful involvement of other interests in planning the park and its management, including provisions for cooperation between the Authority, Queensland state agencies, local governments, and user groups.
5) The boundaries of the park are clearly defined, so the area of jurisdiction of the Authority is also clearly defined.

The park's only significant legislative problem concerns federal legislation that provides a process to resolve appeals against administrative decisions of federal agencies. Other than the courts, appellants have no other recourse for appealing the Authority's decisions. In several cases, resolving these appeals has been a costly process for all involved parties.

The Authority's strong legislative backing and its overriding conservation mandate ensure that it is not captive to a single sector of interest. Because the Authority is removed from the detail of day-to-day park management, it is seen as an independent arbiter and decision maker. It has been successful in securing cooperation with Queensland's state agencies, and has developed sound working relationships with commercial and recreational fishing groups, tourism operators, scientists, and conservationists.

The Great Barrier Reef Marine Park is not a national park, but a multiple-use protected area that is managed using two mechanisms: regulation and zoning (IUCN Commission on National Parks and Protected Areas 1984). Regulation controls *how* activities are carried out. Examples include gear restrictions in fisheries, catch limits, and water quality standards for effluent outfalls. Because they can be changed relatively quickly, regulations are a powerful management tool.

Zoning specifies *where* various activities can be carried out in the park (Figure 8-3). It allows separation of areas with conflicting uses. Levels of protection within the park vary from zones with few restrictions to zones in which almost no human activities are permitted. The only activities prohibited throughout the park are oil exploration, mining (other than for approved research purposes), littering, spearfishing with scuba, and the taking of large individuals of certain species of fishes.

The plans that have been developed thus far specify three major categories of zones:

1) Preservation Zones and Scientific Research Zones, in which the only human activity permitted is strictly controlled scientific research;
2) Marine National Park Zones, in which the major uses permitted are scientific, educational, and recreational; and
3) General Use Zones, in which uses are held at levels that do not jeopardize the ecosystem or its major elements. Commercial and recreational fishing are generally permitted, although bottom trawling is prohibited in one of these two categories.

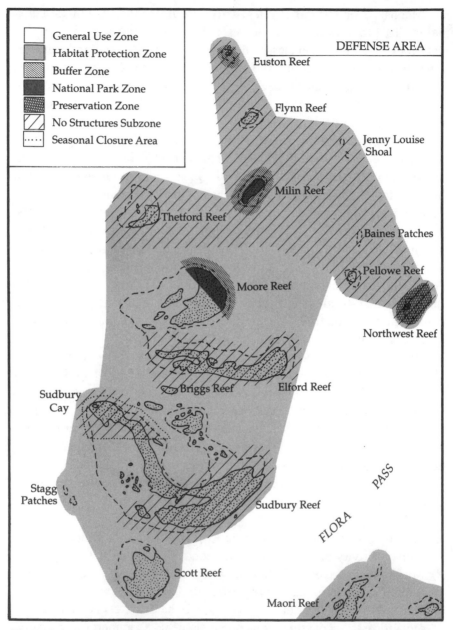

Figure 8-3. Integrated area management zoning map. *The Great Barrier Reef Marine Park (Australia) employs zoning as a potent means of sustaining uses and maintaining biological diversity. Coordinated research, monitoring, education, and public participation are integral to decision making about zoning and regulation.*

The zones are fixed during the life of a zoning plan (generally seven years). They are complemented by smaller areas that give special protection at certain times, for example, animal breeding or nesting sites.

Because tourist use has increased dramatically, the existing zoning system, which focuses on fishing, is proving inadequate. There is increasing competition among tourists for particular sites, something now being addressed through rezoning. A new zoning scheme has been introduced to ensure that the whole GBR does not become dotted with tourist and other structures, while allowing careful development on reefs that are suitable for that purpose.

The Great Barrier Reef Marine Park's multiple-use planning and management approach could well be a model for comparable areas around the world. The following factors have been central to this success:

1) The Authority is backed by legislation with real "teeth." Some of its most powerful sections have never been invoked, but have certainly provided a valuable backing to negotiations.

2) The park is large enough to be managed as a complete marine ecosystem. This provides the Authority with the means to assign both protective measures and multiple uses across the park, which provides flexibility in management options that would not exist in an area focused solely on protection.

3) The Authority manages for the full range of park uses, recognizing the high connectivity of marine ecosystems and the need to keep impacts below levels that will alter the GBR's essential ecological characteristics and processes. The many potential conflicts among park users require the Authority to consult with users and industries to fashion regulations and zoning decisions.

4) Continuing public awareness of, and meaningful participation in, the Authority's decisions have gained the public's support. This is vital to maintaining backing from political parties that decide on legislation governing the Authority's operations, and should be considered by any similar management agency in a democratic society.

5) At the same time, the Authority has remained adaptable to the park's changing patterns of use. The revision of regulations as needed and the regular revision of zoning plans provide mechanisms for adapting

management regimes to suit shifting emphasis in park use and to reflect increasing scientific understanding of the processes that sustain the Great Barrier Reef ecosystem.

The Authority is presently coordinating the preparation of a strategic plan covering the whole of the GBR World Heritage Area, including all the waters, reefs, islands, ports, and harbors. This strategy has involved all of the major interest groups in the area—more than 60—in agreeing on a 25-year vision and 5-year objectives for the area to which each organization commits itself in its corporate plan. This could be the first such exercise covering a large area that is subject to marine and terrestrial multiple jurisdictions anywhere in the world.

Lessons from Existing Institutions and Instruments

More people and organizations are working to ensure the health of the sea than ever before, and some are remarkably effective. In examining existing institutions and instruments to see what does and what does not work, some principles emerge, which, in turn, have led to thematic and specific recommendations (see Chapter 9) on ways to save, study, and use the diversity of marine life sustainably. Insights and a comprehensive course of action are needed because, despite scattered successes, existing efforts are not adequate; marine biodiversity is decreasing.

Commitment to Marine Conservation

The most important lesson is that, at all levels, there is an urgent need to deepen the commitment to conserving the sea. Every major marine treaty has weaknesses that undermine its environmental aims; every major IGO leans too far toward accommodating economic interests or has insufficient authority to implement its environmental goals. The need for economic development does not relieve local institutions, nations, and international organizations of the responsibility for maintaining the diversity and integrity of the living sea. And although this is universally true, from the poorest regions to the richest, inadequate commitment is least forgivable in rich nations that persist in pushing unsustainable marine economic activities. Industrialized nations have a special responsibility to set an example for responsible environmental stewardship within their own waters, to cease damaging the marine resources of poorer nations, and, indeed, to help them to save, study, and use their seas on an ecologically sustainable basis.

Improving Understanding of Marine Populations, Species, and Ecosystems

There is urgent need for increased knowledge about the sea, both from traditional lore and from modern science; as pressures mount, we are decreasingly able to maintain what we do not understand. Incorporating traditional knowledge in decision making is especially important where scientific information is lacking, as it is almost everywhere where there are no major marine laboratories. There needs to be substantial strengthening and coordination of scientific inventory, research, and monitoring to tell governments, IGOs and NGOs which marine components and processes require attention early enough to make a difference. The need for comprehensive, long-term data sets, which can provide the historic context for observed events or trends, is particularly great, as are regional and global networks to gather and share information. Countries such as Costa Rica have proven that there can be effective, mutually rewarding cooperation between industrialized and developing nations in strengthening the biological diversity knowledge base.

Boundaries and Appropriate Spatial Scale for Action

Because the division of the world into nations and subnational governing regions does not reflect marine patterns and processes, we cannot conserve marine life unless our efforts transcend legal boundaries to reflect the workings of the sea. There are precedents for governmental cooperation in conserving marine life, but virtually all have limited scope and/or effectiveness. The necessary cooperation can be achieved only if governments reexamine longstanding views about sovereignty, have enough trust in their neighbors to work together, and fashion effective institutions that address the physical and biological realities of the sea.

National and local institutions have crucial roles to play—they have the most knowledge of the needs of their people—but most do not operate at a geographic scale large enough to be effective at managing marine species. Global governmental institutions and instruments are also crucial—indeed, they are the most appropriate scale to provide framework conventions and to deal with global issues such as satellite-based oceanographic research, oil transportation, introduction of alien species, and atmospheric change—but their scale, in many cases, is *too* large to deal with the issues affecting discrete species populations and biogeographic provinces. In many cases, regional institutions that reflect marine biogeographic provinces and common interests seem to be the most appropriate scale for managing marine species and ecosystems. Whichever level is most effective on a given issue, however, its efforts must be integrated with efforts at other levels to maximize the advantages of local, national, regional, and global institutions and instruments, and to minimize their disadvantages.

Local cooperation from the people most directly involved in activities that affect marine biodiversity ultimately proves to be vital to any conservation effort, but local and national interests can force compromises that critically weaken regional or global agreements, either by forcing governments to remove strong conservation language or by undermining implementation. Longstanding traditions, such as dietary traditions or ways of disposing of wastes, cannot take precedence over the need to maintain the health of the sea; customs born when the world had 500 million people must be adjusted now that there are 5.5 billion of us. No matter how desirable and important they are, activities such as international trade must *not* be given precedence over maintaining life on our planet. Solving the exquisitely difficult challenges of forging equitable relations between poor and rich nations, of finding the correct balance between the needs of particular nations and the world, and of ensuring that fulfilling today's needs does not preclude tomorrow's, is essential for marine conservation.

Balancing Economic Sector and Citizen Input to Government Decision Making

In addition to geographic divisions, people are divided into economic sectors that further fragment decision making within jurisdictions. Local and national governing bodies are usually influenced disproportionately by economic sectors that affect the sea, such as agriculture, forestry, energy, housing, manufacturing, shipping, and large-scale fishing, which disenfranchises many small-scale users of the sea and the general public. Local, national, and international NGOs play a vital role as advocates for the public interest by counterbalancing parochial sectoral interests, thereby keeping decision-making processes honest. They are also instrumental in gathering and communicating information to decision makers. Governments that can limit the influence of major economic interests and that can secure the help of citizens by keeping them informed and actively encouraging their input to decision making are most likely to fashion enduring systems of marine resource management.

Increasing the Role of IUCN and Scientific Research Organizations

IUCN, its commissions, specialist groups, and member organizations are uniquely well-situated to help governments to find solutions to thorny marine conservation problems. Scientific organizations such as SCOPE and IUBS can also play a larger role in helping governments to examine and understand the dimensions of the environmental crisis. It is particularly important that such conservation and scientific organizations improve their communications with national governments and with one another: Better communication is a powerful antidote to fragmented decision making, which favors the short-term interests of particular sectors. Lending and granting institutions should go beyond their traditional national development focus to assist regional and global conservation and scientific organizations that can help to fulfill these integrative functions.

International Financing of Local and National Institutions

The first rule that bilateral and multilateral funders should observe is to do no harm; the second is to do good. Adhering to discredited ideas about economic development, particularly strengthening particular sectors without due attention to externalities, has made funders part of the problem. Now they need to reorganize themselves and reshape their policies so that they become part of the solution. One way is to diminish staffing in areas of traditional concentration—that is, traditional economics and engineering—which have contributed to the current situation, and adding substantial numbers of ecological economists and conservation biologists, whose world views are more likely to emphasize the importance of biological diversity. The GEF's mandate to benefit the environment creates a model for all foreign aid and IGO financial assistance programs, although its insufficient size and temporary tenure currently limit its effectiveness.

Integrated Area Management

Governments can take a major step toward comprehensive protection, study, and sustainable use of the world's seas by undertaking IAM in their own waters and terrestrial watersheds that affect them. By cutting across geographic and economic divisions, IAM has the potential to be *the* most effective means of ending gridlock caused by fragmented decision making. When combined with the LME approach, IAM diminishes the harmful impact of artificial borders that divide continuous species ranges and ecosystems. IAM oriented around LMEs is particularly effective when it includes areas where economic uses are carefully managed and other areas that are strictly protected. In the coastal zone, IAM needs overriding authority to ensure that land-based activities throughout drainage basins do not harm marine resources. IAM in offshore areas within territorial seas is generally simplest because the number of anthropogenic factors influencing marine resources is diminished but authorities are strong. To seaward, in EEZs and on the high seas, anthropogenic factors are similarly diminished but authorities for controlling activities are weaker, and might need to be strengthened before IAM can work effectively. As with all efforts that affect the public, public support and cooperation are essential to the success of IAM.

CHAPTER NINE

Recommendations for Implementing *The Strategy*

IT IS evident that the health of the sea and the well-being of all who depend upon it are by no means secure because humankind has not:

1) defined the factors that determine the biological diversity, integrity, and productivity of the marine realm; and
2) provided guidelines and undertaken practical actions to ensure the protection, study, and sustainable use of marine ecosystems and resources.

Conserving life in the sea has been difficult because our land-based perspective has spawned many erroneous perceptions about the oceans, which has led to defects in the legal and operational regimes that human societies have developed. The challenge we face becomes clear upon examining maps of the world, on which three things often stand out:

1) the division of land into a patchwork of nations;
2) landmasses dividing a world ocean whose most obvious features are lines that divide East and West, North and South; and
3) an expansion of area with increasing distance from the equator, so that tropical regions are deemphasized, as occurs in the widely used Mercator projection.

Such maps both reflect and reenforce people's perceptions. Unfortunately, the world view that they reveal threatens the health of the sea. The division of the land into nations makes it far easier to deal with things that stay in one place than organisms and pollutants that freely cross national borders, as occurs in the sea. The depiction of featureless oceans crisscrossed by straight lines highlights ignorance about the physically and biologically diverse ecosystems that cover more than two-thirds of the globe. And, deemphasizing the area of the tropics diminishes its apparent importance, while, in truth, tropical ecosystems are generally the most biologically diverse ones, where there is the most to lose. To conserve life in the sea for its own sake and for its benefits to humankind, we must begin to see the world differently.

The health of the sea depends on the health of the entire biosphere; threats to life in the sea are a very large subset of those that threaten the biosphere as a whole. As a result, actions to alleviate pressures on the environment in general—for example, efforts to reduce overpopulation, desperate poverty, resource consumption, and production of wastes—will almost always help marine species and ecosystems and the people who depend on them. Of course, other actions must be tailored to the distinctive nature of environmental problems in the marine realm.

Many of these actions can be carried out within nations or even locally. But the marine realm is so much more interconnected than the terrestrial realm that many efforts, though locally or nationally based, require cooperation among nations if they are to have any chance of succeeding. *The overwhelming fact of marine conservation is that the sea does not respect the imaginary lines that we put on maps.* Rather, it is replete with real lines: island arcs, barrier reefs, currents, boundaries of water masses, migration routes, larval trajectories, the edges of species distributions. Some are reasonably fixed; others are fluid. To maintain the sea's wealth, we must learn what these lines are and respect them at least as much as we do legal boundaries that predominate on land.

A successful strategy for conserving marine biodiversity must have certain general characteristics:

1) It must be cross-sectoral, embracing all categories of marine ecosystems and species, all types of human use, and all sources of threats.
2) It must govern actions on land as well as in the sea, since much of the damaging impact on marine systems originates on land.
3) It must be capable of decentralization on a regional, national, and subnational basis.
4) It must have flexibility to address priorities that will inevitably vary from one location to another.

To have any chance of succeeding, people are going to have to bridge something even larger than an ocean: the gulf of resources, understanding, and trust that divide people within and among nations. To conserve something as broad and deep as the sea requires a broadening of our vision, a deepening of our understanding, and, most important of all, a commitment to concerted action for as long as necessary to do the job.

The following strategic recommendations for implementing *The Strategy* are offered with these points in mind. They come, verbatim or with modification, from many sources, including the authors and reviewers of this document, *Agenda 21* (UN 1992), the *Global Biodiversity Strategy* (WRI et al. 1992), *Caring for the Earth: a Strategy for Sustainable Living* (IUCN et al. 1991), the UNCED Global Forum NGO treaties drafted by the Oceans Working Group in Rio, the Globescope Conference on Latin America and the Caribbean, working papers at the IVth International Congress on National Parks and Protected Areas in Caracas, *Preserving Ecological Systems: The Agenda for Long-Term Research and Development* (Draggan et al. 1987), *Diversity of Oceanic Life: An Evaluative Review* (Peterson 1992), *Research Priorities for Conservation Biology* (Soulé and Kohm 1989), *The Diversity of Life* (Wilson 1992), and others. Some of the recommendations are quite specific. They indicate not only what should be done, but who should do it and when. Others are broader. An early goal of activities that implement *The Strategy* should be to review these recommendations and make them much more specific; broad prescriptions alone cannot save the sea.

General Recommendations

Nations should strive to reduce and then stop population growth in all areas where it threatens marine biological diversity, including coastal zones and areas (such as drainage basins) that affect them, by making population stabilization in these areas a national priority, eliminating pro-natalist policies, and establishing family-planning programs to educate people about birth-control options and to provide them the means to prevent unwanted births. Industrialized nations, which have far more environmental impact per capita than developing nations, have a special responsibility to take these steps and to support developing nations to achieve population stabilization.

Industrialized nations along coasts or whose runoff drains into the sea should strive to reduce consumption and wasteful use of marine resources and land-based resources that affect marine biological diversity, in accord with the widely accepted imperative to reduce, reuse, and recycle. For example, to reduce marine oil pollution, physical alteration of coastal wetlands, and sea level rise due to global warming, industrialized nations should increase energy efficiency in the most cost-effective ways consistent with maintaining basic amenities and quality of life, and should remove impediments to increased use of environmentally benign renewable energy resources within their countries and in developing nations. For their part, developing nations should not repeat the North's wasteful path to prosperity, but should adopt technologies and economic strategies that can improve economic and social conditions sustainably without diminishing marine biological diversity.

Nations should strive to achieve an open, nondiscriminatory, equitable, and environmentally sound multilateral trading system that will enable all countries—in particular, developing countries—to improve their economic structures and the standard of living of their populations, so that they have the means to maintain biological diversity in the sea, in freshwaters, and on land.

Nations should sign, ratify, and implement the Convention on Biological Diversity.

Nations should ratify UNCLOS III, and should invest far greater effort to implementing its provisions to protect marine areas and improve management of high seas fisheries.

The UN General Assembly should disseminate, discuss, and, where appropriate, adopt the alternative global environmental and development treaties drafted by environmental NGOs at the Global Forum at UNCED.

The UN should strengthen coordination among its relevant organizations with major marine and coastal responsibilities, including their regional and subregional components, and between those organizations and other UN organizations, institutions, and specialized agencies dealing with development, trade, and other related economic issues. A good beginning would be to establish a marine coordinating committee that includes agencies such as UNEP, UNESCO, FAO, IMO, the International Oceanographic Commission (IOC), the International Atomic

Energy Agency (IAEA), World Health Organization, and the World Meteorological Organization (WMO).

UN agencies and international lending agencies should take no actions that encourage the proliferation of technologies that contribute to the degradation of marine ecosystems or the depletion of marine species below sustainable levels. For example, UN agencies should not promote monofilament gill nets except where they will not take significant numbers of nontarget species.

UNEP, IUCN, WRI, WWF, and other organizations, as appropriate, should produce biennial reports on the status of biological diversity in the sea, in fresh waters, and on land, highlighting ecosystems and taxa of special concern.

IUCN, with cooperation from major funding organizations, should consider publishing an annual "State of the Seas" report that would include an overview of the state of the marine environment, accounts of exciting new scientific findings, descriptions of problems, disasters, actions, and successes in particular areas, and record progress toward attainment of targets such as those set forth in *Agenda 21*, *Caring for the Earth*, and this document.

UNEP, IUCN, FAO, and IMO should convene a global conference to exchange experience in the field of IAM in coastal zones and the sea.

The UN should establish 1995 to 2004 as the International Biological Diversity Decade, to give the highest possible visibility to the need for immediate and farsighted action to protect the world's genetic, species, and ecosystem diversity, including that of the sea. All relevant UN organizations should be full participants in IBDD activities.

The UN should establish a Global Biodiversity Forum to assist the Secretary General in implementing the *Global Biodiversity Strategy* and *The Strategy*, by providing guidance on priorities for the protection, study, and sustainable and equitable use of biological diversity. The forum should consist of leading thinkers, opinion shapers, and decision makers in NGOs, industries, and governments, reflecting the diversity of the UN itself.

The UN should establish and host a periodic (every three to five years) conference on the health of oceans, coastal seas, and estuaries that would bring together government officials, NGOs, scientists, and user groups to discuss progress toward sustainable use, study, and protection of the sea.

The UN should establish an Early Warning Monitoring Network whereby timely information on immediate threats to marine biodiversity is provided to governments, companies, NGOs, and individuals who can act to avert those threats.

With the support and assistance of UNEP, and resources from GEF and other funders, IUCN, CMC, and WWF should hasten the establishment and implementation of an International Marine Conservation Network, which would link decision makers in international governmental organizations, NGOs, and governments in coastal nations worldwide to strengthen and increase communications on marine biological diversity issues, especially among developing nations, and to facilitate implementation of the recommendations in *The Strategy*. Network participants should meet at each IUCN triennial general assembly to strategize on ways to effect the recommendations and improve the Network's functioning.

The International Marine Conservation Network, when established, should assemble a list of subsidy laws and policies that have the worst effects on marine biological diversity, to be published in the World Resources Institute's *World Resources* biennial report.

With funding from GEF and other sources as necessary, and the assistance of participants in the International Marine Conservation Network, eligible nations should assemble national marine biological diversity conservation strategies to serve as blueprints for protecting and ensuring sustainable use of marine species and ecosystems within their waters.

IUCN should consider establishing a Commission on Conservation of the Sea to serve as the focal point of a wide spectrum of activities ranging from science and law to public education.

The Contracting Parties to GATT should consider the impact of trade measures on the environment and formulate amendments to GATT necessary to balance and integrate the objectives of environment and trade policies.

The Contracting Parties to GATT should expedite convening the Working Group on Environmental Measures and amend regulations to encourage sustainable natural resource management and internalization of resource and environmental costs.

Nations should adopt EIA procedures for all government projects or governmentally permitted projects that have significant actual or potential effects on marine or coastal ecosystems. These procedures ensure the participation of the public and of other government agencies, and should offer feasible alternatives to the proposed activities.

Nations should develop economic incentives to apply clean technologies and other means consistent with the internalization of environmental costs, such as the polluter-pays principle, so as to avoid degradation of the marine environment.

Nations that import marine products should create and enforce controls for their industries to ensure that production or extraction activities in other nations that supply these imports neither violate local, national, or international laws where the products originate nor constitute ecologically destructive or economically nonsustainable practices.

Nations should develop ecotourism as one of their tools for conserving biological diversity and as a powerful instrument for sustainable development, as long as they can ensure that its socioeconomic benefits extend to local people, and that tourism development does not result in biodiversity loss or cultural conflict.

Nations in which marine-oriented tourist operations are headquartered should encourage and assist other countries in developing EIA procedures that cover tourist activities, and should ensure that they neither violate local, national, or international laws where activities are occurring nor constitute ecologically destructive or economically nonsustainable practices.

Coastal nations should establish or strengthen coordinating mechanisms for integrated management of coastal and marine areas and their resources, at both the local and national levels. Such mechanisms should include consultation with the academic and private sectors, NGOs, local communities, resource user groups, and indigenous people.

Making Use of Marine Species Sustainable

International

Government agencies and NGOs should refer to the proposed IUCN policy on "Sustainable Use of Wild Species" as the basis for developing national policies for managing marine species sustainably.

The East-West Center and Ocean Voice International should convene an international conference or group of workshops to provide for the UN the necessary documentation and language for a global convention to ban intrinsically destructive and unsustainable fishery techniques such as dynamite fishing, poison fishing, and muro ami.

IUCN, FAO, and UNEP should hold an international forum on inherently vulnerable marine species (particularly long-lived species with late reproductive maturation, including certain corals, mollusks, bony fishes and seabirds, most sharks and marine mammals, and all sea turtles) that are taken commercially, to delimit the special circumstances under which they might be exploited sustainably.

In all of their fisheries activities, FAO and national fisheries agencies should consider "miscellaneous marine products," such as corals and shells, as "serious" fisheries requiring as much monitoring and management as food fishes and shellfishes.

IUCN should, by 1996, issue a *Red Data Book for Fishes* that would compile available information on marine species that are inherently vulnerable to overexploitation, such as blue, lemon (*Negaprion brevirostris*), white, basking (*Cetorhinus maximus*) and hammerhead sharks (*Sphyrna* spp.), Atlantic (*Hippoglossus hippoglossus*) and Pacific halibut, orange roughy, jewfish (*Epinephelus itajara*) and Nassau grouper (*E. striatus*).

The UN should declare 1995 as the Year of the Sustainable Fishers, and should establish and fund 300 awards for Sustainable Fishers, which would go to small-scale fishers from member states who set an example for the sustainable use of living marine resources.

ICES, FAO, the World Bank, and the regional development banks should jointly develop guidelines for sustainable mariculture (including effects on local economies, coastal ecosystems, wild stocks, and diseases) to help public and private funding agencies in making funding decisions.

FAO and UNEP should establish a decade-long program to identify breeding areas of commercial marine species of international importance and establish criteria for their protection from all threats, including overexploitation.

FAO and IUCN should assemble available information on the local and regional physical and biological effects of trawling on the seabed, and should convene an international workshop on the impact of bottom trawling to explore ways to limit its impact.

FAO and IUCN should convene an international workshop on the effectiveness of bycatch reduction devices as a means of diminishing bycatch from bottom trawls.

The UN should work to encourage the creation and implementation of cooperative management regimes for species that move between EEZs of different nations (straddling stocks) or between EEZs and international waters, to remove incentives for individual nations to deplete populations below their optimum sustainable yields. It should convene, as soon as possible, a global conference on implementation of the Law of the Sea on straddling fish stocks and highly migratory fish stocks. The conference should assess existing problems related to the conservation and management of such stocks, and formulate appropriate recommendations. The work and the results of the conference should be fully consistent with the provisions of UNCLOS III, in particular the rights and obligations of coastal states and states fishing on the high seas.

FAO, the World Bank, and the regional development banks should hold two international symposia on fisheries and coastal development, one on government subsidies and the other on tax policies, both of which would examine the effects on local economies and sustainability of artisanal and commercial (wild and mariculture) fisheries, by 1996.

FAO and nations' fisheries research agencies should work to develop materials for fish traps and nets that would degrade within one year under the conditions where traps and nets are most likely to be lost to avoid the sea's growing burden of "ghost" traps and nets. Moreover, fisheries management agencies should mandate that traps and nets be made with such materials for all fishing operations within their waters.

National/Local

Coastal countries should have a management agency with responsibility and authority to protect living marine resources and manage them on a sustainable basis.

Agencies that undertake national and local planning, including EIA, for projects that could affect fisheries and other living marine resources should consult and coordinate with national and local agencies responsible for living marine resources and with organizations that represent fishers and other affected user groups. Nations should ensure that small-scale artisanal fishers are given ample opportunity to participate in coastal and marine development and management programs, including creation of marine and coastal IAM plans.

Nations should recognize and encourage the knowledge and living marine resource management methods of indigenous and traditional peoples, and ensure the opportunity for these groups to share in the economic benefits derived from the use of such traditional methods and knowledge.

Governments should take no actions that subsidize activities—including wetlands destruction, pollution or overcapitalization of fisheries infrastructure—that diminish the productivity, diversity, and sustainable use of living marine resources.

Nations should fully implement General Assembly Resolution 46/215 on large-scale pelagic driftnet fishing.

Nations should recognize the responsibility of IWC for the conservation and management of whale stocks and accept the authority of IWC agreements, including any catch limits or moratoria on whaling, as legal obligations, whether or not they are IWC members.

Nations should monitor and control fishing activities by vessels flying their flags on the high seas to ensure compliance with applicable conservation and management rules, including full, detailed, accurate, and timely reporting of catches and effort. Nations should take effective action, consistent with international law, to deter the reflagging of vessels by their nationals as a means of avoiding compliance with applicable conservation and management rules for fishing activities on the high seas.

Nations should halt the taking of coral reef species for aquaria, curios, and construction materials except where it can be demonstrated that these practices are undertaken in an ecologically sustainable manner.

National governments, individually and jointly, should promulgate fisheries management plans for sustainable use that:

1) incorporate the multispecies approach within large marine ecosystems;
2) incorporate mechanisms to minimize bycatch;
3) take into account effects of exploitation on genetic diversity within target species;
4) completely separate stock assessment and allocation decisions; and
5) incorporate the advice of fishers, conservationists, and members of the general public who use fishery products.

Stopping and Reversing Physical Alteration

International

UNEP should convene, as soon as practicable, a global conference on the protection of the marine environment from land-based activities that physically alter or pollute marine ecosystems.

The UN and all member states should consider international mechanisms to protect hydrothermal vent ecosystems as resources of international importance that should not be degraded by mining for metallic sulfides, disposal of waste, or any other human use.

International lending agencies should take no actions that lead to the destruction, fragmentation, or simplification of mangrove forests, and should encourage only those uses of mangrove forests that can be sustained in perpetuity. Similar actions should be considered for other coastal ecosystems, including seagrass beds, salt marshes, and fringing reefs.

National/Local

Nations should do everything possible to ensure that development activities that divert fresh water from rivers maintain the amount, quality, and timing of flow into estuarine and marine ecosystems to minimize adverse effects on their species and ecological processes.

National and local governments should establish and enforce policies to curtail wetlands disturbance, including logging, draining, diking, dredging, and filling, in relatively undisturbed coastal and estuarine wetland ecosystems.

Nations should ensure that development activities that subject marine wetlands larger than 5 hectares (12.4 acres) to irreversible or irretrievable change should be accompanied by rigorous EIA that details the full spectrum of wetlands values and should be subjected to the stringent standards of proof of the need for such a project. Moreover, these projects should not be carried out until a national wetlands inventory has been completed so the cumulative effect of the project on wetlands loss can be determined.

Nations and localities should ensure that beach renourishment projects are timed to minimize impacts to shore-nesting species and adjacent marine ecosystems.

Preventing Pollution

International

UN members should draft and ratify a convention to discourage transboundary pollution in marine waters. The convention should identify liability procedures so that nations whose interests are harmed by transboundary pollution can be compensated for losses and cleanup.

UNEP should:

1) undertake a study of regional and worldwide accumulation of toxics in marine species and their habitats within EEZs and in international waters;

2) establish a list of persistent toxic substances that should not be discharged into the marine environment; and

3) encourage the drafting of a treaty among member states to ban the production or disposal of persistent toxic chemicals that accumulate in marine species and their habitats on a regional or global basis.

UNEP, through its Regional Seas Programme, should work toward regional agreements to limit land-based sources of marine pollution, with due attention to point sources and special attention to non-point sources.

IMO should assess the state of marine pollution in areas of congested shipping, such as heavily used international straits, with a view to ensuring compliance with international regulations, particularly those related to illegal discharges from ships, in accordance with the provisions of Part III of UNCLOS III.

IMO should establish standards for fittings for equipment that transfers oil and hazardous materials between ships and other facilities, and nations should ensure that their fleets install equipment with standardized fittings.

IMO and national governments should adopt improved training and certification requirements for people who produce and transport oil, an international treaty requiring that all new ships that carry oil or other hazardous materials have double hulls, and regulations requiring crew levels sufficient to meet normal and emergency operation needs for oil tankers and vessels carrying other hazardous substances.

IMO should sponsor an international legal regime for oil pollution damage from ships with enough liability to deter oil pollution.

IMO should accelerate efforts to conclude negotiations on a hazardous and noxious substances liability convention.

IMO should fund legal and economic studies that investigate ways that local communities having fishing or marine park stakes in the coastal zone be compensated for pollution-related damage to those resources.

IMO should reclassify petroleum from the cargo category Type III, which requires moderate safeguards during handling, to Type I, which is used for substances that have substantial impact well beyond the accident scene, and therefore require maximum protection.

With the assistance of UNEP, IMO, and the multilateral development banks, nations singly or in regional consortia should establish or strengthen spill-response centers that would provide a rapid, effective response to oil or chemical spills.

IMO should require tankers to install voyage data recorders, similar to airliners' "black boxes," which would indicate acceleration forces, stresses at various points on the hull, roll and pitch angles of the vessel, draft, and rudder angles and the times at which rudder movement took place, all of which would aid in assessing the causes of spills of oil and other hazardous substances. The black boxes should also record the times at which ballast tanks are filled and emptied, and other activities that affect the health of marine species and ecosystems.

IMO and member nations of the London Dumping Convention should adopt amendments to the Convention that ban the dumping of radioactive and industrial wastes.

IMO, UNEP, IOC, and IAEA should draft a code on transportation of nuclear fuel in flasks on ships, and shipments of such materials, particularly plutonium, should be subject to the most stringent safeguards to ensure that they do not enter the environment.

IMO should revise and update the IMO Code of Safety for Nuclear Merchant Ships and improve implementation of the revised code.

FAO, agricultural organizations, and nations should promote the use of environmentally less harmful pesticides and fertilizers and alternative methods for pest control, and nations should prohibit pesticides known to harm estuarine or marine species and ecosystems.

The GEF should support appropriate regional, national governmental, or non-governmental organizations to monitor marine pollution within GEF-eligible countries.

The World Bank, UNEP, WMO, and IOC should encourage and assist member nations in studies on the impact of conventional air pollutants and global atmospheric changes on marine species and ecosystems, and disseminate existing analyses to nations that are likely to be harmed by these pollutants and their effects.

UNEP should work with the World Bank and other international organizations and national governments to establish methods of assessing the cumulative effects of environmental degradation, especially pollution and physical alteration, from each proposed watershed, coastal, or marine development project involving international assistance or review.

International lending institutions, both public and private, should make special efforts to support programs that reduce the production of waste, encourage reuse, stimulate recycling, use "natural" methods of waste treatment, and diversify the portfolio of waste management options available to communities so that pollution control can be made flexible, economically viable, and environmentally less damaging.

National/Local

Nations should monitor marine ecosystems and river systems that empty into marine waters for persistent toxic substances and regulate toxics with a view to achieving zero discharge levels.

Nations should develop and implement plans for reducing non-point sources of pollution of the marine environment.

Nations should eliminate agricultural policies that promote excessive uniformity of crops and crop varieties, which encourage the overuse of chemical fertilizers and pesticides that end up polluting estuaries and coastal waters.

Nations should build and maintain sewage treatment facilities for their waste waters in accordance with national policies and capacities, with the help of international development agencies if necessary, to have at least primary treatment of municipal sewage discharged to rivers, estuaries, and the sea, and should establish or improve regulatory and monitoring programs to control sewage discharge.

Nations should rigorously monitor and enforce provisions of MARPOL for ships sailing under their flags, owned by their citizens, or that use port facilities in their national waters.

Nations should require tug escorts for all single-boiler, single-engine, or single-screw tank vessels carrying oil or other hazardous products in waterways designated as high risk by individual nations.

Nations should require mandatory participation in vessel traffic service systems in high-risk or congested areas for all oil tankers and other hazardous-product carriers, and require that all such vessels possess and operate an onboard navigation system.

Nations should be aware of the fact that tank barges and other small oil- and hazardous material-transporting vessels are responsible for a large share of spills, and should be subjected to the same stringent regulations as tankers.

Nations should adopt a zero-tolerance policy for illegal drug use and for alcohol abuse on all vessels involved in the transport of oil and hazardous substances.

Nations should mandate oil and hazardous-substances spill-response training for all tanker, barge, and tug crews, and anyone involved in the marine stages of oil and hazardous-substances transportation.

Nations should establish contingency plans to deal with oil spills in environmentally sensitive areas. Drills should be conducted periodically to test the feasibility of contingency plans, and to improve and upgrade them.

Nations should develop and require the use of methods of natural resource valuation that fully incorporate non-market values, as well as market values, in assessing damages resulting from spills of oil and hazardous materials.

Nations should establish strong civil and criminal sanctions for owners of ships, offshore oil and gas facilities, and coastal refineries for spilling oil and other hazardous substances spills, establish adequate environmental resource agency staffing to ensure compliance with spill planning requirements, and aggressively pursue legal action against violators.

Nations should not allow their industries to export waste products for disposal in the waters of other nations.

Nations should ensure that initiatives to protect coastal and marine areas, such as marine parks, sanctuaries, and fishery reserves, consider fully the upstream activities that affect these areas, and extend buffer zones or areas of special management as far up watershed areas as is needed to protect and maintain sustainable use of coastal and marine resources, through such cooperative mechanisms as the UNESCO biosphere reserve approach.

Government agencies should target environmental education efforts on business and industry groups whose activities affect the sea. For example, well-designed education programs for commercial fishers, in combination with incentive programs to return used and unwanted gear to port, can reduce discards of gill nets and trawls at sea. The same approach can help to reduce operational discharges of oil by tanker crews and physical damage to coral reefs by recreational divers.

Nations should take measures to reduce water pollution caused by organotin compounds used in anti-fouling paints, including phase-out of their use, as necessary.

Nations should offer a combination of economic incentives and penalties to encourage the proper disposal and recycling of motor oil and other petroleum-based motor lubricants that are presently poured down the drains of most nations.

Ending Free Rides for Alien Species

International

The UN should declare that the establishment of interoceanic canals that do not serve as complete barriers to the introduction of nonnative marine species is contrary to sustainable development; nations should not permit such canals to be built; and international financial and technical assistance programs should require the establishment of effective barriers as essential conditions for assisting any proposed interoceanic canals.

As a replacement for the voluntary guidelines on ballast water management ratified by the IMO in 1991, IMO and ICES should convene a working group of maritime nations to draft an international treaty to stop the unintended introduction of alien species into marine, estuarine, harbor, and freshwater ecosystems in ships' ballast water, and nations should sign, ratify, and enforce the treaty.

National/Local

Nations should strictly regulate the transfer of marine species and genetic resources and their release into the wild. Such transfers should be allowed only if no native species seems suitable for the purpose, if clear and well-defined benefits to humans or natural communities can be expected, and if it can be demonstrated clearly that the species will not harm populations of native species or the ecosystem into which it is introduced or into which it can spread.

Nations should adopt procedures for implementing recommendations in IUCN's position statement *Translocation of Living Organisms*, and ICES's *The Code of Practice to Reduce the Risks for Adverse Effects Arising from the Introduction and Transfer of Marine Species.*

Nations should establish education programs, such as the Hawaii (USA) Audubon Society's 1991–92 Alien Species Program, targeted at importers and distributors of aquaculture and pet species to minimize the release of alien marine species.

Linking Marine Concerns with Global Atmospheric Solutions

International

Nations should push for a new, binding climate convention to replace the one signed at UNCED. Industrialized nations should set specific target dates and levels for stabilizing and reducing all greenhouse gas emissions and atmospheric additions from land use, with the aim of capping them at 1990 levels and reducing them by 25 percent by 2005.

The UN should seek and establish an international agreement to ban any purposeful efforts to alter the ocean by any means, such as through fertilization, to increase removal of atmospheric carbon dioxide, unless such proposed action is studied in detail by the IUBS and SCOPE and approved by the Security Council.

IOC and other relevant international scientific organizations, with the support of countries having the resources and expertise, should carry out analysis, assessment, and systematic observation of the role of oceans as a carbon sink.

National/Local

Nations should sign and adhere to the Copenhagen amendments to the Montreal Protocol on Substances That Deplete the Ozone Layer by negotiating a sharply accelerated phaseout of chlorofluorocarbons and other ozone-depleting substances.

Nations and local governments should develop CO_2-reduction plans that would include significant financial incentives for local communities to reduce net flux of CO_2 and other greenhouse gases into the atmosphere.

Protecting and Recovering Depleted Populations and Species

International

UN agencies and international lending institutions should take no actions that encourage the proliferation of technologies that lead to the degradation of marine ecosystems or the depletion of marine species below sustainable levels.

FAO should train people in its fisheries observer program to identify and maintain statistics on incidentally caught marine mammals, seabirds, and sea turtles.

IUCN's SSC should set up specialist groups for any taxa of marine organisms at risk, including bony fishes, invertebrates, plants, and microorganisms, and should increase efforts to identify species in these groups that are in jeopardy. SSC should also establish a new interdisciplinary specialist group for coordinating efforts for conserving marine biological diversity.

IUCN's SSC should prepare or adopt action plans for all marine species considered endangered or vulnerable. Special priority should be given to species including the vaquita, northern right whale, Mediterranean monk seal, West Indian manatee (*Trichechus manatus*), short-tailed albatross, and Kemp's ridley sea turtle.

The IWC and IUCN should hold an international symposium on small cetaceans, to establish their status and strategies for conserving species and their genetic diversity.

National/Local

Nations should establish and implement legal authorities to protect and recover species and populations in danger of extinction.

Nations should sign, without reservations, and become active participants in CITES.

Protecting, Managing, and Restoring Marine Ecosystems

International

Although the prevention of ecosystem degradation is clearly the best way to sustain resource values, because many marine and estuarine ecosystems have already been severely altered, the GEF and other funding sources should assist IUCN to establish a new Commission on Ecosystem Restoration to provide technical guidance and help with securing funding for nations seeking to restore the sustainability of their coastal waters, fresh waters, and lands.

GEF, with the help of IUCN, should use geographic information systems and coastal surveys to help eligible nations map all marine ecosystems dominated by macroscopic structure-forming species, including coral, oyster and worm reefs, kelp and seagrass beds, and mangrove forests, which provide essential habitat to myriad other species.

IUCN's CNPPA should develop a marine biogeographic scheme based on patterns of endemism that can be used by nations and international institutions to help establish a global system of marine protected and special management areas. Using this biogeographic scheme, IUCN, in cooperation with the GEF, should establish the framework for a Global Network of Marine Parks by the 2002 World Parks Congress, which should have protection and sustainable use of marine areas as its main theme. Moreover, IUCN and UNESCO's Man and the Biosphere (MAB) Programme should assist nations in establishing marine and estuarine biosphere reserves representing all of the major marine bio-geographic regions featuring protection zones, sustainable use zones, and buffer zones as a means of conserving representative samples of the world's marine biota.

National/Local

Nations should recognize that certain places within each marine biogeo-graphic province merit protection as critical marine areas, including areas of especially high species diversity or endemism, and areas of special importance as sources of recruitment, as nursery grounds, and as stopover points and corridors for migratory species, and should estab-lish these as protected or special management areas, which can be incorporated into the Global Network of Marine Parks outlined above.

Nations should vigorously implement the Ramsar Convention to promote conservation of marine wetlands and freshwater wetlands of impor-tance to marine ecosystems.

Strengthening the Knowledge Base

International

Industrialized nations should establish national institutes for the environ-ment (NIEs) as a mechanism for providing funding for areas of scientific research, monitoring, and training relevant to environmental quality, including maintaining marine biological diversity, and the World Bank should provide substantial long-term assistance to developing nations having adequate infrastructures and human resources to establish their own NIEs. An integral part of these NIEs should be national biodiversity centers, which would coordinate and fund (among other tasks) national marine biodiversity inventories (including inventories of species of po-tential economic value), to provide information to decision makers in industry, government, and the conservation and scientific communities, perhaps using as models Costa Rica's InBio or the Canadian Centre for Biodiversity. The Association of Systematics Collections should provide assistance to these institutions to ensure that voucher specimens are deposited and maintained effectively within host countries and that interested researchers can have free access to them.

UNEP, with funding from the GEF and other funding sources and cooperation from IOC, IUBS, SCOPE, SCOR, ICES, and research institutions worldwide, should establish an integrated global system of regional research and monitoring centers to establish and run a comprehensive network of marine research and monitoring sites where long-term, baseline data on ecological processes can be accumulated and where research essential for the management of similar areas can be conducted. One special emphasis would be to gain far better understanding of variation at different spatial and temporal scales and its influence on our species. Such a system, which would integrate land-based, sea-based, and satellite observation, is essential if humankind is to gain a comprehensive understanding of marine processes and human impacts in time to respond to and prevent undesirable changes to the sea.

Industrialized nations should reverse the steep decline in support for biological systematics research, by substantially increasing and sustaining long term funding for research, training, and collections for traditional and molecular taxonomy of marine organisms. They should offer special economic incentives for people to enter these fields. Moreover, they should make special efforts to attract students from developing countries to increase the pool of qualified scientists for the international marine biodiversity centers mentioned below.

GEF should fund the establishment and maintenance of six to eight international marine biodiversity centers in developing countries in Latin America/Caribbean, Africa, and Australasia/Oceania. These centers, which could be located at or associated with the abovementioned regional research and monitoring centers, would serve two functions: as regional collections and repositories of information for research on marine systematics (traditional and molecular), biogeography, ecology, and sustainable fisheries biology worldwide, and as donors of small grants to marine biodiversity researchers within their regions. These institutions would also provide the personnel and facilities to carry out national marine biodiversity inventories (see above) for nations that do not have the resources to do their own ones.

GEF should fund a consortium of organizations led by IUCN to carry out a global marine wetlands survey that would use remote sensing information and ground-truthing to determine the status and trends of marine wetlands nation-by-nation.

GEF and UNESCO should establish a program for assisting eligible coastal nations in setting up elementary- and secondary school pilot programs in marine resources and conservation, and if successful, should give these programs five-year funding commitments.

Governmental and private science funding agencies in industrialized nations should fund fellowship programs that would support research and training that integrate marine science, conservation biology, and related fields for graduate students and recent graduates from developing countries. These studies could be carried out either in industrialized countries or in those developing countries whose facilities and personnel permit this kind of intensive training.

National/Local

Government and private science funding agencies should target funds for development of natural and social science fields critical to marine conservation, including marine conservation biology, marine landscape (seascape) ecology, marine restoration ecology, use of remote sensing in inventorying marine ecosystems, marine environmental resource valuation, and ecological economics.

Nations should encourage and welcome foreign scientific researchers, monitoring, and inventory efforts in their waters so long as they adhere to national and local laws and provide host nations with both raw data and information in final form, as a means of improving marine resource management.

Nations should establish biodiversity curricula for all primary and secondary schools to increase awareness about biodiversity and the need for its conservation, and should establish or strengthen national or subnational institutions providing information on the conservation and potential values of biodiversity.

Scientists undertaking marine research in foreign developing countries should unfailingly involve institutions, scientists, and students from host countries as active participants in their research, thereby improving the knowledge base in the host country and increasing the commitment to marine research and conservation.

Marine research institutions and scientists in industrialized countries should seek and provide financial support to qualified students from developing countries to assist their research at the researchers' home institutions, thereby developing a pool of qualified scientists committed to studying and conserving marine species and ecosystems.

Nations should design national conservation databases so that they can be linked with one another and with the World Conservation Monitoring Centre.

Providing Environmentally Sound Financial and Technical Assistance

Bilateral and multilateral development agencies and private lenders should not provide any funding for coastal or marine development projects or projects in watersheds significantly affecting coastal or marine ecosystems until there is thorough EIA of the short- and long-term environmental, social and economic effects on marine organisms, ecosystems, and those who depend on them.

To determine the accuracy of environmental impact assessments and whether there is need for modification of projects after they have begun, all projects funded by bilateral and multilateral development agencies and private lenders should include comprehensive environmental monitoring programs.

International financial institutions, including both private sector lenders and foundations and public sector donors and lenders, should develop guidelines or review procedures to ensure that they do not fund activities that will degrade the biological diversity and integrity of marine waters, and so prevent sustainable use of marine resources.

International and national public sector financial institutions should earmark funds for the development and operation of cooperative regional programs to conserve and sustainably use the sea. A good model would be for the GEF to fund cooperative fisheries conservation commissions within LMEs.

Involving Citizens in Decision Making

International

The charters and policies of international agencies and the texts of international treaties, agreements, and programs should be modified as necessary to affirm the rights of individuals and NGOs to participate in decisions affecting environmental quality and the use of natural resources, including those of the sea.

National/Local

Citizens concerned with conserving marine biological diversity should organize to share their views with other sectors of society. They should seek and expect assistance from the IUCN and other established NGOs in their plans to establish, administer, and fund their efforts.

Nations should ensure the rights of individuals and NGOs to participate in decisions affecting environmental quality and the use of resources, including those of the sea, by enacting and/or enforcing appropriate laws institutionalizing such participation. By serving on government delegations and government advisory bodies, by testifying before legislatures, and by providing written testimony and comments, citizens can contribute their expertise and views for the deliberations of policy makers.

National government decision makers, including administrators, legislators, and the courts, should ensure that citizens have access to information relevant to decision-making processes that affect marine biological diversity. Citizens' rights to such information should be guaranteed by law.

National and local governments should encourage citizen oversight of domestic and international laws and treaties by providing funding or tax incentives for funding their organizations. Such tax laws have been crucial to the success of NGOs.

National and local governments and private institutions should encourage citizen participation and oversight in major decisions by private industry and other private organizations whose actions may affect marine biological diversity—for example, through environmental audits and other procedures—to demonstrate their compliance with applicable law and their efforts to minimize adverse environmental effects by their operations.

National and local governments should ensure that citizens can seek judicial review or other methods to ensure the enforcement of laws affecting the conservation of marine biological diversity.

National government agencies charged with conserving marine living resources should enlist the assistance of NGOs to support effective government conservation programs and to secure funding from national legislatures and international funding agencies for implementing that work.

Groups of citizens concerned about marine conservation should communicate with other organizations sharing their interests through such mechanisms as the International Marine Conservation Network to facilitate the implementation of recommendations in *The Strategy*, to compare experiences, and to collaborate in achieving their mutual objectives.

Changing the Burden of Proof

Leading IGOs (including the UN and the MDBs), NGOs, and nations should exert their influence to ensure that policies that encourage the increased use of marine species and ecosystems until there is irrefutable evidence of unacceptable adverse impacts are changed so that the burden is placed on the user to demonstrate reasonably that increased use levels will be sustainable and not have significant adverse impacts on target or associated species or ecosystems. The precautionary principle should govern the way we treat the sea.

APPENDIX A

Acronyms

ATS	Antarctic Treaty System
CARICOMP	Caribbean Coastal Marine Productivity program
CBF	Chesapeake Bay Foundation
CCAMLR	Convention on the Conservation of Antarctic Marine Living Resources
CFC	chlorofluorocarbon
CITES	Convention on International Trade in Endangered Species of Wild Fauna and Flora
CMA	critical marine area
CMC	Center for Marine Conservation
CNPPA	IUCN Commission on National Parks and Protected Areas
CZM	coastal zone management
DMS	dimethyl sulfide
EDF	Environmental Defense Fund
EEZ	exclusive economic zone
EIA	environmental impact assessment
EMAP	US EPA Environmental Monitoring and Assessment Program
EPA	Environmental Protection Agency (USA)
FAO	UN Food and Agriculture Organization
GATT	General Agreement on Tariffs and Trade
GBR	Great Barrier Reef
GEF	Global Environment Facility
GESAMP	Group of Experts on the Scientific Aspects of Marine Pollution
GIS	geographic information systems
GNP	gross national product
IAEA	International Atomic Energy Agency
IAM	integrated area management
IATTC	InterAmerican Tropical Tuna Commission
ICCAT	International Commission for the Conservation of Atlantic Tuna
ICES	International Council for the Exploration of the Sea
ICNAF	International Commission for the Northwest Atlantic Fisheries

ICLARM	International Center for Living Aquatic Resource Management
ICRW	International Convention for the Regulation of Whaling
IFQ	individual fishing quota
IGO	International governmental organization
IMO	International Maritime Organization
INPFC	International North Pacific Fisheries Commission
IOC	International Oceanographic Commission
IPCC	Intergovernmental Panel on Climate Change
IPHC	International Pacific Halibut Commission
ITQ	individual transferrable quota
IUBS	International Union of Biological Sciences
IUCN	World Conservation Union (formerly International Union for Conservation of Nature and Natural Resources)
IWC	International Whaling Commission
LDC	London Dumping Convention
LME	large marine ecosystem
LOS	Law of the Sea
MAB	Man and the Biosphere
MARPOL	International Convention for the Prevention of Pollution from Ships
MDB	multilateral development bank
MEPC	Marine Environment Protection Committee
MMS	US Department of the Interior Minerals Management Service
MPA	marine protected area
MSY	maximum sustainable yield
NAFO	North Atlantic Fisheries Organization
NASCO	North Atlantic Salmon Conservation Organization
NEPA	National Environmental Policy Act (USA)
NGO	non-governmental organization
NIE	national institutes for the environment
NMFS	U.S. National Oceanic and Atmospheric Administration National Marine Fisheries Service
NOAA	U.S. National Oceanic and Atmospheric Administration
ORSTROM	Office de la Recherche Scientific et Technique d'Outre-Mer
OVI	Ocean Voice International
PCBs	polychlorinated biphenyls
SCOPE	Scientific Committee on Problems of the Environment
SPAW	Protocol on Specially Protected Areas and Wildlife of the Wider Caribbean Region
SSC	IUCN Species Survival Commission
TAC	total allowable catch

TED	turtle excluder device
UN	United Nations
UNCED	United Nations Conference on Environment and Development
UNCLOS	United Nations Convention on the Law of the Sea
UNDP	United Nations Development Programme
UNEP	United Nations Environment Programme
UNESCO	United Nations Educational, Scientific, and Cultural Organization
WWF	World Wide Fund for Nature or World Wildlife Fund

Institutions Mentioned in the Text

Association of Systematics
    Collections
730 11th Street, N.W.
Washington, D.C. 20001 (USA)
(202) 347-2850 (telephone)
(202) 347-0072 (fax)

Center for Marine Conservation
1725 DeSales Street, N.W.
Washington, D.C. 20036 (USA)
(202) 429-5609 (telephone)
(202) 872-0619 (fax)

Chesapeake Bay Foundation
162 Prince George Street
Annapolis, Maryland 21401 (USA)
(301) 268-8816 (telephone)
(301) 268-6687 (fax)

Commission on National Parks and
    Protected Areas
(see IUCN)

Consultative Group on Biological
    Diversity
1290 Avenue of the Americas
Room 3450
New York, New York 10104 (USA)
(212) 373-4200 (telephone)
(212) 315-0996 (fax)

Cousteau Society, Inc.
870 Greenbriar Circle
Chesapeake, Virginia 23320 (USA)
(804) 523-9335 (telephone)
(804) 523-2747 (fax)

East West Center
1777 East West Road
Honolulu, Hawaii 96848 (USA)
(808) 944-7111 (telephone)
(808) 944-7970 (fax)

Environmental Defense Fund, Inc.
257 Park Avenue South
New York, New York 10010 (USA)
(212) 505-2100 (telephone)
(212) 505-2375 (fax)

European Community
200 rue de la Loi
1049 Bruxelles (Belgium)

General Agreement on Tariffs and
    Trade (GATT)
154 rue de Lausanne
1211 Geneva 21 (Switzerland)

Global Environment Facility
(c/o World Bank)

Great Barrier Reef Marine Park
Authority
G.P.O. Box 791
Canberra City, A.C.T. 2601
(Australia)
(062) 47 0211 (telephone)
(062) 47 5761 (fax)

Greenpeace Council
Keizersgracht 176
1016 Dw Amsterdam (Netherlands)
20-253-6555 (telephone)
20-523-6500 (fax)

Group of Experts on Problems of
Marine Pollution (GESAMP)
Administrative Secretary
Marine Environment Division
International Maritime Organization
(IMO)
4 Albert Embankment
London SE1 7SR (UK)

Hawaii Audubon Society
212 Merchant Street, Suite 320
Honolulu, Hawaii 96813 (USA)
(808) 522-5566

Instituto Nacional de Biodiversidad
(INBio)
3100 Santa Domingo
Heredia (Costa Rica)
(506) 36-7690 (telephone)
(506) 36-2818 (fax)

InterAmerican Tropical Tuna
Commission (IATTC)
Scripps Institution of Oceanography
9500 Gilman Drive
La Jolla, California 92093-0210
(USA)

International Atomic Energy Agency
Wagramerstrasse 5
POB 100
1400 Wien (Austria)

International Commission for the
Conservation of Atlantic Tunas
(ICCAT)
Principe de Vergara 17-7
28001 Madrid (Spain)
431-03-29 (telephone)
576-19-68 (fax)

International Maritime Organization
(IMO)
4 Albert Embankment
London SE1 7SR (UK)
71-587-3123 (telephone)
71-587-3210 (fax)

International North Pacific Fisheries
Commission
6640 Northwest Marine Drive
Vancouver, British Columbia,
V6T 1X2 (Canada)
(604) 228-1128 (telephone)
(604) 228-1135 (fax)

International Pacific Halibut
Commission
P.O. Box 95009
Seattle, Washington 98145-2009
(USA)
(206) 634-1838 (telephone)
(206)632-2983 (fax)

International Whaling Commission
The Red House, Station Road
Histon, Cambridge CB44NP (UK)
02 23 23 3971 (telephone)
02 23 23 2876 (fax)

IUCN (World Conservation Union)
Rue Mouvernay 28
CH-1196 Gland (Switzerland)
22-999-0001 (telephone)
22-999-0002 (fax)

Marine Conservation Society
9, Gloucester Road
Ross-on-Wye
Herefordshire HR9 5BU (UK)
98-96-60-17

National Marine Fisheries Service
NOAA
East West Highway
Silver Spring, Maryland 20910 (USA)
(301) 713-2239 (telephone)
(301) 713-2258 (fax)

National Oceanic and Atmospheric
    Administration (NOAA)
U.S. Department of Commerce
14th and Constitution Avenue, NW
Washington, D.C. 20230 (USA)
(301) 443-8910

National Research Council
National Academy of Sciences
2101 Constitution Avenue, N.W.
Washington, D.C. 20418 (USA)
(202) 334-2000 (telephone)
(202) 334-1597 (fax)

National Science Foundation
1800 G Street, NW
Washington, D.C. 20550 (USA)
(202) 357-5000 (telephone)
(202) 357-7745 (fax)

National Wildlife Federation
1400 Sixteenth Street, NW
Washington, D.C. 20036-2266 (USA)
(202) 797-6800 (telephone)
(202) 797-6646 (fax)

Northwest Atlantic Fisheries
    Organization (NAFO)
P.O. Box 638
Dartmouth, Nova Scotia, B24 3Y9
    (Canada)

Ocean Voice International
2883 Otterson Drive
Ottawa, Ontario KIV 7B2 (Canada)
(613) 990-8818

Office of Technology Assessment
United States Congress
Washington, D.C. 20510-8025 (USA)
(202) 228-6098 (telephone)
(202) 228-6208 (fax)

Office of Oceanic Research Programs
NOAA
1335 East West Highway
Silver Spring, Maryland 20901 (USA)
(301) 713-2448 (telephone)
(301) 713-0799 (fax)

Partners of the Americas
2001 S Street NW
Washington, D.C. 20009 (USA)

Species Survival Commission of
    IUCN (SSC)
c/o Chicago Zoological Society
Brookfield, Illinois 60513 (USA)
(708) 485-0263 (telephone)
(708) 485-3532 (fax)

United Nations Development
    Programme (UNDP)
One United Nations Plaza
New York, New York 10017 (USA)
(212) 906-5328 (telephone)
(212) 906-5364 (fax)

United Nations Food and Agriculture
Organization (FAO)
Via delle Terme di Caracalla
00100
Roma (Italy)

United Nations Environment
Programme (UNEP)
P.O. Box 30552
Nairobi (Kenya)

U.S. National Cancer Institute
9000 Rockville Pike
Bethesda, Maryland 20814 (USA)
(301) 496-4000

Woods Hole Research Center
13 Church Street
Woods Hole, Massachusetts 02543
(USA)
(508) 540-9900 (telephone)
(508) 540-9700 (fax)

World Bank
1818 H Street, NW
Washington, D.C. 20433 (USA)
(202) 477-1234 (telephone)
(202) 477-6391 (fax)

World Conservation Monitoring
Centre
219 Huntingdon Road
Cambridge CB3 ODL (UK)
02 23 277 314 (telephone)
02 23 277 136 (fax)

World Conservation Union
(*see* IUCN)

World Meteorological Organization
41 avenue Guiseppe Motta
Casa Postale 2300
1211 Geneva 2 (Switzerland)

World Resources Institute
1709 New York Avenue, NW
Washington, D.C. 20006 (USA)
(202) 638-6300 (telephone)
(202) 638-0036 (fax)

World Wildlife Fund-U.S.
1250 24th Street, NW
Washington, D.C. 20037 (USA)
(202) 293-4800 (telephone)
(202) 293-9211 (fax)

World Wide Fund for Nature
World Conservation Center
Avenue du Mont-Blanc
1196 Gland (Switzerland)
22-36-49-111

Legal Citations

Chapter 7. Tools for Conserving Marine Biological Diversity

Expanding the Knowledge Base

Draft Global Plan of Action, FAO/UNEP project no. 0502–78/02 (FAO Rome 1981).

Information Transfer

Convention for the Protection and Development of the Marine Environment of the Wider Caribbean Region (known as the Cartagena Convention), March 24, 1983, 22 I.L.M. 227 (1983).

Protocol concerning Cooperation in Combatting Oil Spills in the Wider Caribbean Region, March 24, 1983, reprinted in 22 I.L.M. 227 (1983).

Protocol on Specially Protected Areas and Wildlife of the Wider Caribbean Region (SPAW), January 18, 1990, reprinted in *2 New Directions in the Law of the Sea*, Doc. J. 36, at 3 (K. R. Simmonds loose-leaf ed. September 1990).

Environmental Impact Assessment

National Environmental Policy Act of 1969 (NEPA), 44 U.S.C. section 4331–4332 (1992).

Regulating Threats

Stockholm Declaration on the Human Environment, UN Doc. A/CONF. 48/14, reprinted in 11 I.L.M. 1416 (1972).

World Charter for Nature, October 28, 1982, UN General Assembly Resolution 37/7, 37 UN GAOR, Supp. No. 51 at 17–18, U.N. Doc. A/37/51 (1982).

Convention on International Trade in Endangered Species of Wild Fauna and Flora (CITES), March 6, 1973, 27 U.S.T. 1087, T.I.A.S. No. 8249, 993 U.N.T.S. 243, reprinted in 12 I.L.M. 1085 (1973).

InterAmerican Tropical Tuna Commission (IATTC), was created InterAmerican Tropical Tuna Convention, May 31, 1949, 1 U.S.T. 230, 80 U.N.T.S. 3.

United Nations Convention on the Law of the Sea (UNCLOS III), opened for signature December 10, 1982, UN Doc. A/CONF.62/122, reprinted in United Nations, *Official Text of the United Nations Convention on the Law of the Sea with Annexes and Index*, UN Sales No. E.83.V.5. (1983), 21 I.L.M. (1982).

The Convention on the Prevention of Marine Pollution by Dumping of Wastes and Other Matter, done December 29, 1972, 26 U.S.T. 2403, T.I.A.S. No. 8165, 1046 U.N.T.S. 120, 11 I.L.M. 1291 (known as the London Dumping Convention).

Convention for the Prevention of Marine Pollution by Dumping from Ships and Aircraft, done February 15, 1972, 932 U.N.T.S. 3 (known as the Oslo Convention).

United Nations Environment Programme, created by General Assembly Resolution, G.A. Res. 2997 (XXVII), 26 U.N. GAOR, Supp. 30, pp. 43–45 (1972).

Many of the conventions that have resulted from the UNEP Regional Sea Programme are reprinted in *2 New Directions in the Law of the Sea* (K. R. Simmonds loose-leaf ed. September 1986); Sand, *Marine Environment in the United Nations Environment Programme* (1988).

United States Clean Water Act, 33 U.S.C. sections 466 et seq. (1992).

International Maritime Organization (IMO), previously known as the Inter-Governmental Consultative Organization, March 6, 1948, 9 U.S.T. 621, T.I.A.S. No. 4044, 289 U.N.T.S. 48.

International Convention for the Prevention of Pollution from Ships (known as the MARPOL Convention), November 2, 1973, 12 I.L.M. 1319.

General Agreement on Tariffs and Trade (GATT), opened for signature October 30, 1947, 61 Stat. A3, T.I.A.S. No. 1700, 55 U.N.T.S. 194.

Chapter 8. Existing Marine Institutions and Instruments

International Organizations and Programs

Law of the Sea

United Nations Convention on the Law of the Sea (UNCLOS III), opened for signature December 10, 1982, UN Doc. A/CONF.62/122, reprinted in *United Nations, Official Text of the United Nations Convention on the Law of the Sea with Annexes and Index*, UN Sales No. E.83.V.5. (1983), 21 I.L.M. (1982).

Antarctic Treaties

Convention on the Regulation of Antarctic Mineral Resource Activities, June 2, 1988, reprinted 27 I.L.M. 860.

Protocol on Environmental Protection to the Antarctic Treaty, June 21, 1991, reprinted 30 I.L.M. 1461.

Convention on the Conservation of Antarctic Marine Living Resources (CCAMLR), May 20, 1980, 33 U.S.T. 3476, T.I.A.S. No. 10, 240.

Convention on the Conservation of Antarctic Seals, June 1, 1972, 29 U.S.T. 441, T.I.A.S. No. 8826.

A general discussion of all Antarctic materials is contained within *1 Antarctica and International Law: A Collection of Inter-state and National Documents* (W. Bush ed. 1982).

Fishery Treaties

International Commission for the Northwest Atlantic Fisheries (ICNAF), created by the International Convention for the Northwest Atlantic Fisheries, done February 8, 1949, 1 U.S.T. 477, T.I.A.S. No. 2089, 157 U.N.T.S. 158.

International North Pacific Fisheries Commission (INPFC), created by the International Convention for the High Seas Fisheries of the North Pacific Ocean, May 9, 1952, 4 U.S.T. 380, T.I.A.S. No. 2786, 205 U.N.T.S. 65.

International Pacific Halibut Commission (IPHC), created by the Convention for the Preservation of the Halibut Fishery of the North Pacific Ocean and Bering Sea, March 2, 1953, 5 U.S.T. 5, T.I.A.S. 2900, 222 U.N.T.S. 77.

The MARPOL Convention

International Convention for the Prevention of Pollution from Ships (known as the MARPOL Convention), November 2, 1973, 12 I.L.M. 1319 and the Protocol of 1978 Relating to the International Convention for the Prevention of Pollution from Ships, 1973, done at London, February 17, 1978, 17 I.L.M. 546.

London Dumping Convention

The Convention on the Prevention of Marine Pollution by Dumping of Wastes and Other Matter, done December 29, 1972, 26 U.S.T. 2403, T.I.A.S. No. 8165, 1046 U.N.T.S. 120, 11 I.L.M. 1291 (known as the London Dumping Convention).

International Whaling Commission

International Convention for the Regulation of Whaling, December 2, 1946, 62 Stat. (2) 1716, T.I.A.S. No. 1849, 161 U.N.T.S. 72.

United Nations Food and Agriculture Organization

United Nations Food and Agriculture Organization, citations to the various conventions and conferences establishing the pertinent programs of the

FAO can be found in *Basic Texts of the Food and Agriculture Organization of the United Nations* (1989 ed.).

United Nations International Maritime Organization

International Maritime Organization (IMO), previously known as the Inter-Governmental Consultative Organization, March 6, 1948, 9 U.S.T. 621, T.I.A.S. No. 4044, 289 U.N.T.S. 48.

National and Local Organizations and Programs

Federal Management of Marine Fisheries, USA

Magnuson Fishery Conservation and Management Act, 16 U.S.C. section 1801 et seq. (1992).

Great Barrier Reef Marine Park, Australia

Great Barrier Reef Marine Park Authority, created by the Great Barrier Reef Marine Park Act of 1975 (Cth.).

Endangered Marine Animal Species

The following is a compilation of marine animals listed under the IUCN Red List of Threatened Animals, US Endangered Species Act (ESA), and the Convention on International Trade in Endangered Species of Wild Fauna and Flora (CITES). The species range from inhabitants of the open ocean through coastal and estuarine species to dwellers in intertidal salt marshes. Some are obligate marine inhabitants for all or part of the year; others occur in both marine and freshwater or terrestrial realms.

The list includes only those species whose status is sufficiently known to have justified listing under the institutions' two categories of highest risk. The actual list of species in danger of extinction is likely to be considerably larger. Moreover, IUCN and CITES have additional listing categories; species in these categories are not included below. Under IUCN, endangered (E) species are at risk of extinction; vulnerable (V) is the next highest level of risk. The highest category under the US ESA is endangered (E); species likely to become endangered are listed as threatened (T). Under CITES, Appendix I (1) also represents the highest level of risk for species that enter international trade; Appendix II (2) has two meanings, one of which is the next highest level of risk.

SCIENTIFIC NAME	COMMON NAME	IUCN	ESA	CITES
Mammals				
Pontoporia blainvillei	franciscana (La Plata dolphin)			2
Cephalorhynchus commersonii	Commerson's dolphin			2
Cephalorhynchus eutropia	black dolphin			2
Cephalorhynchus heavisidii	Heaviside's dolphin			2
Cephalorhynchus hectori	Hector's dolphin	V		2

Scientific Name	Common Name	IUCN	ESA	CITES
Orcaella brevirostris	Irrawaddy dolphin			2
Sotalia spp.	tucuxis			1
Sousa spp.	hump-backed dolphins			1
Phocoena phocoena	harbor porpoise			2
Phocoena sinus	vaquita (cochito, Gulf of California harbor porpoise)	E	E	1
Neophocaena phocaenoides	finless porpoise			1
Physeter macrocephalus (=*catodon*)	sperm whale		E	1
Delphinapterus leucas	white whale			2
Monodon monoceros	narwhal			2
Hyperoodon spp.	bottlenose whale			1
Hyperoodon ampullatus	northern bottlenose whale	V		
Berardius spp.	beaked (=fourtooth) whales			1
Balaenoptera acutorostrata	minke whale			1/2
Balaenoptera borealis	sei whale		E	1
Balaenoptera edeni	Bryde's whale			1
Balaenoptera musculus	blue whale	E	E	1
Balaenoptera physalus	fin(back) whale	V	E	1
Megaptera novaeangliae	humpback whale	V	E	1
Eubalaena (=*Balaena*) *glacialis*	northern right whale	E	E	1
Eubalaena (=*Balaena*) *australis*	southern right whale	V		1
Balaena mysticetus	bowhead whale	V	E	1
Caperea marginata	pygmy right whale			1
Eschrichtius robustus (=*glacus*)	gray whale		E	1
Ursus (=*Thalarctos*) *maritimus*	polar bear	V		2
Enhydra lutris nereis	southern sea otter		T	1
Lutra felina	marine otter	V	E	1
Lutra longicaudis (incl. *platensis*)	long-tailed otter		E	1
Lutra longicaudis longicaudis	La Plata otter	V		
Lutra lutra	European otter			1
Lutra lutra lutra	European otter	V		
Lutra perspicillata	smooth-coated otter			2

Scientific Name	Common Name	IUCN	ESA	CITES
Lutra provocax	southern river otter	V	E	1
Lutra sumatrana	hairy-nosed otter			2
Arctocephalus philippii	Juan Fernandez fur seal	V		2
Arctocephalus townsendi	Guadalupe fur seal	V	T	1
other *Arctocephalus* spp.	fur seals			2
Eumetopias jubata	Steller's sea lion		T	
Monachus monachus	Mediterranean monk seal	E	E	1
Monachus schauinslandi	Hawaiian monk seal	E	E	1
Phoca hispida saimensis	Saimaa ringed seal	E		
Phoca vitulina stejnegeri	Kuril (harbor) seal	V		
Mirounga angustirostris	northern elephant seal			2
Mirounga leonina	southern elephant seal			2
Dugong dugon	dugong	V	E	1/2
Trichechus inunguis	Amazonian manatee	V	E	1
Trichechus manatus	West Indian (Florida) manatee	V	E	1
Trichechus senegalensis	West African manatee	V	T	2
Dipodomys heermanni morroensis	Morro Bay kangaroo rat	E	E	
Peromyscus polionotus allophrys	Choctawhatchee beach mouse	E	E	
Peromyscus polionotus ammobates	Alabama beach mouse	E	E	
Peromyscus polionotus niveiventris	southeastern beach mouse		T	
Peromyscus polionotus phasma	Anastasia Island beach mouse	E	E	
Peromyscus polionotus trissyllepsis	Perdido Key beach mouse	E	E	
Reithrodontomys raviventris	salt marsh harvest mouse	E	E	
Microtus pennsylvanicus dukecampbelli	Florida salt marsh vole		E	
Pseudomys praeconis	Shark Bay mouse		E	1
Xeromys myoides	false water rat		E	1

Birds

Spheniscus demersus	jackass penguin			2
Spheniscus humboldti	Peruvian penguin			1

Scientific Name	Common Name	IUCN	ESA	CITES
Spheniscus mendiculus	Galápagos penguin		E	
Megadyptes antipodes	yellow-eyed penguin	V		
Diomedea albatrus	short-tailed albatross		E	1
Diomedea amsterdamensis	Amsterdam albatross	E		
Pterodroma axillaris	Chatham Island petrel	V		
Pterodroma cahow	cahow (Bermuda petrel)	E	E	
Pterodroma defilippiana	Defilippe's petrel	V		
Pterodroma madeira	freira	E		
Pterodroma phaeopygia sandwichensis	Hawaiian dark-rumped petrel		E	
Pterodroma pycrofti	Pycroft's petrel	V		
Procellaria parkinsoni	black petrel	V		
Puffinus auricularis (formerly *newelli*)	Newell's Townsend's shearwater ('A'o)		T	
Puffinus newelli	Newell's shearwater	V		
Puffinus auricularis	Townsend's shearwater	E		
Puffinus creatopus	pink-footed shearwater	V		
Pelecanus crispus	Dalmation pelican	E		1
Pelecanus occidentalis	brown pelican		E	
Sula abbotti	Abbott's booby	E	E	1
Fregata andrewsi	Andrew's (Christmas) frigatebird	E	E	1
Egretta eulophotes	Chinese egret	E	E	
Leptoptilos javanicus	lesser adjutant	V		
Mycteria cinerea	milky stork	V		1
Geronticus eremita	northern bald ibis	E		1
Eudocimus ruber	scarlet ibis			2
Platalea leucorodia	white spoonbill			2
Platalea minor	black-faced spoonbill	E		
Phoenicopteridae spp.	flamingoes			2
Branta rufficolis	red-breasted goose			2
Cygnus melanocoryphus	black-necked swan			2
Dendrocygna arborea	West Indian whistling duck			2
Oxyura leucocephala	white-headed duck	V		2
Marmaronetta angustirostris	marbled teal	V		
Anas aucklandica aucklandica	Auckland Island flightless teal			2

Scientific Name	Common Name	IUCN	ESA	CITES
Anas aucklandica chlorotis	New Zealand brown teal			2
Anas aucklandica nesiotis	Campbell Island flightless teal		E	1
Anas bernieri	Madagascar teal	V		2
Anas laysanensis	Laysan teal		E	1
Aythya baeri	Baer's pochard	V		
Haliaeetus albicilla	white-tailed fish-eagle		E	
Haliaeetus albicilla groenlandicus	Greenland white-tailed eagle		E	
Haliaeetus leucocephalus	bald eagle		E/T	1
Haliaeetus pelagicus	Steller's sea eagle			2
Haliaeetus sanfordi	Solomon's sea eagle			2
Haliaeetus vociferoides	Madagascar fish-eagle	E		2
Rallus longirostris levipes	light-footed clapper rail		E	
Rallus longirostris obsoletus	California clapper rail		E	
Gallirallus australis hectori	buff weka			2
Haematopus chathamensis	Chatham black oystercatcher	E		
Himantopus mexicanus (=himantopus) knudseni	Hawaiian stilt (Ae'o)		E	
Charadrius melodus	piping plover		E/T	
Thinornis novaeseelandiae	New Zealand shore plover	E	E	
Tringa guttifer	Nordmann's greenshank		E	1
Numenius borealis	Eskimo curlew	E	E	1
Numenius tenuirostris	slender-billed curlew			1
Larus audouninii	Audouin's gull		E	
Larus relictus	relict gull		E	1
Sterna albostriata	black-fronted tern	V		
Sterna antillarum	least tern		E	
Sterna dougallii dougallii	roseate tern		E/T	
Sterna virgata	Kerguelen tern	V		
Synthliboramphus wumizusume	Japanese murrelet	V		
Ammodramus (=Ammospiza) maritimus mirabilis	Cape Sable seaside sparrow		E	

Scientific Name	Common Name	IUCN	ESA	CITES
Reptiles				
Chelonia mydas (incl. *agassizi*)	green sea turtle	E	T/E	1
Eretmochelys imbricata	hawksbill sea (carey) turtle	E	E	1
Lepidochelys kempi	Kemp's (Atlantic) ridley sea turtle	E	E	1
Lepidochelys olivacea	olive (Pacific) ridley sea turtle	E	T/E	1
Caretta caretta	loggerhead sea turtle	V	T	1
Dermochelys coriacea	leatherback sea turtle	E	E	1
Amblyrhynchus cristatus	Galápagos marine iguana			2
Nerodia fasciata taeniata	Atlantic salt marsh snake		T	
Alligator mississippiensis	American alligator		T	2
Caiman latirostris	broad-snouted caiman	E	E	1
Crocodylus acutus	American crocodile	E	E	1
Crocodylus cataphractus	African slender-snouted crocodile		E	1/2
Crocodylus palustris	marsh crocodile	V		1
Crocodylus palustris palustris	mugger crocodile		E	
Crocodylus palustris kimbula	Ceylon mugger crocodile		E	
Crocodylus porosus	saltwater (estuarine) crocodile	V	E	1/2
Crocodylus niloticus	Nile crocodile	V	T/E	1/2
Crocodylus rhombifer	Cuban crocodile	E	E	1
Fishes				
Latimeria chalumnae	coelacanth	V		1
Acipenser brevirostrum	shortnose sturgeon	V	E	1
Acipenser oxyrhynchus	Atlantic sturgeon	V		2
Acipenser sturio	common sturgeon		E	1
Hypomesus transpacificus	Delta smelt	V		
Prototroctes maraena	Australian grayling	V		
Coregonus canadensis	Atlantic whitefish	E		
Coregonus oxyrhynchus	houting	E		
Oncorhynchus clarki	cutthroat trout	V		
Oncorhynchus tshawytscha	chinook salmon		T	
Stenodus leucichthys leucichthys	beloribitsa	V		

Scientific Name	Common Name	IUCN	ESA	CITES
Nothobranchius sp.	Caprivi killifish	E		
Valencia hispanica	Valencia toothcarp	E		
Menidia conchorum	key silverside	V		
Pseudomugil mellis	honey blue-eye	V		
Gasterosteus aculeatus santaeannae	Santa Anna stickleback	E		
Hippocampus capensis	Knysna seahorse	V		
Syngnathus watermayeri	river pipefish	V		
Myoxocephalus quadricornis	fourhorn sculpin	V		
Totoaba (=*Cynoscion*) *macdonaldi*	totoaba (seatrout, weakfish)	E	E	1
Taenioides jacksoni	bearded goby	V		

Invertebrates

Scientific Name	Common Name	IUCN	ESA	CITES
Nematostella vectensis	starlet sea anemone	V		
Antipatharia spp.	black corals			2
Scleractinia	stony corals			2
Milleporidae spp.	fire corals			2
Stylasteridae spp.	lace corals			2
Coenothecalia spp.	blue corals			2
Tubiporidae spp.	organpipe corals			2
Hippopus hippopus	horse's hoof clam			2
Hippopus porcellanus	China clam			2
Tridacna crocea	crocus clam			2
Tridacna derasa	southern giant clam	V		2
Tridacna gigas	giant clam	V		2
Tridacna maxima	small giant clam			2
Tridacna squamosa	scaly clam			2

Literature Cited

Ahmad, Yusuf J. and George K. Sammy (1985). *Guidelines to Environmental Impact Assessment in Developing Countries.* Sponsored by UN Environment Programme. Hodder and Stoughton, London (UK).

Alcala, Angel C. (1988). Effects of marine reserves on coral fish abundances and yields of Philippine coral reefs. *Ambio* 17:194–99.

Alexander, Vera (1992). Arctic marine ecosystems. Pp. 221–32 in Robert L. Peters and Thomas E. Lovejoy, eds. *Global Warming and Biological Diversity.* Yale University Press, New Haven, Connecticut (USA).

Anderson, Lee G. (1986). *The Economics of Fishery Management.* The Johns Hopkins University Press, Baltimore, Maryland (USA).

Andrew, N.L. and J.G. Pepperell (1992). The bycatch of shrimp trawl fisheries. *Oceanography and Marine Biology: An Annual Review* 30:527–65.

Angel, Martin V. (1992). Managing biodiversity in the oceans. Pp. 23–59 in Melvin N. A. Peterson, ed. *Diversity of Oceanic Life: An Evaluative Review.* Center for Strategic and International Studies, Washington, D.C. (USA).

Anonymous (1991). *Report of the Study Group on Ecosystem Effects of Fishing Activities.* Lowestoft, 11–15 March 1991. ICES C.M. 1991/G:7, Session Y.

Atkins, Natasha and Burr Heneman (1987). The dangers of gill netting to seabirds. *American Birds* 41(5):1395–1403.

Ayling, Tony and Geoffrey J. Cox (1987). *Collins Guide to the Sea Fishes of New Zealand.* William Collins Publishers, Ltd., Auckland (New Zealand).

Bailey, Robert G. (1989). Explanatory supplement to ecoregions map of the continents. *Environmental Conservation* 16(4):307–10.

Balazs, George H. (1981). Growth rates of immature green turtles in the Hawaiian archipelago. Pp. 117–26 in Karen A. Bjorndal, ed. *Biology and Conservation of Sea Turtles.* Smithsonian Institution Press, Washington, D.C. (USA).

Ballantine, W. J. (1991). *Marine Reserves for New Zealand.* University of Auckland, Leigh Laboratory Bulletin No. 25, Auckland (New Zealand).

Bally, R. and C. L. Griffiths (1989). Effects of human trampling on an exposed rocky shore. *International Journal of Environmental Studies* 34(1/2):115–25.

Barber, Richard T. and Francisco P. Chavez (1983). Biological consequences of El Niño. *Science* 222:1203–10.

Beddington, J. R. (1984). The response of multispecies systems to perturbations. Pp. 209–25 in R. M. May, ed. *Exploitation of Marine Communities.* Report of the Dahlem Workshop on Exploitation of Marine Communities. Springer-Verlag, Berlin (Germany).

Bhaskar, S. (1981). Preliminary report on the status and distribution of sea turtles in Indian waters. *Indian Forester* November 1981:707–11.

Birnie, Patricia (1985). *International Regulation of Whaling*. Oceana Publications, Inc., New York, New York (USA).

Boekschoten, G. J. and Maya Wijsman-Best (1981). *Pocillopora* in the Miocene reef at Baixo, Porto Santo (Eastern Atlantic). *Proceedings of the Koninklijke Nederlandse Akademie van Wetenschapen Series B: Palaeontology, Geology, Physics and Chemistry* 84(1):13–20.

Bohnsack, J. A. (1990). *The Potential of Marine Fishery Reserves for Reef Fish Management in the U.S. Southern Atlantic*. NOAA Technical Memorandum NMFS-SEFC-261. NOAA/NMFS Southeast Fisheries Center, Miami, Florida (USA).

Boicourt, William C. (1982). Estuarine larval retention on two scales. Pp. 445–57 in Victor S. Kennedy, ed. *Estuarine Comparisons*. Academic Press, New York, New York (USA).

Bold, Harold C. and Michael J. Wynne (1985). *Introduction to the Algae: Structure and Reproduction, Second Edition*. Prentice-Hall, Inc., Englewood Cliffs, New Jersey (USA).

Bricklemyer, Eugene C., Jr., Suzanne Iudicello, and Hans J. Hartmann (1989). Discarded catch in U.S. commercial marine fisheries. Pp. 259–95 in W. J. Chandler and L. Labate, eds. *Audubon Wildlife Report 1989/1990*. Academic Press, New York, New York (USA).

Briggs, John C. (1969). The sea-level Panama canal: Potential biological catastrophe. *BioScience* 19(1):44–47.

Briggs, John C. (1974). *Marine Zoogeography*. McGraw-Hill, New York, New York (USA).

Broecker, Wallace S. (1987). The biggest chill. *Natural History* 96(10):74–82.

Broecker, Wallace S. and George H. Denton (1989). The role of ocean-atmosphere reorganizations in glacial cycles. *Geochimica et Cosmochimica Acta* 53:2465–2501

Brown, B. and R. P. Dunne (1988). The environmental impact of coral mining on coral reefs in the Maldives. *Environmental Conservation* 15(2):159–65.

Brown, Lester R. et al. (1989). *State of the World*. W. W. Norton and Company, New York, New York (USA).

Brownel, Robert L., Jr., Katherine Ralls, and William F. Perrin (1989). The plight of the "forgotten" whales. *Oceanus* 32(1):5–11.

Brusca, Richard C. (1980). *Common Intertidal Invertebrates of the Gulf of California, Second Edition*. University of Arizona Press, Tucson, Arizona (USA).

Caddy, J.F. and M. Savini (1989). *Report of the Second GFCM Technical Consul-*

tation on Red Coral in the Mediterranean. Torre del Greco, Italy. September 27–30, 1988. FAO, Rome (Italy).

Caldwell, Lynton Keith (1990). *International Environmental Policy: Emergence and Dimensions*. Duke University Press, Durham, North Carolina (USA).

Cambares, G. and H. Lima (1990). Leatherbacks disappearing from the BVI. *Marine Turtle Newsletter* 49:4–7.

Canada Department of Fisheries and Oceans (1991). *Scientific Review of North Pacific Driftnet Fisheries*. Sidney, British Columbia, June 11–14, 1991. Unpublished report, 3 vols., available from: DFO Canada, P.O. Box 6000, Sidney, British Columbia V8L 4B2 (Canada).

Carlton, James T. (1979). *History, Biogeography, and Ecology of the Introduced Marine and Estuarine Invertebrates of the Pacific Coast of North America*. Ph.D. dissertation, University of California, Davis, California (USA).

Carlton, James T. (1985). Transoceanic and interoceanic dispersal of coastal marine organisms: The biology of ballast water. *Oceanography and Marine Biology, an Annual Review* 23:313–71.

Carlton, James T. (1987). Mechanisms and patterns of transoceanic marine biological invasions in the Pacific Ocean. *Bulletin of Marine Science* 41:467–99.

Carlton, James T. (1989). Man's role in changing the face of the ocean: Biological invasions and implications for conservation of near-shore environments. *Conservation Biology* 3(3):265–73.

Carlton, James T. and J. A. Scanlon (1985). Progression and dispersal of an introduced alga: *Codium fragile* sp. *tomentosoides* (Chlorophyta) on the Atlantic coast of North America. *Botanica Marina* 28:155–65.

Carlton, James T., J. K. Thompson, L. E. Schemel, and F. H. Nichols (1990). Remarkable invasion of San Francisco Bay (California, USA) by the Asian clam *Potamocorbula amurensis*. Introduction and dispersal. *Marine Ecology Progress Series* 66:81–94.

Carlton, James T., Geerat J. Vermeij, David R. Lindberg, Debby A. Carlton, and Elizabeth Dudley (1991). The first historical extinction of a marine invertebrate in an ocean basin: The demise of the eelgrass limpet *Lottia alveus*. *Biological Bulletin* 180(1):72–80.

Carpenter, Richard A. and James E. Maragos (1989). *South Pacific Regional Environment Programme (SPREP): How to Assess Environmental Impacts on Tropical Islands and Coastal Areas*. Environment and Policy Institute, East-West Center, Honolulu, Hawaii (USA).

Carr, Archie (1987). Impact of nondegradable marine debris on the ecology and survival outlook of sea turtles. *Marine Pollution Bulletin* 18(6B):352–56.

Castro, Gonzalo, J. P. Myers, and Allen R. Place (1989). Assimilation efficiency of sanderlings (*Calidris alba*) feeding on horseshoe crab (*Limulus polyphemus*) eggs. *Physiological Zoology* 62(3):716–31.

Caton, A., K. McLoughlin, and M. J. Williams (1989). *Southern Bluefin Tuna: Scientific Background to the Debate*. Bulletin No. 3, Department of Primary Industries and Energy, Bureau of Rural Resources, Australian Government Publishing Service, Canberra (Australia).

Chabreck, Robert H. (1988). *Coastal Marshes: Ecology and Wildlife Management*. University of Minnesota Press, Minneapolis, Minnesota (USA).

Chambers, J. R. (1991). Coastal degradation and fish population losses. *Marine Recreational Fisheries* 14:45–51.

Charlson, R. E., J. Langner, and H. Rodhe (1990). Sulfate aerosol and climate. *Nature* 348:22.

Christy, Francis T., Jr. (1982). *Territorial Use of Rights in Marine Fisheries*. FAO Fisheries Technical Paper No. 227.

Clark, Colin W. (1973). Profit maximization and the extinction of animal species. *Journal of Political Economy* 81(4):950–61.

Clark, C. W. and R. Lamberson (1982). An economic history and analysis of pelagic whaling. *Marine Policy* 6:103–20.

Clark, R. B. (1989). *Marine Pollution*. Clarendon Press, Oxford (UK).

Clarke, Andrew (1992). Is there a latitudinal diversity cline in the sea? *Trends in Ecology and Evolution* 7(9):286–87.

Collette, Bruce B. and Cornelia E. Nauen (1984). *FAO Species Catalogue, Vol. 2. Scombrids of the World*. Fisheries Synopsis Series No. 125.

Colwell, Rita R. and Russell Hill (1992). Microbial diversity. Pp. 100–106 in Melvin N. A. Peterson, ed. *Diversity of Oceanic Life: An Evaluative Review*. Center for Strategic and International Studies, Washington, D.C. (USA).

Corpuz, V. T., P. Casteneda, and J. C. Sy (1983). Muro ami. *Fisheries Newsletter of the Philippine Bureau of Fisheries and Aquatic Science* 12(1):2–13.

Cortes, J. N. and M. J. Risk (1985). Reef under siltation stress: Cahuita, Costa Rica. *Bulletin of Marine Science* 35:339.

Costanza, R., ed. (1991). *Ecological Economics: The Science and Management of Sustainability*. Columbia University Press, New York, New York (USA).

Council on Environmental Quality (1993). *Incorporating Biodiversity Considerations Into Environmental Impact Analysis Under the National Environmental Policy Act*. Council on Environmental Quality, Washington, D.C. (USA).

Courtenay, W. R., Jr. and J. R. Stauffer, Jr., eds. (1984). *Distribution, Biology, and Management of Exotic Fishes*. The Johns Hopkins University Press, Baltimore, Maryland (USA).

Crouse, Deborah T., Larry B. Crowder, and Hal Caswell (1987). A stage-based population model for loggerhead sea turtles and implications for conservation. *Ecology* 68(5):1412–23.

Croxall, J. P., P. G. H. Evans, and R. W. Schreiber (1984). *Status and Conservation of the World's Seabirds*. International Council for Bird Preservation, Cambridge (UK).

Cury, Philippe (1991). Les contraintes biologiques liées à une gestion des ressources instables. Pp. 506–18 in Philippe Cury and Claude Roy, eds. *Pêcheries Ouest-Africaines (Variabilité, Instabilité et Changement)*. OR-STROM, Paris (France).

Darwin, Charles R. (1842). *The Structure and Distribution of Coral Reefs*. Smith, Elder and Co., London (UK).

Davis, A. R. and A. J. Butler (1989). Direct observations of larval dispersal in the colonial ascidian *Podoclavella moluccensis* Sluiter: Evidence for closed populations. *Journal of Experimental Marine Biology and Ecology* 127: 189–203.

Davis, C. O. (1982). The San Francisco Bay ecosystem: A retrospective overview. Pp. 17–37 in W. J. Kockelman, T. J. Conomos, and A. E. Levitan, eds. *San Francisco Bay: Use and Protection*. Pacific Division, American Association for the Advancement of Science, San Francisco, California (USA).

Davis, Kimberely (1985). International management of cetaceans under the new Law of the Sea Convention. *Boston University International Law Journal* 3:477–518.

Davis, W. J. (1990). Global aspects of marine pollution policy: The need for a new international convention. *Marine Policy* 14(3):191–97.

Davis, W. J., and J. M. VanDyke (1990). Dumping of decommissioned nuclear submarines at sea: A technical and legal analysis. *Marine Policy* 14(6): 467–76.

Day, David (1981). *The Doomsday Book of Animals: A Natural History of Vanished Species*. Viking Press, New York, New York (USA).

Dayton, Paul K. (1975). Experimental studies of algal canopy interactions in a sea otter-dominated kelp community at Amchitka Island, Alaska. *Fishery Bulletin* 73(2):230–37.

Dayton, Paul K. (1989). Interdecadal variation in an Antarctic sponge and its predators from oceanographic climate shifts. *Science* 245:1484–86.

Debenham, P. and L. K. Younger (1991). *Cleaning North America's Beaches: 1990 Beach Cleanup Results*. Center for Marine Conservation, Washington, D.C. (USA).

Denny, Mark W. (1988). *Biology and the Mechanics of the Wave-Swept Environment*. Princeton University Press, Princeton, New Jersey (USA).

Derr, M. (1992). Raiders of the reef. *Audubon* 94(2):48–56.

di Castri, F. and T. Younes (1990). Ecosystem function of biological diversity. *Biology International*, Special Issue 22:1–20.

Dixon, J. A. (1989). Coastal resources: Assessing alternatives. Coastal Area Management in South East Asia: Policies, Management Strategies and Case Studies. *ICLARM Conference Proceedings* 19:153–62.

Doty, M. S. (1961). *Acanthophora*, a possible invader of the marine flora of Hawaii. *Pacific Science* 15:547–52.

Doumenge, F., S. Edwards, and R. J. Planck (1990). *Evaluation of the Impact of the Proposed Airport at Shiraho Reef.* Report to the IUCN–The World Conservation Union, November 1990, Perth, Australia.

Draggan, S., J. J. Cohrssen, and R. E. Morrison, eds. (1987). *Preserving Ecological Systems: The Agenda for Long-Term Research and Development.* Praeger, New York, New York (USA).

Drake, James A., Harold A. Mooney, F. di Castri, R. H. Groves, F. J. Kruger, M. Rejmanek, and M. Williamson, eds. (1989). *Biological Invasions: A Global Perspective.* SCOPE report 37. John Wiley and Sons, Chichester (UK).

Druehl, L., M. Cackette, and J. M. D'Auria (1988). Geographical and temporal distribution of Iodine-131 in the brown seaweed *Fucus* subsequent to the Chernobyl incident. *Marine Biology* 98(1):125–29.

Earll, R. C. (1992). Commonsense and the precautionary principle: An environmentalist's perspective. *Marine Pollution Bulletin* 24(4):182–86.

Edeson, W. R. and J. F. Pulvenis (1983). *The Legal Regime of Fisheries in the Caribbean Region. Lecture Notes on Coastal and Estuarine Studies, Vol. 7.* Springer-Verlag, Berlin (Germany).

Ehrenfeld, David (1978). *The Arrogance of Humanism.* Oxford University Press, New York, New York (USA).

Ehrlich, P. R. and P. H. Raven (1964). Butterflies and plants: A study in coevolution. *Evolution* 18:586–608.

El-Sayed, Sayed Z. (1985). Plankton of the Antarctic seas. Pp. 135–53 in W.N. Bonner and D.W.H. Walton, eds. *Key Environments: Antarctica.* Pergamon Press, Oxford (UK).

Elton, Charles S. (1958). *The Ecology of Invasions by Animals and Plants.* Methuen, London (UK).

Emlet, R. B., L. R. McEdward, and Richard R. Strathmann (1987). Echinoderm larval ecology viewed from the egg. Pp. 55–136 in M. Jangoux and J. M. Lawrence, eds. *Echinoderm Studies.* A.A. Balkema, Rotterdam (The Netherlands).

Engle, J. A. (1979). *Ecology and Growth of Juvenile California Spiny Lobster, Panulirus interruptus (Randall).* Ph.D. dissertation, University of Southern California, Los Angeles, California (USA).

Eton, P. (1985). Tenure and taboo: Customary rights and conservation in the South Pacific. Pp. 164–75 in *Third South Pacific National Parks and Reserves Conference: Conference Report Vol. 2.* SPREP, Noumea (New Caledonia).

Fagerstrom, J. A. (1991). Reef-building guilds and a checklist for determining guild membership. *Coral Reefs* 10:47–52.

FAO (United Nations Food and Agriculture Organization) (1985). *Review of the Regular Programme 1984–85.* FAO, Rome (Italy).

FAO (United Nations Food and Agriculture Organization) (1991). *Statistics Series No. 68: Catches and Landings*. FAO, Rome (Italy).

Fisheries Agency of Japan, Canadian Department of Fisheries and Oceans, U.S. National Marine Fisheries Service, and the U.S. Fish and Wildlife Service (1991). *Final Report of 1990 Observations of the Japanese High Seas Squid Driftnet Fishery in the North Pacific Ocean*. Joint report of the Fisheries Agency of Japan, Canadian Department of Fisheries and Oceans, U.S. National Marine Fisheries Service, and the U.S. Fish and Wildlife Service. Washington, D.C. (USA).

Fonseca, M. S., W. J. Kenworthy, and G. W. Thayer (1991). Seagrass beds: Nursery for coastal species. *Marine Recreational Fisheries* 14:141–47.

Forman, Richard T. T. and Michel Godron (1986). *Landscape Ecology*. John Wiley and Sons, New York, New York (USA).

Foster, M. S. and D.C. Barilotti (1990). An approach to determining the ecological effects of seaweed harvesting: A summary. *Hydrobiologia* 204/205:15–16.

Foster, M. S., A. P. DeVogelaere, C. Harrold, J. S. Pearse, and A. B. Thum (1988). Causes of spatial and temporal patterns in rocky intertidal communities of central and northern California. *Memoirs of the California Academy of Sciences* 9:1–45.

Fot'yanova L. I. and M. Ya Serova (1987). The late Miocene climatic optimum in the Northeast Pacific province. *International Geology Review* 29(5):515–28.

Fowler, C. W. (1987). *A Review of Seal and Sea Lion Entanglement in Marine Fishing Debris*. Paper presented at the North Pacific Rim Fishermen's Conference on Marine Debris. October 13–16, 1987. Kailua-Kona, Hawaii (USA).

Freeman, A. Myrick III (1982). *Air and Water Pollution Control: A Benefit-Cost Assessment*. John Wiley and Sons, New York, New York (USA).

Futuyma, Douglas J. (1979). *Evolutionary Biology*. Sinauer Associates, Sunderland, Massachusetts (USA).

Gadgil, Madhav (1987). Diversity: Cultural and biological. *Trends in Ecology and Evolution* 2(12):369–73.

Gámez, Rodrigo (1991). Biodiversity conservation through facilitation of its sustainable use: Costa Rica's National Biodiversity Institute. *Trends in Ecology and Evolution* 6(12):377–78.

GESAMP (Joint Group of Experts on the Scientific Aspects of Marine Pollution) (1987). *Land/Sea Boundary Flux of Contaminants: Contributions from Rivers*. GESAMP Reports and Studies No. 32.

GESAMP (Joint Group of Experts on the Scientific Aspects of Marine Pollution) (1990). *The State of the Marine Environment*. Blackwell Scientific Publications, Oxford (UK).

Glantz, M. H. (1984). *Report of the Working Group on Societal Implications of*

Varying Fishery Resources: Report of the Expert Consultation to Examine Changes in Abundance and Species Composition of Neritic Fish Resources, San Jose, Costa Rica, April 18–29, 1983. FAO Fishery Report No. 291, Vol. 1.

Glynn, P. W. and W. H. de Weerdt (1991). Elimination of two reef-building hydrocorals following the 1982–83 El Niño warming event. *Science* 253(5015):69–71.

Glynn, Peter W. and Joshua S. Feingold (1992). Hydrocoral species not extinct. *Science* 257(5078):1845.

Gomez, E. D., A. C. Alcala, and A. C. San Diego (1981). Status of Philippines coral reefs-1981. *Proceedings of the 4th International Coral Reef Symposium, Manila* 1:275–82.

Goodwin, C. Lynn, and Warren Shaul (1984). *Age, Recruitment and Growth of the Geoduck Clam* (Panope generosa *Gould) in Puget Sound, Washington*. Washington State Department of Fisheries Progress Report No. 215, Olympia, Washington (USA).

Goodyear, C. Philip (1990). *Status of Red Snapper Stocks of the Gulf of Mexico*. U.S. Department of Commerce, National Marine Fisheries Service, Southeast Fisheries Center, Miami Laboratory Contribution CRD89/90–05, Miami, Florida (USA).

Gordon, H. S. (1954). The economic theory of a common-property resource: The fishery. *Journal of Political Economy* 62:124–42.

Gordon, Rue E., ed. (1993). *1993 Conservation Directory, 38th Edition*. National Wildlife Federation, Washington, D.C. (USA).

Gould, Steven Jay (1989). *Wonderful Life: The Burgess Shale and the Nature of History*. W. W. Norton and Company, New York, New York (USA).

Graham, M. (1955). Effect of trawling on animals in the sea bed. *Deep Sea Research* 3 (Suppl.):1–6.

Grassle, J. Frederick (1986). The ecology of deep-sea hydrothermal vent communities. *Advances in Marine Biology* 23:301–62.

Grassle, J. Frederick (1989). Species diversity in deep-sea communities. *Trends in Ecology and Evolution* 4:12–15.

Grassle, J. Frederick (1991). Deep-sea benthic biodiversity. *BioScience* 41:464–69.

Grassle, J. F., P. Laserre, A. D. McIntyre, and G. C. Ray (1991). Marine biodiversity and ecosystem function. *Biology International*, Special Issue 23:1–19.

Grassle, J. Frederick and Nancy J. Maciolek (1992). Deep-sea species richness: Regional and local diversity estimates from quantitative bottom samples. *American Naturalist* 139(2):313–41.

Grassle, Judith P. and J. Frederick Grassle (1976). Sibling species in the marine pollution indicator *Capitella* (Polychaeta). *Science* 192(4239):567–69.

Gray, Mark Allan (1990). The United Nations Environment Programme: An assessment. *Environmental Law* 20:291–319.

Grigg, Richard W. (1984). Resource management of precious corals: A review and application to shallow reef building corals. *Marine Ecology* 5(1):57–74.

Grimes, C. B. (1987). Reproductive Biology of Lutjanidae: A review. Pp.239–94 in J.J. Polovina and S. Ralston, eds., *Tropical Snappers and Groupers: Biology and Fisheries Management*. Westview Press, Boulder, Colorado (USA).

Groombridge, Brian (1992). *Global Biodiversity: Status of the Earth's Living Resources*. Chapman and Hall, London (UK).

Groves, R. H. and J. J. Burdon, eds. (1986). *Ecology of Biological Invasions: An Australian Perspective*. Cambridge University Press, London (UK).

Gulland, J. A. and S. Garcia (1984). Observed patterns in multispecies fisheries. Pp. 155–90 in R. M. May, ed. *Exploitation of Marine Communities*. Report of the Dahlem Workshop on Exploitation of Marine Communities. Springer-Verlag, Berlin (Germany).

Häder, D.-P., R. C. Worrest, and H. D. Kumar (1989). Aquatic ecosystems. Pp. 39–48 in J.C. van der Leun and M. Tevini, eds. *Environmental Effects Panel Report*. United Nations Environment Programme, Nairobi (Kenya).

Häder, D.-P., R. C. Worrest, and H. D. Kumar (1991). Aquatic ecosystems. Pp. 33–40 in J.C. van der Leun and M. Tevini, eds. *Environmental Effects of Ozone Depletion: 1991 Update*. United Nations Environment Programme, Nairobi (Kenya).

Halim, Y. (1990). Manipulation of hydrological cycles. Pp. 231–319 in *Technical Annexes to the Report on the State of the Marine Environment*. UNEP Regional Seas Reports and Studies No. 114/1.

Halverson, M. and D. F. Martin (1980). Studies of cytoclysis of *Chattonella subsalsa*. *Florida Science* 43(1):35.

Hamilton, L. S. and S. C. Snedaker, eds. (1984). *The Mangrove Management Handbook*. IUCN/UNESCO/UNEP/EWC, Honolulu, Hawaii (USA).

Hanna, Susan S. (1990). The eighteenth-century English commons: A model for ocean management. *Ocean and Shoreline Management* 14:155–72.

Harden Jones, F. R. (1968). *Fish Migration*. St. Martin's Press, New York, New York (USA).

Hardin, Garrett (1968). The tragedy of the commons. *Science* 162(3859): 1243–48.

Hay, Mark E. (1984). Patterns of fish and urchin grazing on Caribbean coral reefs: Are previous results typical? *Ecology* 65(2):446–54.

Hayden, B. P., G. C. Ray, and R. Dolan (1984). Classification of coastal and marine environments. *Environmental Conservation* 11:199–207.

Hedgecock, D. (1986). Is gene flow from pelagic larval dispersal important in the adaptation and evolution of marine invertebrates? *Bulletin of Marine Science* 39(2):550–64.

Heneman, Burr (1992). Troubled habitats. *The Observer (Point Reyes Bird Observatories)* Winter 1991–92:12–15.

Herrnkind, William F. (1980). Spiny lobsters: Patterns of movement. Pp. 349–407 in J. S. Cobb and B. F. Phillips, eds. *Biology and Management of Lobsters, Vol. 1.* Academic Press, New York, New York (USA).

Hicklin, P. W. (1987). The migration of shorebirds in the Bay of Fundy. *Wilson Bulletin* 99:540–70.

Highsmith, Raymond C. (1985). Floating and algal rafting as potential dispersal mechanisms in brooding invertebrates. *Marine Ecology Progress Series* 25(2):169–79.

Hildebrand, Henry H. (1982). A historical review of the status of sea turtle populations in the western Gulf of Mexico. Pp. 447–54 in Karen A. Bjorndal, ed. *Biology and Conservation of Sea Turtles.* Smithsonian Institution Press, Washington, D.C. (USA).

Hoffman, J. S., D. Keyes, and J. G. Titus (1983). *Projecting Future Sea Level Rise: Methodology, Estimates to the Year 2100, and Research Needs.* U.S. Environmental Protection Agency, EPA 230–09–007, Washington, D.C. (USA).

Hounshell, Paul B. and Gerry M. Madrazo (1990). Why marine education? Pp. 1–5 in Gerry M. Madrazo and Paul B. Hounshell, eds. *Oceanography for Landlocked Classrooms.* National Association of Biology Teachers, Reston, Virginia (USA).

Hurme, Arthur K. and Edward J. Pullen (1988). Biological effects of marine sand mining and fill placement for beach replenishment: Lessons for other uses. *Marine Mining* 7:123–36.

Hutchings, P. A. and B. L. Wu (1987). Coral reefs of Hainan Island, South China Sea. *Marine Pollution Bulletin* 18(1):25–26.

IPCC (Intergovernmental Panel on Climate Change) (1990). *Climate Change: The IPCC Scientific Assessment.* J. T. Houghton, G. J. Jenkins, and J. J. Ephraums, eds. Cambridge University Press, London (UK).

IUCN (International Union for Conservation of Nature and Natural Resources) (1976). *Proceedings of an International Conference on Marine Parks and Reserves, Tokyo, 12–14 May 1975.* IUCN Publications New Series 37, Gland (Switzerland).

IUCN (World Conservation Union) (1988). *IUCN Action Plan: Dolphins, Porpoises and Whales—Action Plan for Conservation of Biological Diversity 1988–1992.* W.F. Perrin, ed. International Union for Conservation of Nature and Natural Resources, Gland (Switzerland).

IUCN (World Conservation Union) (1991). *Directory of Members.* IUCN, Gland (Switzerland).

IUCN Commission on National Parks and Protected Areas (1984). Categories,

Objectives, and Criteria for Protected Areas. Pp. 47–53 in Jeffrey A. McNeely and Kenton R. Miller, eds. *National Parks, Conservation and Development: The Role of Protected Areas in Sustaining Society.* Smithsonian Institution Press, Washington, D.C. (USA).

IUCN, UNEP, and WWF (International Union for the Conservation of Nature and Natural Resources, United Nations Environment Programme, and World Wildlife Fund) (1980). *World Conservation Strategy: Living Resources Conservation for Sustainable Development.* IUCN, Morges (Switzerland).

IUCN, UNEP, and WWF (World Conservation Union, United Nations Environment Programme, and World Wide Fund for Nature) (1991). *Caring for the Earth: A Strategy for Sustainable Living.* IUCN, Gland (Switzerland).

IWC (International Whaling Commission) (1991). *Report of the Workshop on Mortality of Cetaceans in Passive Fishing Nets and Traps.* IWC Meeting Document Sc/43/Rep 1. Unpublished report, available from: IWC, The Red House, 135 Station Road, Histon, Cambridge CB4 4NP (UK).

Jablonski, David (1986). Larval ecology and macroevolution in marine invertebrates. *Bulletin of Marine Science* 39(2):565–87.

Jablonski, David and Richard A. Lutz (1983). Larval ecology of marine benthic invertebrates: Paleobiological implications. *Biological Reviews* 58:21–89.

Jeftig, L. (1987). The Mediterranean Action Plan. *Chemosphere* 16(2–3):N1–N3.

Jickells, Timothy D., Roy Carpenter, and Peter S. Liss (1990). Marine environment. Pp. 313–34 in Billie Lee Turner II, ed. *The Earth as Transformed by Human Action.* Cambridge University Press, New York, New York (USA).

Johannes, R. E. (1978). Traditional marine conservation methods in Oceania and their demise. *Annual Review of Ecology and Systematics* 9:349–64.

Johannes, R. E. (1979). Reproductive strategies of coastal marine fishes in the tropics. *Environmental Biology of Fishes* 3:65–84.

Johannes, R. E. (1982). Traditional conservation methods and protected marine areas in Oceania. *Ambio* 11(5):258–61.

Johnston, D. M. (1965). *The International Law of Fisheries.* Yale University Press, New Haven, Connecticut (USA).

Jokiel, P. L. (1990). Long-distance dispersal by rafting: Reemergence of an old hypothesis. *Endeavour* 14(2):66–73.

Jokiel, P. L. and S. L. Coles (1990). Response of Hawaiian and other Indo-Pacific reef corals to elevated temperatures. *Coral Reefs* 8(4):155–62.

Joseph, James (1983). International tuna management revisited. Pp. 123–50 in Brian J. Rothchild, ed. *Global Fisheries Perspectives for the 1980s.* Springer-Verlag, New York, New York (USA).

Josupeit, Helga (1991). Sponges: World production and trade. *Infofish International* 2/91: 21–27.

Karr, James R. (1991). Biological integrity: A long-neglected aspect of water resources management. *Ecological Applications* 1(1):66–84.

Kelleher, Graeme and Richard Kenchington (1991). *Guidelines for Establishing Marine Protected Areas*. Great Barrier Reef Marine Park Authority, Townsville (Australia) and IUCN, Gland (Switzerland).

Kerr, Richard A. (1992). Unmasking a shifty climate system (includes Did the Great Salinity Anomaly cool the Atlantic?). *Science* 255:1508–10.

Kidron, Michael and Ronald Segal (1987). *The New State of the World Atlas*. Simon and Schuster, New York, New York (USA).

King, F. Wayne (1981). Historical review of the decline of the green turtle and the hawksbill. Pp. 183–88 in Karen A. Bjorndal, ed. *Biology and Conservation of Sea Turtles*. Smithsonian Institution Press, Washington, D.C. (USA).

King, Judith E. (1983). *Seals of the World, Second Edition*. Cornell University Press, Ithaca, New York (USA).

Kirkman, Hugh (1992). Large-scale restoration of seagrass meadows. Pp. 111–40 in Gordon W. Thayer, ed. *Restoring the Nation's Marine Environment*. Maryland Sea Grant, College Park, Maryland (USA).

Kitching, R. L., ed. (1986). *The Ecology of Exotic Animals and Plants: Some Australian Case Histories*. John Wiley and Sons, New York, New York (USA).

Knowlton, Nancy, Ernesto Weil, Lee A. Weigt, and Héctor M. Guzmán (1992). Sibling species in *Montastraea annularis*, coral bleaching, and coral climate record. *Science* 255(5042):330–33.

Küchler, A. W. (1964). *Potential Natural Vegetation of the Conterminous United States*. American Geographical Society Special Publication No. 36, New York, New York (USA).

Kusler, Jon and Mary Kentula (1989). Executive Summary in Jon Kusler and Mary Kentula, eds. *Wetlands Creation and Restoration. Vol. 1, The Status of the Science*, Ecological Research Series, U.S. Environmental Protection Agency, Washington, D.C. (USA).

Lande, Russell (1988). Genetics and demography in biological conservation. *Science* 241:1455–60.

Larkin, P. A. (1977). An epitaph for the concept of maximum sustained yield. *Transactions of the American Fishery Society* 106(1):1–11.

Laws, E. A. and D. G. Redalje (1982). Sewage diversion effects on the water column of a subtropical estuary. *Marine Environmental Research* 6(4): 265–79.

Lawson, Rowena and Michael Robinson (1983). Artisanal fisheries in West Africa. *Marine Policy* 7(4):279–90.

Leatherman, Stephen P. (1990). Environmental implications of shore protection strategies along open coasts (with a focus on the United States). Pp. 197–207 in J. G. Titus, ed. *Changing Climate and the Coasts*. UNEP, Nairobi (Kenya).

Leatherman, Stephen P. (1991). Impact of climate-induced sea level rise on coastal areas. Pp. 170–79 in R. L. Wyman, ed. *Global Climate Change and Life on Earth*. Routledge, Chapman and Hall, New York, New York (USA).

Legrande, M. R., R. J. Delmas, and R. J. Charlson (1988). Climate forcing implications from Vostok ice core sulphate data. *Nature* 334:418–20.

Lehman, Scott J. and Lloyd D. Keigwin (1992). Sudden changes in North Atlantic circulation during the last deglaciation. *Nature* 356:757–62.

Leigh, Egbert G. Jr., Robert T. Paine, James F. Quinn and Thomas H. Suchanek (1986). Wave energy and intertidal productivity. *Proceedings of the National Academy of Sciences, USA* 84:1314–18.

Leopold, Aldo (1949). *A Sand County Almanac, With Essays on Conservation from Round River*. Oxford University Press, New York, New York (USA).

Lessios, H. A., D. R. Robertson, and J. D. Cubit (1984). The spread of *Diadema* mass mortality through the Caribbean. *Science* 226:335–37.

Lewis, Jerry G., Norman F. Stanley, and G. Gordon Guist (1988). Commercial production and applications of algal hydrocolloids. Pp. 205–36 in Lembi, Carole A. and J. Robert Waaland, eds. *Algae and Human Affairs*. Cambridge University Press, New York, New York (USA).

Lien, Jon and Robert Graham, eds. (1985). Marine Parks and Conservation: Challenge and Promise. National and Provincial Parks Association of Canada, Henderson Park Book Series No. 10.

Lindstrom, Sandra C. and Paul W. Gabrielson, eds. (1990). *Proceedings of the Thirteenth International Seaweed Symposium held in Vancouver, Canada, August 13–18, 1989*. Kluwer Academic Publishers, London (UK).

Little, E. J., Jr. (1977). Observations on recruitment of postlarval spiny lobsters, *Panulirus argus*, to the south Florida coast. *Florida Marine Research Publications* 29:1–358

Lopez E., G. A. Gill, G. Camprasse, S. Camprasse, and F. Lallier (1989). Soudure sans transition (octecassimilation) entre l'os maxillaire humain et un implant dentaire compact en calite naturelle d'invertèbres marins. *Comptes-Rendus Hebdomadaires des Séances de l'Académie des Sciences* 309(6):203–10.

Lovejoy, Thomas (1980). A projection of species extinctions. Pp. 328–32 in Council on Environmental Quality and US Department of State. *The Global 2000 Report to the President, Vol. 2, The Technical Report*. U.S. Government Printing Office, Washington, D.C. (USA).

Lugo, Ariel E. (1990). *Mangroves of the Pacific Islands: Research Opportunities*. US Department of Agriculture/Forest Service, Pacific Southwest Research Station. General Technical Report PSW-118.

Lukowski, A. B. (1983). DDT residues in the tissues and eggs of three species of penguins from breeding colonies at Admiralty Bay (King George Island, South Shetland Islands). *Polish Polar Research* 4(1/4):127–34.

MacNeill, J., P. Winsemius and T. Yakushiji (1991). *Beyond Interdependence:*

The Meshing of the World's Economy and the Earth's Ecology. Oxford University Press, Oxford (UK).

Magraw, Daniel Barstow (1990). Legal treatment of developing countries: Differential, contextual, and absolute norms. *Colorado Journal of International Environmental Law and Policy* 1:69.

Manomet Bird Observatory (1986). *The International Shorebird Surveys Newsletter*, March 1986.

Maragos, James E. (1992). Restoring coral reefs with emphasis on Pacific reefs. Pp. 141–222 in Gordon W. Thayer, ed. *Restoring the Nation's Marine Environment*. Maryland Sea Grant, College Park, Maryland (USA).

Margulis, Lynn and Karlene V. Schwartz (1988). *Five Kingdoms: An Illustrated Guide to the Phyla of Life on Earth, Second Edition*. W. H. Freeman and Company, New York, New York (USA).

Marshall, David B. (1988). *Status of the Marbled Murrelet in North America: With Special Emphasis on Populations in California, Oregon, and Washington*. US Fish and Wildlife Service Biological Report 88(30). Washington, D.C. (USA).

Martin, Gene S., Jr. and James W. Brennan (1989). Enforcing the International Convention for the Regulation of Whaling: The Pely and Packwood-Magnuson Amendments. *Denver Journal of International Law and Policy* 17:293–95.

Mathews, C. P., V. R. Gouda, Wafa T. Riad, and J. Dashti (1987). Pilot study for the design of a long life fish trap (gargoor) for Kuwait's fisheries. *Kuwait Bulletin of Marine Science* 1987(9):221–43.

May, John (1990). *The Greenpeace Book of Dolphins*. Sterling Publishing Company, New York, New York (USA).

McAllister, Don E. (1991a). Questions about the impacts of trawling. *Sea Wind* 5(2):28–32.

McAllister, Don E. (1991b). Measuring Arctic Ocean shrinkage and warming: The planetary solar mirror and global thermostat. *Sea Wind* 5(4):19–24.

McAllister, Don E., S. P. Plataia, F. W. Schueler, M. E. Baldwin, and D. S. Lee (1986). Ichthyofaunal patterns on a geographic grid. Pp. 18–51 in C. H. Hocutt and E. O. Wiley, eds. *Zoogeography of North American Freshwater Fishes*. John Wiley and Sons, New York, New York (USA).

McClanahan, Timothy R. (1990). Are conservationists fish bigots? *BioScience* 40(1):2.

McCoy, Floyd W. (1988). Floating megalitter in the Eastern Mediterranean. *Marine Pollution Bulletin* 19(1):25–28.

McKinney, Michael L. (1987). Taxonomic selectivity and continuous variation in mass and background extinctions of marine taxa. *Nature* 325:143–45.

McNeely, Jeffrey A., Kenton R. Miller, Walter V. Reid, Russell A. Mittermeier, and Timothy B. Werner (1990). *Conserving the World's Biological Diversity*.

International Union for Conservation of Nature and Natural Resources, World Resources Institute, Conservation International, World Wildlife Fund-US, and the World Bank, Gland (Switzerland).

McNeely, J. A. and David Pitt, eds. (1984). *Culture and Conservation: The Human Dimension in Environmental Planning*. Croom Helm, London (UK).

Mead, James G. and Edward D. Mitchell (1984). Atlantic gray whales. Pp. 33–53 in Mary Lou Jones, Steven L. Swartz, and Stephen Leatherwood, eds. *The Gray Whale* Eschrichtius robustus. Academic Press, Orlando, Florida (USA).

Meffe, Gary K. (1986). Conservation genetics and the management of endangered fishes. *Fisheries* 11(1):14–23.

Meybeck, Michel (1982). Carbon, nitrogen and phosphorus transport by world rivers. *American Journal of Science* 282:401–50.

Miller, G. H. and A. de Vernal (1992). Will greenhouse warming lead to Northern Hemisphere ice-sheet growth? *Nature* 355:244–46.

Minisian, Stanley M., Kenneth C. Balcomb III, and Larry A. Foster (1984). *The World's Whales*. Smithsonian Institution Press, Washington, D.C. (USA).

Mitton, J. B. and M. C. Grant (1984). Associations among protein heterozygosity, growth rate, and developmental homeostasis. *Annual Review of Ecology and Systematics* 15:479–99.

Mooney, H. A. and J. A. Drake, eds. (1986). *Ecology of Biological Invasions of North America and Hawaii*. Ecological Studies No. 58. Springer-Verlag, New York, New York (USA).

Moyer, J. (1989). Reef channels as spawning sites for fishes on the Shiraho Coral Reef, Ishigaki Island, Japan. *Japanese Journal of Ichthyology* 36(3):371–75.

Mrosovsky, N. (1983). *Conserving Sea Turtles*. British Herpetological Society, London (UK).

Mrosovsky, N. and Jane Provancha (1992). Sex ratio of hatchling loggerhead sea turtles: Data and estimates from a five-year study. *Canadian Journal of Zoology* 70:530–38.

Mrosovsky, N. and C. L. Yntema (1980). Temperature dependence of sexual differentiation in sea turtles: Implications for conservation. *Biological Conservation* 18:271–80.

Muller-Karger, F. E., C. R. McClain, T. R. Fisher, W. E. Esaias, and R. Varela (1989). Pigment distributions in the Caribbean Sea: Observations from space. *Progress in Oceanography* 23:23–64.

Munro, J. L. (1989). Fisheries for giant clams (Tridacnidae: Bivalvia) and prospects for stock enhancement. Pp. 541–58 in J. Caddy, ed. *Marine Invertebrate Fisheries: Their Assessment and Management*. John Wiley and Sons, New York, New York (USA).

Munro, J. L., J. D. Parrish, and F. H. Talbot (1987). The biological effects of intensive fishing on coral reef communities. Pp. 41–49 in B. Salvat, ed. *Human Impacts on Coral Reefs: Facts and Recommendations*. Musée

National d'Histoire Naturelle et École Pratique des Hautes Études, Athénée de Tahiti, Centre de l'Environement, Moorea (French Polynesia).

Murawski, Steven A. (1991). Can we manage our multispecies resources? *Fisheries* 16(5):5–13.

Murawski, Steven A. and J. S. Idoine (1989). *Multispecies Size Composition: A Conservative Property of Exploited Fishery Systems?* North Atlantic Fisheries Organization, Scientific Council Report, Doc. 89/76.

Muzik, K. (1985). Dying coral reefs of the Ryukyu Archipelago (Japan). *Proceedings of the 5th International Coral Reef Congress, Tahiti* 6:483–89.

Myers, J. P. (1983). Conservation of migrating shorebirds: Staging areas, geographic bottlenecks and regional movements. *American Birds* 37:23–25.

Myers, J. P. (1986). Sex and gluttony on Delaware Bay. *Natural History* 95:68–77.

Myers, J. P., R. I. G. Morrison, P. Z. Antas, B. A. Harrington, T. E. Lovejoy, M. Sallaberry, S. E. Senner, and A. Tarak (1987). Conservation strategy for migratory species. *American Scientist* 75:18–26.

National Academy of Sciences (1971). *Radioactivity in the Marine Environment.* National Academy of Sciences, Washington, D.C. (USA).

National Research Council (1975). Marine litter. Pp. 405–33 in *Assessing Potential Ocean Pollutants. A Report of the Study Panel on Assessing Potential Ocean Pollutants to the Ocean Affairs Board.* Commission on Natural Resources, National Research Council, National Academy of Sciences, Washington, D.C. (USA).

National Research Council (1980). *Research Priorities in Tropical Biology.* National Academy of Sciences, Washington, D.C. (USA).

National Research Council (1990a). *Decline of Sea Turtles: Causes and Prevention.* National Academy Press, Washington, D.C. (USA).

National Research Council (1990b). *Managing Troubled Waters: The Role of Marine Environmental Monitoring.* National Academy Press, Washington, D.C. (USA).

National Research Council (1990c). *Fulfilling the Promise: Biology Education in the Nation's Schools.* National Academy Press, Washington, D.C. (USA).

Newman, W. (1963). On the introduction of an edible Oriental shrimp (Caridea, Palaemonidae) to San Francisco Bay. *Crustaceana* 5:119–32.

Nichols, Frederic H., James C. Cloern, Samuel N. Luoma, and David H. Peterson (1986). The modification of an estuary. *Science* 231:567–73.

Nichols, F. H. and J. K. Thompson (1985). Persistence of an introduced mudflat community in South San Francisco Bay, California. *Marine Ecology Progress Series* 24:83–97.

Nicolaidou, A., M. A. Pancucci, and A. Zenetos (1989). The impacts of dumping coarse metalliferous waste on the benthos in Evoikos Gulf, Greece. *Marine Pollution Bulletin* 20(1):28–33.

Norse, Elliott A. (1987). Habitat diversity and genetic variability: Are they necessary ecosystem properties? Pp. 93–113 in S. Draggan, J. J. Cohrssen, and R. E. Morrison, eds. *Preserving Ecological Systems: The Agenda for Long-Term Research and Development.* Praeger, New York, New York (USA).

Norse, Elliott A. (1991). Conserving the neglected 71%: Marine biological diversity. *Species (IUCN Species Survival Commission)* 16:16–18.

Norse, E. A. and V. Fox-Norse (1977). Studies on portunid crabs from the Eastern Pacific. II. The unusual distribution of *Euphylax dovii. Marine Biology* 40:374–75.

Norse, Elliott A. and Roger E. McManus (1980). Ecology and living resources— biological diversity. Pp. 31–80 in *The Eleventh Annual Report of the Council on Environmental Quality.* U.S. Government Printing Office, Washington, D.C. (USA).

Norse, Elliott A., Kenneth L. Rosenbaum, David S. Wilcove, Bruce A. Wilcox, William H. Romme, David W. Johnston, and Martha L. Stout (1986). *Conserving Biological Diversity in Our National Forests.* The Wilderness Society, Washington, D.C. (USA).

Novak, Milan, James A. Baker, Martyn E. Obbard, and Bruce Malloch, eds. (1987). *Wild Furbearer Management and Conservation in North America.* Ontario Ministry of Natural Resources, Toronto (Canada).

Nunny, R. S. and P. C. H. Chillingworth (1986). *Marine Dredging for Sand and Gravel.* Department of the Environment, Minerals Division, London (UK).

Office of Technology Assessment (1987). *Technologies to Maintain Biological Diversity,* OTA-F-330. U.S. Government Printing Office, Washington, D.C. (USA).

Ogden, John C. and Elisabeth H. Gladfelter, eds. (1986). Caribbean coastal marine productivity. *UNESCO Reports in Marine Science* 41. UNESCO, Paris (France).

Ogden, J. C. and R. Wicklund, eds. (1988). *Mass Bleaching of Coral Reefs in the Caribbean: A Research Strategy.* National Undersea Research Program, Research Report 88–2. NOAA, Office of Undersea Research, Rockville, Maryland (USA).

O'Hara, Kathryn J., Suzanne Iudicello, and Rose Bierce (1987). *A Citizen's Guide to Plastics in the Ocean: More Than a Litter Problem.* Center for Marine Conservation, Washington, D.C. (USA).

Olivera, Baldomero M., Jean Rivier, Craig Clark, Cecilia A. Ramilo, Gloria P. Corpuz, Fe C. Abogadie, E. Edward Mena, Scott R. Woodward, David R. Hillyard, and Lourdes J. Cruz (1990). Diversity in *Conus* neuropeptides. *Science* 249:257–63.

Omori, Makoto, Christopher P. Norman, and Hiroshi Tamakawa (1992). Biodiversity: Human impacts through fisheries and transportation. Pp. 63–75 in

Melvin N. A. Peterson, ed. *Diversity of Oceanic Life: An Evaluative Review.* Center for Strategic and International Studies, Washington, D.C. (USA).

Ono, R. Dana, James D. Williams, and Anne Wagner (1983). *Vanishing Fishes of North America.* Stone Wall Press, Washington, D.C. (USA).

Orians, Gordon H. (1993). Endangered at what level? *Ecological Applications* 3(2):206–208.

Paine, Robert T. (1969). A note on trophic complexity and community stability. *American Naturalist* 103(929):91–93.

Paine, Robert T. (1980). Food webs: Linkage, interaction strength and community infrastructure. *Journal of Animal Ecology* 49:667–85.

Palumbi, S. L. and A. C. Wilson (1990). Mitochondrial DNA diversity in the sea urchins *Strongylocentrotus purpuratus* and *S. droebachiensis.* *Evolution* 44(2):403–15.

Panayotou, Theodore (1982). Management concepts for small-scale fisheries: Economic and social aspects. *FAO Fisheries Technical Paper* No. 228.

Partners of the Americas (1988). *Natural Resource Directory: Latin America and the Caribbean, First Edition.* Washington, D.C. (USA).

Patterson, G. M. L. and C. D. Smith (1991). Novel pharmaceuticals from blue-green algae. *International Marine Biotechnology Conference, Baltimore, Maryland,* Paper S109.

Pellegrin, Gilmore J., Jr., Shelby B. Drummond, and Robert S. Ford, Jr. (1981). *The Incidental Catch of Fish by the Northern Gulf of Mexico Shrimp Fleet.* NOAA/NMFS Southeast Fisheries Center, Miami, Florida (USA).

Pernetta, J. C. and P. J. Hughes, eds. (1990). *Implications of Expected Climate Changes in the South Pacific Region: An Overview.* UNEP Regional Seas Reports and Studies No. 128. Nairobi (Kenya).

Perrin, William F. (1989). *Dolphins, Porpoises, and Whales: An Action Plan for the Conservation of Biological Diversity: 1988–1992.* IUCN, Gland (Switzerland).

Peters, Robert Henry (1983). *The Ecological Implications of Body Size.* Cambridge Studies in Ecology, Cambridge University Press, Cambridge (UK).

Peterson, Melvin N. A., ed. (1992). *Diversity of Oceanic Life: An Evaluative Review.* Center for Strategic and International Studies, Washington, D.C. (USA).

Pimm, Stuart L. (1980). Food web design and the effects of species deletion. *Oikos* 35:139–49.

Polar Research Board (1985). *Glaciers, Ice Sheets and Sea Level: Effect of a CO_2-induced Climatic Change.* Report of a workshop held in Seattle, Washington (USA), September 13–15, 1984. U.S. Department of Energy/ER/60235–1.

Posey, M. (1988). Community changes associated with the spread of an introduced seagrass *Zostera japonica. Ecology* 69:974–83.

Povey, Anna and Michael J. Keough (1991). Effects of trampling on plant and animal populations on rocky shores. *Oikos* 61:355–68.

Powers, J. E., C. P. Goodyear, and G. P. Scott (1987). *The Potential Effect of Shrimp Fleet Bycatch on Fisheries Production of Selected Fish Stocks in The Gulf of Mexico*. NOAA/NMFS Southeast Fisheries Center, Contribution No. CRD87/88–06.

Pratt, Harold L., Jr., and John G. Casey (1990). Shark reproductive strategies as a limiting factor in indirected fisheries, with a review of Holden's method of estimating growth-parameter. Pp. 97–107 in Pratt, Harold L., Jr., Samuel H. Gruber, and Toru Taniuchi, eds. *Elasmobranchs as Living Resources: Advances in the Biology, Ecology, Systematics, and the Status of the Fisheries*. NOAA Technical Report NMFS 90.

Premnatha, M., K. Chandra, S. K. Bajpai, and K. Kathiresan (1992). A survey of some Indian marine plants for antiviral activity. *Botanica Marina* 35:321–24.

Preston, M. R. (1989). Marine pollution. Pp. 53–196 in J. P. Riley, ed. *Chemical Oceanography Vol. 9*. Academic Press, London (UK).

Primavera, J. Honculada (1991). Intensive prawn farming in the Philippines: Ecological, social and economic implications. *Ambio* 20(1):28–33.

Pritchard, P. C. H. (1979). *Encyclopedia of Turtles*. T. F. H. Publications, Neptune, New Jersey (USA).

Pritchard, Peter C. H. (1990). Kemp's ridleys are rarer than we thought. *Marine Turtle Newsletter* 49:1–3.

Rabinowitz, Deborah (1978). Early growth of mangrove seedlings in Panama, and an hypothesis concerning the relationship of dispersal and zonation. *Journal of Biogeography* 5:113–33.

Race, M. S. (1982). Competitive displacement and predation between introduced and native mud snails. *Oecologia* 54:337–47.

Randall, J. E. (1987). Introduction of marine fishes to the Hawaiian Islands. *Bulletin of Marine Science* 41:490–502.

Ray, G. Carleton (1976). Critical marine habitats. Pp. 15–59 in *Proceedings of an International Conference on Marine Parks and Reserves, Tokyo, Japan, 12–14 May 1974*. IUCN Publications New Series, No. 37.

Ray, G. Carleton (1988). Ecological diversity in coastal zones and oceans. Pp. 36–50 in E. O. Wilson, ed. *Biodiversity*. National Academy Press, Washington, D.C. (USA).

Ray, G. Carleton (1991). Coastal-zone biodiversity patterns. *BioScience* 41(7):490–98.

Reeves, J. E. and G. S. DiDonato (1972). *Effects of Trawling in Discovery Bay, Washington*. Washington Department of Fisheries, Management and Research Division, Olympia, Washington (USA).

Reid, D. G. (1985). Habitat and zonation patterns of *Littoraria* species (Gastropoda: Littorinidae) in Indo-Pacific mangrove forests. *Biological Journal of the Linnaean Society* 26(1):39–68.

Reid, Walter V. and Mark C. Trexler (1991). *Drowning the National Heritage: Climate Change and U.S. Coastal Biodiversity.* World Resources Institute, Washington, D.C. (USA).

Reynolds, John E., III and Daniel K. Odell (1991). *Manatees and Dugongs.* Facts on File, New York, New York (USA).

Riemann, Bo and Erik Hoffmann (1991). Ecological consequences of dredging and bottom trawling in the Limfjord, Denmark. *Marine Ecology Progress Series* 69:171–78.

Rinehart, K. L. (1991). Pharmaceutical agents from marine sources. *International Marine Biotechnology Conference, Baltimore, Maryland,* Paper S87.

Robison, Bruce H. and Karen Wishner (1990). Biological research needs for submersible access to the greatest ocean depth. *Marine Technology Society Journal* 24(2):34–37.

Rosenblatt, R. (1967). The zoogeographic relationships of the marine shore fishes of tropical America. *Studies in Tropical Oceanography, University of Miami* 5:579–87.

Ross, James Perran (1981). Historical decline of loggerhead, ridley and leatherback sea turtles. Pp. 189–95 in Karen A. Bjorndal, ed. *Biology and Conservation of Sea Turtles.* Smithsonian Institution Press, Washington, D.C. (USA).

Roux, François X., Danial Brasnu, Bernard Loty, Bernard George, and Geneviève Guillemin (1988). Madreporic coral: A new bone graft substitute for cranial surgery. *Journal of Neurosurgery* 69(1):510–13.

Rozengurt, M. A. (1991). Alteration of freshwater inflows. *Marine Recreational Fisheries* 14:73–80.

Ruddle, Kenneth (1986). *No Common Property Problem: Village Fisheries in Japanese Coastal Waters.* Paper presented at Workshop on Ecological Management of Common Property Resources, IVth International Congress of Ecology, Syracuse, New York (USA).

Rugh, David J. (1984). Census of gray whales at Unimak Pass, Alaska: November–December 1977–1979. Pp. 225–48 in Mary Lou Jones, Steven L. Swartz, and Stephen Leatherwood, eds. *The Gray Whale* Eschrichtius robustus. Academic Press, Orlando, Florida (USA).

Saenger, P., E. J. Hegerl, and D. D. S. Davie, eds. (1983). Global Status of Mangrove Ecosystems. *The Environmentalist* 3, Supplement 3.

Sainsbury, K. J. (1988). The ecological basis for multispecies fisheries, and management of a demersal fishery in tropical Australia. Pp. 349–82 in J. A. Gulland, ed. *Fish Population Dynamics, Second Edition.* John Wiley and Sons, New York, New York (USA).

Sakai, T. (1976). Notes from the carcinological fauna of Japan (IV). *Researches on Crustacea* 7:29–40.

Sale, Peter F. (1977). Maintenance of high diversity in coral reef fish communities. *American Naturalist* 111:337–59.

Sale, Peter F., ed (1991). *The Ecology of Fishes on Coral Reefs*. Academic Press, San Diego, California (USA).

Salm, Rodney V. (1987). Coastal zone management planning and marine protected areas. *Parks* 12(1):18–19.

Salm, Rodney V. and John R. Clark (1984). *Marine and Coastal Protected Areas: A Guide for Planners and Managers*. International Union for the Conservation of Nature and Natural Resources, Gland (Switzerland).

Salm, Rodney V. and James A. Dobbin (1987). A coastal zone management strategy for the Sultanate of Oman. Pp. 97–106 in Orville T. Magoon, ed. *Coastal Zone '87: Proceedings of the Fifth Symposium on Coastal and Ocean Management, Vol. 1*. American Society of Civil Engineers, New York, New York (USA).

Salm, Rodney V. and James A. Dobbin (1989). Coastal zone management, planning, and implementation in the Sultanate of Oman. Pp. 72–78 in Orville T. Magoon, ed. *Coastal Zone '89: Proceedings of the Sixth Symposium on Coastal Zone Management, Vol. 1*. American Society of Civil Engineers, New York, New York (USA).

Sætersdal, Gunmar (1980). A review of past management of some pelagic stocks and its effectiveness. *Rapport et Procès-Verbaux des Réunions Conseil International pour l'Exploration de la Mer* 177:505–12.

Savina, Gail C. and Alan T. White (1986). A tale of two islands: Some lessons for marine resource management. *Environmental Conservation* 13(2): 107–13.

SC-CAMLR-VII (Scientific Committee for the Conservation of the Antarctic Marine Living Resources) (1988). *Report of the Seventh Meeting of the Scientific Committee*. 24–31 October 1988, Hobart (Australia).

Schaeff, C., S. Kraus, M. Brown, J. Perkins, R. Payne, D. Gaskin, P. Boag, and B. White (1991). Preliminary analysis of mitochondrial DNA variation within and between the right whale species *Eubalaena glaciensis* and *Eubalaena australis*. Pp. 217–24 in A.R. Hoelzel, ed., *Genetic Ecology of Whales and Dolphins Report of the International Whaling Commission, Special Issue 13*. Cambridge (UK).

Scheltema, R. S. (1989). Planktonic and non-planktonic development among prosobranch gastropods and its relationship to the geographic range of species. Pp. 183–88 in J. S. Ryland and P. A. Tyler, eds., *Reproduction, Genetics and Distribution of Marine Organisms*. Olsen and Olsen, Fredensborg (Denmark).

Schneider, Stephen H. (1989). The greenhouse effect: Science and policy. *Science* 243(4892):771–81.

Schonewald-Cox, C. M., S. M. Chambers, B. MacBryde, and L. Thomas (1983).

Genetics and Conservation. Benjamin/Cummings Publishing Company, Menlo Park, California (USA).

Seliger, H. H., J. A. Boggs, R. B. Rivkin, W. H. Bigley, and K. R. H. Aspden (1982). The transport of oyster larvae in an estuary. *Marine Biology* 71: 57–72.

Shaffer, M.L. (1981). Minimum population sizes for species conservation. *Bio-Science* 31:131–34.

Shepherd, J. G. (1988). Fish stock assessments and their data requirements. Chapter 2 in J.A. Gulland, ed. *Fish Population Dynamics, Second Edition.* John Wiley and Sons, New York, New York (USA).

Sherman, Kenneth (1990). Productivity, perturbations, and options for biomass yields in Large Marine Ecosystems. Pp. 206–19 in Kenneth Sherman, Lewis M. Alexander, and Barry D. Gold, eds. *Large Marine Ecosystems: Patterns, Process, and Yields.* American Association for the Advancement of Science, Washington, D.C. (USA).

Sherman, Kenneth, Lewis M. Alexander, and Barry D. Gold (1990). Large Marine Ecosystems. Pp. vii–xi in Kenneth Sherman, Lewis M. Alexander, and Barry D. Gold, eds. *Large Marine Ecosystems: Patterns, Process, and Yields.* American Association for the Advancement of Science, Washington, D.C. (USA).

Shomura, R. S. and M. L. Godfrey, eds. (1990). *Proceedings of the Second International Conference on Marine Debris, 2–7 April 1989, Honolulu, Hawaii.* U.S. Department of Commerce. NOAA Technical Memorandum NMFS, NOAA-TM-NMFS-SWFSC-154.

Shomura, R. S. and H. O. Yoshida, eds. (1985). *Proceedings of the Workshop on the Fate and Impact of Marine Debris, 27–29 November 1984, Honolulu, Hawaii.* U.S. Department of Commerce. NOAA Technical Memorandum, NMFS, NOAA-TM-NMFS-SWFC-54.

Simenstad, Charles A., James A. Estes, and Karl W. Kenyon (1978). Aleuts, sea otters and alternate stable-state communities. *Science* 200:403–11.

Simon, Joseph L. and Daniel M. Dauer (1977). Reestablishment of a benthic community following natural defaunation. Pp. 139–54 in Bruce C. Coull, ed. *Ecology of Marine Benthos.* University of South Carolina Press, Columbia, South Carolina (USA).

Sinclair, Michael (1988). *Marine Populations: An Essay on Population Regulation and Speciation.* Washington Sea Grant Program, University of Washington Press, Seattle, Washington (USA).

Skjak-Braek, Gudmund, Thorleif Anthonsen, and Paul A. Sandford (1989). *Chitin and Chitosan. Sources, Chemistry, Biochemistry, Physical Properties and Applications.* Elsevier, London (UK).

Smith, G. (1984). The United Nations and the environment: Sometimes a great notion? *Texas International Law Journal* 19:335.

Smith, Joel B. (1990). From global to regional climate changes: Relative knowns and unknowns about global warming. *Fisheries* 15(6):2–6.

Smith, R. C., B. B. Prézelin, K. S. Baker, R. R. Bidigare, N. P. Boucher, T. Coley, D. Karentz, S. MacIntyre, H. A. Matlick, D. Menzies, M. Ondrusek, Z. Wan, and K. J. Waters (1992). Ozone depletion: Ultraviolet radiation and phytoplankton biology in Antarctic waters. *Science* 255:952–59.

Smith, S.H. (1988). Cruise ships: A serious threat to coral reefs and associated organisms. *Ocean and Shoreline Management* 11(3):231–48.

Smith, Thomas J. III, Kevin G. Boto, Stewart D. Frusher, and Raymond L. Giddins (1991). Keystone species and mangrove forest dynamics: The influence of burrowing by crabs on soil nutrient status and forest productivity. *Estuarine, Coastal, and Shelf Science* 33:419–32.

Soekarno, R. (1989). Comparative studies on the status of Indonesian corals. *Netherlands Journal of Sea Research* 23(2):215–22.

Sohn, Louis (1992). From the Hills of Tennessee to the Forests of Brazil: A Short History of International Environmental Law. Pp. 1–32 in *International Environmental Law: Recent Developments and Implications*. American Bar Association, Section of International Law and Practice, Chicago, Illinois (USA).

Soulé, Michael E. (1991). Conservation: Tactics for a constant crisis. *Science* 253(5021):744–50.

Soulé, Michael E. and Kathryn A. Kohm (1989). *Research Priorities for Conservation Biology*. Island Press, Washington, D.C. (USA).

Southward, A. J. (1991). Forty years of changes in species composition and population density of barnacles on a rocky shore near Plymouth. *Journal of Marine Biological Association, United Kingdom* 71:495–513.

Spanier, Ehud and Bella S. Galil (1991). Lessepsian migration: A continuous biogeographical process. *Endeavour, New Series* 15(3):102–106.

Springer, V. G. (1982). Pacific plate biogeography, with special reference to shorefishes. *Smithsonian Contributions in Zoology* 367:1–182.

Stavins, Robert N. (1990). *Innovative Policies for Sustainable Development in the 1990's: Economic Incentives for Environmental Protection*. Resources for the Future, Discussion Paper No. QE90–11.

Steele, John H. (1985). A comparison of terrestrial and marine ecological systems. *Nature* 313(6001):355–58.

Steele, John H. (1991). Marine functional diversity. *BioScience* 41(7):470–74.

Steele, John H., S. Carpenter, J. Cohen, P. Dayton, and R. Ricklefs (1989). *Comparison of Terrestrial and Marine Ecological Systems*. Report of a Workshop, Santa Fe, New Mexico. National Science Foundation Blue Paper (USA).

Stern, Paul C., Oran R. Young, and Daniel Druckman, eds. (1992). *Global Environmental Change: Understanding the Human Dimensions*. National Academy Press, Washington, D.C. (USA).

Stolarski, Richard, Rumen Bojkov, Lane Bishop, Christos Zerefos, Johannes Staehelin, and Joseph Zawodny (1992). Measured trends in stratospheric ozone. *Science* 256:342–49.

Stone, C. P. and D. B. Stone, eds. (1989). *Conservation Biology in Hawaii*. University of Hawaii Cooperative National Park Resources Studies Unit, Honolulu, Hawaii (USA).

Strathmann, Richard R. (1974). The spread of sibling larvae of sedentary marine invertebrates. *American Naturalist* 108(959):29–44.

Strathmann, Richard R. (1985). Feeding and nonfeeding larval development and life-history evolution in marine invertebrates. *Annual Review of Ecology and Systematics* 16:339–61.

Strathmann, Richard R. (1990). Why life histories evolve differently in the sea. *American Zoologist* 30:197–207.

Sutinen, J. G. and J. R. Gauvin (1987). Assessing compliance with fishery regulations. *Maritimes* 33(1):10–12.

Tharpes, Yvonne L. (1989). International environmental law: Turning the tide on marine pollution. *University of Miami Inter-American Law Review* 20: 579–614.

Thayer, Gordon W. (1992). Overview of the status and directions of the science of restoration. Pp. 1–6 in Gordon W. Thayer, ed. *Restoring the Nation's Marine Environment*. Maryland Sea Grant, College Park, Maryland (USA).

Thomson, Donald A., Lloyd T. Findley, and Alex N. Kerstich (1987). *Reef Fishes of the Sea of Cortez*. University of Arizona Press, Tucson, Arizona (USA).

Thomson, Keith S. (1991). *Living Fossil: The Story of the Coelacanth*. W. W. Norton and Company, New York, New York (USA).

Thorne-Miller, Boyce and John Catena (1991). *The Living Ocean: Understanding and Protecting Marine Biodiversity*. Island Press, Washington, D.C. (USA).

Titus, J. G. (1990). Greenhouse effect, sea level rise and land use. *Land Use Policy* 7(2):138–53.

Titus, James G. (1991). Greenhouse effect and coastal wetland policy: How Americans could abandon an area the size of Massachusetts at minimum cost. *Environmental Management* 15(1):39–58.

Townsend, Richard (1990). *Shipping Safety and America's Coasts*. Center for Marine Conservation, Washington, D.C. (USA).

Troadec, Jean-Paul and Francis T. Christy, Jr. (1990). *Temporarily Out of Stock: A Diagnosis and a Strategy for International Cooperation on Fishery Research*. Report of the Study on International Fisheries Research (World Bank, FAO, UNDP, EC), Washington, D.C. (USA).

Trzyna, Thaddeus C., ed. (1989). *World Directory of Environmental Organizations: A Handbook of National and International Organizations and Programs—Governmental and Non-Governmental—Concerned with Protecting the Earth's Resources*. Third Edition. California Institute of Public

Affairs in cooperation with the Sierra Club and IUCN-The World Conservation Union, Claremont, California (USA).

Tseng, C. K. (1981). Commercial cultivation. Pp. 680–725 in C. S. Lobban and M. J. Wynne, eds. *Biology of Seaweeds*. University of California Press, Berkeley, California (USA).

Tunnicliffe, Verena (1991). The biology of hydrothermal vents: Ecology and evolution. *Oceanography and Marine Biology Annual Review* 29:319–407.

Turekian, Karl K., J. Kirk Cochran, D. P. Kharkar, Robert M. Cerrato, J. Rimas Vaišnys, Howard L. Sanders, J. Frederick Grassle, and John A. Allen (1975). Slow growth rate of a deep-sea clam determined by ^{228}Ra chronology. *Proceedings of the National Academy of Sciences* 72(7):2829–32.

Turner, Kerry (1991). Economics and wetland management. *Ambio* 22(2):59–63.

Udvardy, M. D. F. (1975). *A Classification of the Biogeographical Provinces of the World*. IUCN Occasional Paper No. 18.

UNEP, IOC, and WMO (United Nations Environment Programme, International Oceanographic Commission, and World Meteorological Organization) (1990). *UNEP-IMO-WMO Meeting of Experts on Long-term Global Monitoring System of Coastal and Near-shore Phenomena Related to Climate change*.

UNEP and IUCN (United Nations Environment Programme and World Conservation Union) (1988/89). *Coral Reefs of the World. Vols. 1–3*. UNEP Nairobi(Kenya)/IUCN, Cambridge (UK) and Gland (Switzerland).

United Nations (1972). *Stockholm Declaration on the Human Environment, adopted by consensus by the Stockholm Conference on the Human Environment on June 16, 1972*. UN Doc. A/CONF.48/14/Rev. 1, Pp. 3–5, Principle 21.

United Nations (1973). *The International Convention for the Prevention of Pollution from Ships of 1973, November 2, 1973*, International Maritime Consultative Organization (IMCO) Doc. MP/CONF/WP.35 (1973) reprinted in 12 I.L.M. 1319 (1973) [MARPOL] and Protocol of 1978.

United Nations (1982). *U. N. General Assembly, Resolution 37/7, 28 October 1982*, 37 GAOR, Supp. No. 51 (A/37/51), at 17–18).

United Nations (1983). *The Law of the Sea: Official Text with Annexes and Index, Final Act, and Introductory Material*. United Nations Publications, New York, New York (USA).

United Nations (1992). *Adoption of Agreements on Environment and Development: Agenda 21, United Nations Conference on Environment and Development*. United Nations Doc. A/CONS.151/26.

United Nations General Assembly (1991). *Resolution 46/215. Large-scale pelagic driftnet fishing and its impacts on the living marine resources of the world's oceans and seas. Resolution adopted by the 46th General Assembly [on the report of the Second Committee (A//RES/46/215)]*.

Urban, D. L., R. V. O'Neill, and H. H. Shugart, Jr. (1987). Landscape Ecology. *BioScience* 37(2):119–27.

van der Elst, R. P. (1979). A proliferation of small sharks in the shore-based Natal sport fishery. *Environmental Biology of Fishes* 4:349–62.

Vandermeulen, J. H. (1982). Some conclusions regarding long-term biological effects of some major oil spills. Pp. 151–67 in R. B. Clark, ed. *Long-term Effects of Oil Pollution on Marine Populations, Communities and Ecosystems*. Proceedings of the Royal Society Discussion Meeting. The Royal Society, London (UK).

Vecchione, M. (1987). Juvenile ecology (cephalopods). Pp. 61–84 in Peter R. Boyle, ed. *Cephalopod Life Cycles. Vol. II*. Academic Press, Orlando, Florida (USA).

Vermeij, Geerat J. (1978). *Biogeography and Adaptation: Patterns of Marine Life*. Harvard University Press, Cambridge, Massachusetts (USA).

Vermeij, Geerat J. (1987). *Evolution and Escalation: An Ecological History of Life*. Princeton University Press, Princeton, New Jersey (USA).

Veron, J. E. N. (1992). Conservation of biodiversity: A critical time for the hermatypic corals of Japan. *Coral Reefs* 11:13–21.

Villier, G. (1979). Recovery of a population of white mussels *Donax serra* at Elands Bay, South Africa, following a mass mortality. *Fisheries Bulletin (South Africa. Sea Fisheries Branch)* 12:69–74.

Vitousek, Peter M., Paul R. Ehrlich, Anne H. Ehrlich, and Pamela A. Matson (1986). Human appropriation of the products of photosynthesis. *BioScience* 36(6):368–73.

Walker, Brian H. (1992). Biodiversity and ecological redundancy. *Conservation Biology* 6(1):18–23.

Walsh, Don (1990). Thirty thousand feet and thirty years later: Some thoughts on the Deepest Presence concept. *Marine Technology Society Journal* 24(2):7–8.

Warrick, R. A. and J. Oerlemans (1990). *Sea Level Rise*. Peer Reviewed Assessment for WGI Plenary, 30 April 1990. Unpublished report, Section 9.

Watson, R. T. and D. L. Albritton (1992). *Scientific Evidence of Ozone Depletion: 1991*. World Meteorological Organization, United Nations Environment Programme, National Aeronautics and Space Administration, National Oceanic and Atmospheric Administration, United Kingdom Department of Environment, Washington, D.C. (USA).

Weinberg, J. R., V. R. Starczak, C. Mueller, G. C. Pesch, and S. M. Lindsay (1990). Divergence between populations of a monogamous polychaete with male parental care: Premating isolation and chromosome variation. *Marine Biology* 107:205–13.

Wells, Sue M., Robert Michael Pyle, and N. Mark Collins (1983). *The IUCN Invertebrate Red Data Book*. IUCN, Gland (Switzerland).

White, A. W. (1980). Recurrence of kills of Atlantic herring (*Clupea harengus harengus*) caused by dinoflagellate toxins transferred through herbivorous zooplankton. *Canadian Journal of Fisheries and Aquatic Science* 37(2):2262–65.

Wieland, Robert (1992). *Why People Catch Too Many Fish: A Discussion of Fishing and Economic Incentives*. Center for Marine Conservation, Washington, D.C. (USA)

Wilson, E. O. (1992). *The Diversity of Life*. Belknap Press of Harvard University Press, Cambridge, Massachusetts (USA).

Wise, John P. (1991). *Federal Conservation and Management of Marine Fisheries in the United States*. Center for Marine Conservation, Washington, D.C. (USA).

Wood, E. and S. Wells (1988). *The Marine Curio Trade: Conservation Issues*. Marine Conservation Society, Ross-on-Wye (UK).

Woodley, Thomas H. and David M. Lavigne (1991). *Incidental Capture of Pinnipeds in Commercial Fishing Gear*. International Marine Mammal Association, Inc., Technical Report. No. 91–01., Guelph, Ontario (Canada).

Woodwell, G. M. (1967). Toxic substances and ecological cycles. *Scientific American* 216(3):24–31.

Woody, J. (1991). Guest editorial: It's time to stop headstarting Kemp's ridley. *Marine Turtle Newsletter* 55:7–8.

World Bank Environment Department (1991). *Environmental Assessment Sourcebook. Vol. I. Policies, Procedures, and Cross-Sectoral Issues. Vol. II. Sectoral Guidelines. Vol. III. Guidelines or Environmental Assessment of Energy and Industry Projects*. World Bank Technical Paper Nos. 139, 140, and 154. The World Bank, Washington, D.C. (USA).

World Commission on Environment and Development (1987). *Our Common Future* (also known as The Brundtland Commission Report). Oxford University Press, New York, New York (USA).

World Conservation Monitoring Centre (1990). *Draft List of Marine and Coastal Protected Areas*. WCMC Protected Areas Data Unit, Cambridge (UK).

Worrest, R. C. (1989). What are the effects of UV-B radiation on marine organisms? Pp. 269–78 in T. Schneider et al., eds. *Atmospheric Ozone Research and Its Policy Implications*. Elsevier, Amsterdam (Netherlands).

WRI, IUCN, and UNEP (World Resources Institute, World Conservation Union, and United Nations Environment Programme) (1992). *Global Biodiversity Strategy*. World Resources Institute, Washington, D.C. (USA).

Yolen, Naida Maris (1991). *Survey of Marine Conservation Activities by U.S.*

Conservation Organizations. Center for Marine Conservation, Washington, D.C. (USA).

Zann, L. P. (1982). Changing technology in subsistence fisheries. Pp. 69–71 in P. Helfrich, ed. *Proceedings of the Seminar/Workshop on Utilization and Management of Inshore Marine Ecosystems of the Tropical Islands, November 24–30, 1979, University of the South Pacific, Suva, Fiji.* Sea Grant Cooperative Report, University of Hawaii, Hawaii (USA).

Glossary

abyssal the deep sea, variously defined as starting at 2,000- to 4,000-meter (6,600- to 13,000-foot) depths, a region of low temperature, high pressure, and absence of sunlight

anadromous a type of life cycle in which fishes mature in the ocean, while spawning occurs in freshwater streams or lakes

alien species (also called introduced, exotic, nonindigenous, or nonnative species) a species that has been transported by human activity, intentional or accidental, into a region where it does not naturally occur

allele any of the different versions of a gene occupying a particular locus (place) on a chromosome

anoxia the absence of oxygen

ballast weight carried by a vessel to improve stability or the act of adding such weight; in modern vessels, ballast is usually water

bathypelagic the dark, deepsea (1,000 to 4,000 meters, or 3,300 to 13,000 feet) water column below the euphotic (or well-lighted) and mesopelagic zones (poorly lighted) but above the abyssopelagic zone

benthic pertaining to the seabed; bottom-dwelling

billfish the swordfish and the members of the family Istiophoridae (e.g., sailfish, marlin), pelagic fishes that have the upper jaw prolonged into a long spear or sword

bioerosion the erosion of material, such as coral rock, that results from the direct action of living organisms such as boring sponges, sipunculid worms, pholadid bivalve mollusks, or sea urchins

biogeochemical cycle the cyclical transformation of compounds of chemical elements by interacting biological, geological, and chemical processes; well-known ones include the carbon and nitrogen cycles

biogeography the study of geographical distributions of organisms, their habitats, and the historical and biological factors that produce them

biological diversity (= biodiversity) the diversity of life, often divided into three hierarchical levels: genetic (diversity within species), species (diversity among species), and ecosystem (diversity among ecosystems)

biological pump a biogeochemical process by which marine organisms take up and transport

CO$_2$ from surface waters into the deep sea

bloom a sharp increase in density of phytoplankton or benthic algae in a given area

blue-green algae *see* cyanobacteria

bottleneck (population) a sharp decrease in population density to very low numbers, often with a corresponding reduction of total genetic variability

bycatch *see* incidental take (also called non-target species)

cetaceans an order of marine mammals distinguished by the absence of visible hind legs, and having a layer of blubber, paddle-like front flippers, and a horizontal tail (includes great whales, dolphins, and porpoises)

chitin a polymer of glucosamine that is a major component of the exoskeletons of arthropods, including shrimp, crabs, and lobsters; used in a variety of industrial and commercial processes

coelacanths a primitive group of bony fishes related to the ancestor of amphibians, reptiles, birds, and mammals; only one extant species

coastal waters marine benthic and pelagic ecosystems having substantial influence from the land

conservation biology the science of conserving biological diversity

continental shelf the edges of continental landmasses, now covered with seawater; generally the most productive part of the sea

coral bleaching a phenomenon occurring when corals under stress expel their mutualistic zooxanthellae or the concentration of photosynthetic pigments is decreased; as a result, the corals' white skeletons show through their tissues, and they appear bleached

cosmopolitan in biogeography, having an extremely broad or global distribution; sperm whales, for example, are cosmopolitan

cyanobacteria (formerly known as blue-green algae) a phylum of single-celled or filamentous plant-like, usually photosynthetic bacteria in which the dominant pigment (c-phycocyanin) imparts a blue, red, or black color

demersal fishes (also called groundfishes), cephalopods, or crustaceans that can swim but spent their time on or near the bottom (e.g., blue crabs, halibut); also eggs laid on or near the seabed by actively swimming species

detritus a major food-source in marine ecosystems, consisting of organic remains of plants and animals, often heavily colonized by bacteria

dredge to remove sediments from canals, rivers, or harbors by means of a floating device (a dredge), or to catch shellfish in the seabed (and the heavy mesh fishing gear for doing so)

diadromous migrating between fresh water and seawater as a regular part of the life cycle, either anadromous (*see*) or catadromous (maturing in fresh waters but reproducing in the sea)

do-no-harm principle *see* precautionary principle

downwelling a process by which surface waters increase in density and sink; strong downwelling occurs mainly off Greenland and Antarctica

driftnet a gill net suspended vertically from floats at a specific depth and left to drift freely

dumping (as defined in London Dumping Convention) any deliberate disposal at sea of wastes or other matter, or any deliberate disposal of vessels or other human-made structures (wastes derived from normal vessel operations or related to the development of seabed mineral resources are not discussed in the Convention)

ecological economics a branch of economics that takes account of ecological principles and examines economic values of nonmarket ecological products and services

ecology the scientific study of the interactions of living things and their environment

ecosystem a community of organisms in their physical environment

ecosystem diversity the diversity among biological communities and their physical settings, characterized by differences in species composition, physical structure, and function; the highest level of biological diversity

El Niño event a regional or global oceanic-atmospheric perturbation whose manifestations range from increased sea surface temperatures in the tropical East Pacific to aberrant rainfall patterns

endemic native to and restricted to a specific geographic area

environmental impact assessment (EIA) process by which the consequences of proposed projects or programs are evaluated as an integral part of planning the project, alternatives are analyzed, and the general public has ample opportunity to comment

estuary an ecosystem in which a river or stream meets ocean waters; characterized by intermediate or variable salinity levels and often by high productivity

eutrophication enrichment of a water body with nutrients, resulting in excessive growth of phytoplankton, seaweeds, or vascular plants, and, often, depletion of oxygen

exclusive economic zone (EEZ) that part of the marine realm seaward of territorial waters within which nations have exclusive fishing rights

ex situ out of the original location (in conservation, often in a laboratory, botanical garden, zoo, or aquarium); the opposite of *in situ*

extant still surviving; opposite of extinct; the sea is now home to only about 300 extant species of lamp shells, only a small fraction of the tens of thousands of lamp shell species known from the fossil record

extinct no longer surviving; opposite of extant; the eelgrass limpet became extinct in the 1930s

fecundity the number of young produced

fitness the genetic contribution to

future generations; the average number of descendants

food web a network of interconnected trophic chains in a community

fouling communities benthic organisms attached to submerged objects of economic importance, such as pilings or boat-bottoms

gene pool the total amount of genetic material within a freely interbreeding population at a given time

genetic diversity the diversity of genes within and among populations of a species; the lowest level of biological diversity

genetic drift random changes in the gene pool not due to selection, mutation, or immigration, sometimes leading to the loss of particular alleles, through unequal reproductive contributions of individuals; non-Darwinian evolution

genotype the hereditary or genetic expression of an individual or a group of individuals

ghost fishing the continued ensnarement of marine animals in lost or discarded traps, nets, or line

glacial maximum the time of maximum glacial advance during an ice age

greenhouse gases gases, such as carbon dioxide and methane, that tend to trap heat radiating from the Earth's surface, thus causing warming in the lower atmosphere

groundfish fishes that live on or near the seabed (e.g., cod, flounders); demersal fishes

guild a group of species having similar functional roles in their ecosystem; the species (e.g., mangrove trees, filter-feeders) are not necessarily closely related

habitat the space in which an organism, population, or species lives

harvest in fisheries, catching fishes, whether they are wild or their life cycles are manipulated in mariculture operations; appropriately used in mariculture but not in capture fisheries

holoplankton species that remain in planktonic form throughout their life cycle

hotspot an area rich in total numbers of species, *or* an area of especially high pollutant (e.g., PCB) concentration

hydrology the scientific study of waters of the Earth, especially in relation to the effect of precipitation and evaporation upon the occurrence and character of water in streams, lakes, and on or below land surface

hypoxia a state of low oxygen concentration relative to the needs of most aerobic species

in situ in the original location; opposite of *ex situ*

inbreeding mating of close relatives, which reduces genetic diversity, often leading to expression of deleterious recessive characteristics and reduction of fitness in the offspring

incidental take (= bycatch or non-target species) the portion of a fishery catch consisting of species other than the target species

integrated area management (IAM) management approach whereby a specific area is zoned and regulated for a variety of uses, including research, species protection, tourism, or fishing, that is compatible with the management goals for the area

intertidal zone (= littoral zone) the zone of overlap between land and sea that is submerged at high tide and exposed at low tide

invertebrate any animal without backbone or spinal column

keystone species a species that influences the ecological composition, structure, or functioning of its community far more than its abundance would suggest

large marine ecosystem (LME) a large marine region that has unique physical and biological characteristics and within which organisms have distinctive reproductive, growth and feeding strategies

lecithotrophic pertains to an organism whose developing embryos gain their nourishment from yolky eggs

long-term research (also used for monitoring) program for acquiring information over ecologically meaningful time period, rather than the one to three years of a typical funding cycle; long-term programs are far more likely to record phenomena that last years or decades

mangrove forest a community of salt-tolerant trees, with associated shrubs or vines and other organisms, that grows in a zone roughly

coinciding with the intertidal zone along protected tropical and subtropical coasts

mariculture controlled cultivation of marine organisms in tanks, ponds, cages, rafts, or other structures

maximum sustainable yield (MSY) the maximum amount of a species or group of species that can be taken without diminishing the future take

meroplankton organisms that are temporarily members of the plankton community as early life history stages, but are not planktonic as adults

mudflat an intertidal ecosystem whose substrate consists predominantly of fine silts, clays, and organic material

multilateral development bank (MDB) an international institution, such as the World Bank or the Asian Development Bank, that has been established by a broad spectrum of national governments to provide grants or loans for projects to improve economic conditions in recipient nations

muro ami fishing technique that uses weighted bags to smash coral reefs, thereby scaring fishes from their hiding places

mutualism a kind of symbiotic relationship, such as the one between corals and zooxanthellae, in which both organisms benefit; the relationship can be obligate or facultative for one or both partners

nekton aquatic organisms, such as mackerels and penguins, that are powerful enough swimmers to move against currents

neritic pertaining to the water column overlying the continental shelf (as opposed to oceanic, which is seaward of the shelf-break)

neuston organisms that occur at or just below the air-sea interface

outfall place where a sewer, drain, or stream discharges

pelagic free-swimming (nektonic) or floating (planktonic) organisms that live exclusively in the water column of the open sea (oceanic pelagic) or continental shelf waters (neritic pelagic), not on the bottom

plankton drifting or slowly swimming organisms that cannot swim against currents

planktotrophic feeding on plankton

pleuston organisms that float on the sea surface

precautionary principle (= do-no-harm principle) a proactive method of dealing with the environment that places the burden of proof on those whose activities could harm the environment; the opposite of the wait-and-see principle

primary production amount of organic material synthesized by organisms from inorganic substrates in a given area in a given period

proximate immediate, near

purse seine large nets used in commercial fisheries encircle a school of fish and are drawn together

recruitment the influx of new members into a population by reproduction or immigration

red tide reddish-brown discoloring of surface water from blooming populations of dinoflagellate phytoplankton; long associated with nutrient pollution, these might be population outbreaks of alien dinoflagellate species

sessile fixed or attached, unable to move

siltation the settling of fine mineral particle matter

smolt the early life history stage of an anadromous salmonid fish that can migrate to and live in the sea

spat the spawn or young of bivalve mollusks

spawn the eggs of certain aquatic organisms or the act of producing such eggs or egg masses

Special Areas designation protected status awarded under MARPOL to some enclosed and semi-enclosed seas with restricted circulation as part of an effort to control noxious liquid, garbage, or oil pollution (e.g., the Baltic, Black, and Red seas)

species diversity the diversity of species in a higher taxon or a particular place; the middle, most familiar level of biological diversity

stock a specific population or group of populations

straddling stock a population of organisms that travels between the exclusive economic zones of two or more countries or (by some definitions) between them and the high seas

stratosphere the layer of the atmosphere (15-50 kilometers or 9-31 miles above the Earth's surface) above the troposphere, in which ozone prevents most ultraviolet radiation from reaching the Earth's surface

supply-side ecology   the ecological concept that differences in the amount and timing of recruitment can determine a species' distribution and abundance

symbiosis   the close relationship of two organisms in proximity, with one benefiting and the other either benefiting (mutualism), not being significantly affected (commensalism), or being harmed (parasitism)

target species   the intended catch of a fishery; opposite of bycatch

taxon (pl. taxa): any group of organisms or populations considered to be sufficiently distinct from other such groups to be treated as a separate unit; in order of increasing inclusiveness, commonly used hierarchical levels of taxa include the subspecies, species, genus (pl. genera), family, order, class, phylum (pl. phyla), and kingdom

trawl   a funnel-shaped fishing net towed behind a vessel

trophic level   feeding level in food chain or pyramid; for example, herbivores (organisms that eat plants) constitute one trophic level

tsunami   a very fast-moving wave, incorrectly called a tidal wave, that is initiated by an underwater disturbance, such as an earthquake, slumping event, or volcanic eruption; tsunamis can cause severe damage to low-lying coastal lands

turbid   exhibiting reduced water clarity because of the presence of suspended matter

upwelling   a process by which water rises from lower depths into the shallows, usually the result of divergence or offshore currents

valuation   the attachment of monetary value to an object through a consideration of both internalized and externalized costs

wait-and-see principle   a reactive method of dealing with the environment that places the burden of proof on those who would conserve; the opposite of the precautionary principle

warm-core ring   a large, isolated, lens-shaped mass of warm water surrounded by cooler water that is created when an eddy or swirl of water breaks off from a warm current

zooxanthellae   microscopic dinoflagellate algae that live mutualistically in the tissues of certain marine invertebrates, including reef-building corals and giant clams

Index

Biological diversity (*continued*)
institutions, 152, 163, 174, 175, 211
judicial review, 306
limits to understanding, 68
North-South issues, 171
physical alteration, 106
pollution, 122
proximate threats to, 88, 150
research/monitoring centers, 302
science, 196, 197
species diversity, 9, 10, 57, 58, 68, 69, 256, 301
sustainability, 304
traditional management, 160
United Nations, 208
Biological invasion, 130
see also Alien species
Biological pump, 27, 28
Biosphere reserve, 219, 297, 301
Samana Bay project (Dominican Republic), 192
Sian Ka'an Biosphere Reserve, 266
Birgus latro, see Crab, coconut
Black Sea, 116, 132
pollution control, 236
Bloom, algal, 124, 126
Norway, 127
Phaeocystis, 125
Southern Ocean, 127
toxic, 6, 125, 132, 135
Bolivia, 218
Bottlenecks, 67, 68, 74
Boundaries, 204, 231
delimitation of marine ecosystems, 169, 204, 230, 269, 281
Great Barrier Reef Marine Park, 272
international cooperation, 176, 224, 263, 277
lack of in the sea, 42, 43, 50, 174
migration, 84, 174
natural, 43, 282
Brachyrhamphus marmoratus, see Murrelet, marbled
Branta bernicla, see Goose, brant
Brazil, 13, 24, 56, 59, 85, 145
Breeding areas, 289
Bristlemouths, genus *Cyclothone*, 19
British Columbia, 44, 134, 185
introduced species, 134
Bryozoan, 14, 21
Bugula neritina, see Bryozoan
Burden of proof, 157, 174, 180–183, 229, 230, 306
CCAMLR, 229
Butterflyfish (family Chaetodontidae), 106
Bycatch, 89, 93–95, 111, 291
cetacean, 94
dolphin, 209

finfish, 94, 264
pinniped, 94
sea turtle, 95
totoaba, 148
Bycatch reduction device, 95, 289

Cadmium, 85, 118, 121
Calidris alba, see Sanderling
Calidris canutus, see Knot, red
Calidris pusilla, see Sandpiper, semi-palmated
California, 70, 149, 177, 181
California Current, 61
Callinectes sapidus, see Crab, blue
Callinectes spp., *see* Crab, blue
Callorhinus ursinus, see Fur seal, northern
Calyptogena magnifica, see Clam, hydrothermal vent
Cameroon, 221
Camptorhynchus labradorius, see Duck, Labrador
Canada, 10, 18, 23, 24, 44, 68, 79, 100, 125, 134, 181, 185, 214, 223, 227, 232, 255
Canadian Centre for Biodiversity, 301
Cancer magister, see Crab, Dungeness
Capital, 164, 170, 180, 188
biotic, 152, 164, 165, 170, 189
Capital liquidation, 165
Carcharodon carcharias, see Shark, white
Caretta caretta, see Sea turtle, loggerhead
Caribbean, 48
biodiversity databases, 257, 302
conservation efforts, 254, 257, 259, 265
corals, 104
endemism, 59
exploitation of algae, 104
sea turtle study, 265
SPAW Protocol, 249
spawning aggregations, 63
sponges, 104
TURFs, 215
waste disposal, 241
Caribbean Coastal Marine Productivity program (CARICOMP), 199, 259
Caribbean Convention, 210
Caribbean Environment Programme, 249
Caribbean Sea, 112, 141, 204, 257, 259
CARIDAT, 257
Carrageenanweed
Chondrus, 24
Eucheuma, 24
Gigartina, 24
Iridaea, 24
Kappaphycus, 24
Cartagena Convention, 249, 259
Caspian Sea, 116
Cayman Islands, 32, 260,

Mussel
  aquaculture, 19
  hydrothermal vent (*Bathymodiolus thermophilus*), 7
  starfish interaction, 30
  toxic algal blooms, 125
  zebra (*Dreissena polymorpha*), 132, 209
Mustela macrodon, see Sea mink

Narmada River (India), 240
Narwhal (*Monodon monoceros*), 100
National institutes for the environment (NIEs), 301
National Seafood Promotion Council (US), 264
Nature Conservancy, The, 198
Nauru Group, 233
Negaprion brevirostris, see Shark, lemon
Netherlands, 260
Netherlands Antilles, 260
Network
  electronic, 189, 198, 240, 257–259, 277, 286, 306
  marine research, 302
New Caledonia (France), 82
New York Bight (USA), 125
New Zealand, 58, 71, 91, 165, 185, 214, 219
Nicaragua, 260
Nile River (Egypt), 116
Nitrogen, 11, 112, 123–127
Non-governmental organizations (NGOs), 253, 278, 305
  concern for the whales, 243
  information transfer, 195
  management, 252
  marine related, 254
  oversight responsibilities, 253
  political advocacy, 191, 253
  public education, 194
  role, 252, 253, 305
  USA, 264
North America
  alien species, 131
  climatic change, 144
  eelgrass beds, 77
  seabirds, 101
North Atlantic Salmon Conservation Organization (NASCO), 231
North Sea, 6, 85, 89, 94, 110, 121, 123, 125, 150, 204
North-South interaction, 170–172, 190, 242, 245
Northwest Atlantic Fisheries Organization (NAFO), 232
Norway, 11, 23, 71, 99, 125, 243
Novaya Zemlya military activities, 177
Nuclear submarines, 177

Nuclear testing
  circulation of radioactive materials, 44
  Greenpeace, 254
  Pacific atolls, 177
Numenius borealis, see Curlew
Nursery grounds, 58, 65–67, 85, 108, 116, 154, 206, 214, 239, 264, 301
Nutrient circulation, 11
Nutrient enrichment, 19, 25, 45, 48, 49, 61, 65, 85, 112, 117, 123–126, 143
Nutrient limitation, 11, 28, 48, 60, 61, 123, 124, 144
Nutrient pollution, 125
Nutrient transport, 108, 116
Nycticorax caledonicus crassirostris, see Heron, Bonin night

Ocean Voice International (OVI), 255, 288
Oceanic circulation, 44, 144
Oceanodroma macrodactyla, see Storm petrel, Guadalupe
Oceanography, 50, 195
Octopod concentration of paralarvae, 65
Oil
  disposal, 297
  field accidents (IXTOC), 119
  properties of, 119
  transport, 277
  see also Pollution, oil
Okhotsk Sea (Russia), 59
Olympic Peninsula (USA), 114
Oman, 269–271
  fisheries, 270
Ommatophoca rossii, see Seal, Ross
Oncorhynchus keta, see Salmon, chum
Oncorhynchus tshawhytscha, see Salmon, chinook
Orang-utan (*Pongo pygmaeus*), 151
Orange roughy (*Hoplostethus atlanticus*), 91, 288
Orcinus orca, see Whale, orca
Organotin, 120, 297
  see also TBT
Oslo Convention, 210
Otter
  marine (*Lutra felina*), 56
  marine, listed under CITES, 53
  sea (*Enhydra lutris*), 53
  sea, exploitation, 83, 97, 98
  sea urchin interaction, 53, 181
Overexploitation, 24, 53, 54, 56, 62, 65, 75, 82, 86, 88–90, 97–106, 110, 114, 150, 154, 159, 161, 165, 180, 189, 213, 214, 219, 226, 229, 230, 288, 289
Ovula ovum, see Egg cowries
Ownership, 156, 159, 160, 196, 212, 213, 215, 219
  adjacent waters, 168